T0329317

Advanced Data Analysis and Modelling in Chemical Engineering

Advanced Data Analysis and Modelling in Chemical Engineering

Denis Constales
Gregory S. Yablonsky
Dagmar R. D'hooge
Joris W. Thybaut
Guy B. Marin

ELSEVIER

AMSTERDAM · BOSTON · HEIDELBERG · LONDON · NEW YORK · OXFORD
PARIS · SAN DIEGO · SAN FRANCISCO · SINGAPORE · SYDNEY · TOKYO

Elsevier
Radarweg 29, PO Box 211, 1000 AE Amsterdam, Netherlands
The Boulevard, Langford Lane, Kidlington, Oxford OX5 1GB, United Kingdom
50 Hampshire Street, 5th Floor, Cambridge, MA 02139, United States

Library of Congress Cataloging-in-Publication Data
A catalog record for this book is available from the Library of Congress

British Library Cataloguing-in-Publication Data
A catalogue record for this book is available from the British Library

ISBN: 978-0-444-59485-3

For information on all Elsevier publications
visit our website at https://www.elsevier.com/

Working together
to grow libraries in
developing countries

www.elsevier.com • www.bookaid.org

Publisher: John Fedor
Acquisition Editor: Kostas Marinakis
Editorial Project Manager: Sarah Watson
Production Project Manager: Anitha Sivaraj
Cover Designer: Maia Cruz Ines

Typeset by SPi Global, India

Contents

Preface xiii

Chapter 1: Introduction 1
   1.1 Chemistry and Mathematics: Why They Need Each Other 1
   1.2 Chemistry and Mathematics: Historical Aspects 2
   1.3 Chemistry and Mathematics: New Trends 3
   1.4 Structure of This Book and Its Building Blocks 5
   References 7

Chapter 2: Chemical Composition and Structure: Linear Algebra 9
   2.1 Introduction 9
   2.2 The Molecular Matrix and Augmented Molecular Matrix 11
      2.2.1 The Molecular Matrix 11
      2.2.2 Application of the Molecular Matrix to Element Balances 12
      2.2.3 Remarks 21
   2.3 The Stoichiometric Matrix 21
      2.3.1 Application of the Stoichiometric Matrix to Reactions 21
      2.3.2 The Augmented Stoichiometric Matrix 24
   2.4 Horiuti Numbers 26
   2.5 Summary 30
   Appendix. The RREF in Python 31
   Nomenclature 33
   Further Reading 34

Chapter 3: Complex Reactions: Kinetics and Mechanisms – Ordinary Differential Equations – Graph Theory 35
   3.1 Primary Analysis of Kinetic Data 35
      3.1.1 Introduction 35
      3.1.2 Types of Reactors for Kinetic Experiments 35
      3.1.3 Requirements of the Experimental Information 36
   3.2 Material Balances: Extracting the Net Rate of Production 36
      3.2.1 Transport in Reactors 37
      3.2.2 Change Due to the Chemical Reaction—Conversion 39
      3.2.3 Ideal Reactors (Homogeneous) 39

3.2.4 Ideal Reactors With Solid Catalyst.. 42
3.2.5 Summary.. 44
3.3 Stoichiometry: Extracting the Reaction Rate From the Net Rate
of Production 45
3.3.1 Complexity and Mechanism: Definition of Elementary Reactions.............. 45
3.3.2 Homogeneous Reactions: Mass-Action Law 46
3.3.3 Heterogeneous Reactions 47
3.3.4 Rate Expressions for Single Reactions: Stoichiometry 49
3.3.5 Relationships Between Reaction Rate and Net Rate of Production 49
3.4 Distinguishing Kinetic Dependences Based on Patterns and Fingerprints 50
3.4.1 Single Reactions 50
3.4.2 Distinguishing Between Single and Multiple Reactions 55
3.4.3 Distinguishing Between Elementary and Complex Reactions 55
3.4.4 Summary of Strategy... 56
3.5 Ordinary Differential Equations.. 58
3.6 Graph Theory in Chemical Kinetics and Chemical Engineering.................... 61
3.6.1 Introduction... 61
3.6.2 Derivation of a Steady-State Equation for a Complex Reaction Using
Graph Theory.. 62
3.6.3 Derivation of Steady-State Kinetic Equations for Multiroute
Mechanisms: Kinetic Coupling 68
3.6.4 Bipartite Graphs of Complex Reactions 73
3.6.5 Graphs for Analyzing Relaxation: General Form of the Characteristic
Polynomial 78
Nomenclature... 80
References... 81

Chapter 4: Physicochemical Principles of Simplification of Complex Models 83
4.1 Introduction... 83
4.2 Physicochemical Assumptions 84
4.2.1 Assumptions on Substances 84
4.2.2 Assumptions on Reactions and Their Parameters 85
4.2.3 Assumptions on Transport-Reaction Characteristics 86
4.2.4 Assumptions on Experimental Procedures.. 87
4.2.5 Combining Assumptions 87
4.3 Mathematical Concepts of Simplification in Chemical Kinetics 88
4.3.1 Introduction... 88
4.3.2 Simplification Based on Abundance... 89
4.3.3 Rate-Limiting Step Approximation... 90
4.3.4 Quasiequilibrium Approximation... 92
4.3.5 Quasi-Steady-State Approximation.. 94
4.3.6 Mathematical Status of the Quasi-Steady-State Approximation.................... 95
4.3.7 Simplifications and Experimental Observations............................... 98
4.3.8 Lumping Analysis.. 100
Nomenclature... 101
References... 102

Chapter 5: Physicochemical Devices and Reactors.. **105**

5.1 Introduction.. 105
5.2 Basic Equations of Diffusion Systems... 106
    5.2.1 Fick's Laws.. 106
    5.2.2 Einstein's Mobility and Stokes Friction 106
    5.2.3 Knudsen Diffusion and Diffusion in Nonporous Solids....................... 107
    5.2.4 Onsager's Approach... 107
    5.2.5 Teorell's Approach 108
    5.2.6 Nonlinear Diffusion Models.. 109
5.3 Temporal Analysis of Products Reactor: A Basic Reactor-Diffusion
    System.. 110
    5.3.1 Introduction.. 110
    5.3.2 Typical TAP Reactor Configurations... 110
    5.3.3 Main Principles of TAP... 114
5.4 TAP Modeling on the Laplace and Time Domain: Theory 115
    5.4.1 The Laplace Transform 115
    5.4.2 Key Results.. 116
5.5 TAP Modeling on the Laplace and Time Domain: Examples........................ 121
    5.5.1 Moment Definitions... 121
    5.5.2 One-Zone Reactor, Diffusion Only... 121
    5.5.3 One-Zone Reactor, Irreversible Adsorption.. 123
    5.5.4 One-Zone Reactor, Reversible Adsorption.. 124
    5.5.5 Three-Zone Reactor... 126
    5.5.6 Reactive Thin Zones.. 126
    5.5.7 Other Thin Zones... 127
5.6 Multiresponse TAP Theory 129
5.7 The Y Procedure: Inverse Problem Solving in TAP Reactors....................... 132
5.8 Two- and Three-Dimensional Modeling... 134
    5.8.1 The Two-Dimensional Case... 135
    5.8.2 The Three-Dimensional Case... 137
    5.8.3 Summary.. 138
    5.8.4 Radial Diffusion in Two and Three Dimensions................................... 139
5.9 TAP Variations 141
    5.9.1 Infinite Exit Zone 141
    5.9.2 Internal Injection... 143
    5.9.3 Nonideal Plug-Flow Reactor via Transfer Matrix 145
5.10 Piecewise Linear Characteristics 146
    5.10.1 Introduction.. 146
    5.10.2 Conical Reactors.. 147
    5.10.3 Surface Diffusion... 147
    5.10.4 Independence of the Final Profile of Catalyst Surface Coverage 148
5.11 First- and Second-Order Nonideality Corrections in the Modeling
    of Thin-Zone TAP Reactors.. 151
    5.11.1 Thin-Zone TAP Reactor 151
    5.11.2 Thin-Sandwiched TAP Reactor Matrix Modeling............................... 152

Nomenclature.. 154
References.. 156

Chapter 6: Thermodynamics.. 159

6.1 Introduction... 159
   6.1.1 Thermodynamic Systems 159
   6.1.2 Thermodynamic State Functions................................... 160
   6.1.3 Steady State and Equilibrium....................................... 163
   6.1.4 Reversibility and Irreversibility 164
   6.1.5 Equilibrium and Principle of Detailed Balance............ 165
6.2 Chemical Equilibrium and Optimum.. 166
   6.2.1 Introduction... 166
   6.2.2 Le Chatelier's Principle.. 166
   6.2.3 Relation Between Equilibrium and Optimum Regime.... 168
   6.2.4 Summary.. 174
6.3 Is It Possible to Overshoot an Equilibrium?.............................. 175
   6.3.1 Introduction... 175
   6.3.2 Mass-Action Law and Overshooting an Equilibrium.... 175
   6.3.3 Example of Equilibrium Overshoot: Isomerization....... 176
   6.3.4 Requirements Related to the Mechanism of a Complex Chemical
         Reaction 181
6.4 Equilibrium Relationships for Nonequilibrium Chemical Dependences 183
   6.4.1 Introduction... 183
   6.4.2 Dual Experiments 184
6.5 Generalization. Symmetry Relations and Principle of Detailed Balance 189
   6.5.1 Introduction... 189
   6.5.2 Symmetry Between Observables and Initial Data......... 189
   6.5.3 Experimental Evidence... 192
   6.5.4 Concluding Remarks 194
6.6 Predicting Kinetic Dependences Based on Symmetry and Balance 194
   6.6.1 Introduction... 194
   6.6.2 Theoretical Model... 195
   6.6.3 Matrix of Experimental Curves.................................... 195
   6.6.4 General Considerations... 198
   6.6.5 Checking Model Completeness..................................... 199
   6.6.6 Concluding Remarks 202
6.7 Symmetry Relations for Nonlinear Reactions 202
   6.7.1 Introduction... 202
   6.7.2 Single Reversible Reactions... 203
   6.7.3 Two-Step Reaction Mechanisms................................... 210
   6.7.4 Physicochemical Meaning of New Invariances and Their Novelty............ 214
   6.7.5 Concluding Remarks 215
Appendix... 216
Nomenclature.. 217
References.. 219

Chapter 7: Stability of Chemical Reaction Systems **221**
    7.1 Stability—General concept 221
        7.1.1 Introduction 221
        7.1.2 Non-Steady-State Models 222
        7.1.3 Local Stability—Rigorous Definition 224
        7.1.4 Analysis of Local Stability 225
        7.1.5 Global Dynamics 230
    7.2 Thermodynamic Lyapunov Functions 231
        7.2.1 Introduction 231
        7.2.2 Dissipativity of Thermodynamic Lyapunov Functions 233
        7.2.3 Chemical Oscillations 235
    7.3 Multiplicity of Steady States in Nonisothermal Systems 236
        7.3.1 Introduction 236
        7.3.2 Mathematical Model of the Nonisothermal CSTR 237
        7.3.3 Types of Stability of the Steady State 239
    7.4 Multiplicity of Steady States in Isothermal Heterogeneous
          Catalytic Systems 242
        7.4.1 Historical Background 242
        7.4.2 The Simplest Catalytic Trigger 244
        7.4.3 Influence of Reaction Reversibility on Hysteresis 246
        7.4.4 Transient Characteristics and "Critical Slowing Down" 247
        7.4.5 General Model of the Adsorption Mechanism 249
    7.5 Chemical Oscillations in Isothermal Systems 251
        7.5.1 Historical Background 251
        7.5.2 Model of Catalytic Chemical Oscillations 252
    7.6 General Procedure for Parametric Analysis 257
    Nomenclature 262
    References 264

Chapter 8: Optimization of Multizone Configurations **267**
    8.1 Reactor Model 267
    8.2 Maximizing the Conversion 269
    8.3 Optimal Positions of Thin Active Zones 270
    8.4 Numerical Experiments in Computing Optimal Active Zone
          Configurations 272
    8.5 Equidistant Configurations of the Active Zones 274
        8.5.1 General Equation 274
        8.5.2 Influence of Da on Conversion Characteristics With Increasing
                Number of Thin Zones 278
        8.5.3 Activity for One Zone: Shadowing or Cooperation? 279
        8.5.4 Comparison of Model Reactors 280
    Nomenclature 283
    Reference 284

Chapter 9: Experimental Data Analysis: Data Processing and Regression **285**
9.1 The Least-Squares Criterion 285
9.2 The Newton-Gauss Algorithm 286
9.3 Search Methods From Optimization Theory 287
    9.3.1 Univariate Search 287
    9.3.2 The Rosenbrock Method 288
    9.3.3 The Gradient Method 290
9.4 The Levenberg-Marquardt Compromise 291
9.5 Initial Estimates 294
    9.5.1 Power-Law Models for the Reaction Rate Equation 294
    9.5.2 Arrhenius Equation 294
    9.5.3 Rate Laws for Heterogeneously Catalyzed Reactions 296
9.6 Properties of the Estimating Vector 299
9.7 Temperature Dependence of the Kinetic Parameters k and K_i 301
9.8 Genetic Algorithms 302
Nomenclature 304
References 306

Chapter 10: Polymers: Design and Production **307**
10.1 Introduction 307
10.2 Microscale Modeling Techniques 310
    10.2.1 Deterministic Modeling Techniques 310
    10.2.2 Stochastic Modeling Techniques 322
10.3 Macroscale Modeling Techniques 328
10.4 Extension Toward Heterogeneous Polymerization in Dispersed Media 330
    10.4.1 Suspension Polymerization 331
    10.4.2 Emulsion Polymerization 334
10.5 Extension Toward Heterogeneous Polymerization With Solid Catalysts 339
10.6 Conclusions 342
Nomenclature 342
Acknowledgments 347
References 347

Chapter 11: Advanced Theoretical Analysis in Chemical Engineering: Computer Algebra and Symbolic Calculations **351**
11.1 Critical Simplification 351
    11.1.1 Model of the Adsorption Mechanism 351
    11.1.2 Transformation of the Langmuir System to a Slow-Fast System 354
    11.1.3 Zero Approximation for Singular Perturbation 354
    11.1.4 First Approximation for Singular Perturbation 355
    11.1.5 Bifurcation Parameters 356
    11.1.6 Calculation of Reaction Rates 357
    11.1.7 Dynamics 361
    11.1.8 An Attempt at Generalization 366

11.2 "Kinetic Dance": One Step Forward—One Step Back 366
    11.2.1 History.. 366
    11.2.2 Experimental Procedures... 367
    11.2.3 Concluding Remarks 373
11.3 Intersections and Coincidences 373
    11.3.1 Introduction... 373
    11.3.2 Two-Step Consecutive Mechanism With Two Irreversible Steps 374
    11.3.3 Two-Step Consecutive Mechanism With One Reversible and One
           Irreversible Step.. 385
    11.3.4 Concluding Remarks 389
Nomenclature.. 390
References... 392

Index 395

Preface

This book is written by Denis Constales, Dagmar D'hooge, Joris Thybaut, and Guy Marin from Ghent University (Belgium) and Gregory Yablonsky from St. Louis University (United States), who has been affiliated with Ghent University as a visiting professor for more than 15 years. During that period, we have been working toward the development of a center of chemico-mathematical collaboration, organizing, among other things, the workshops and conferences related to "Mathematics in Chemical Kinetics and Chemical (Bio) Engineering" (MACKiE). More generally, we have attempted to contribute to the integration of "Science-Technology-Engineering-Mathematics" (STEM).

Our book is a first consolidation of our efforts in this respect. It is about advanced data analysis and modeling in chemical engineering. We dare to use the term "advanced," as we aim to help the reader to increase the level of understanding of the physicochemical phenomena relevant for the field of chemical engineering using effective mathematical models and tools. This does not mean that these models have to be complex, but rather that the complexity of a mathematical model and the corresponding tools have to be optimal. Also, we would like to stress that, in our opinion, the role of mathematics is not limited to that of servant of all masters, a tireless and beautiful Cinderella, but that it provides an inexhaustible source of powerful concepts, both theoretical and experimental.

Our Team

Denis Constales is associate professor at the Department of Mathematical Analysis in the faculty of Engineering and Architecture of Ghent University. He obtained his doctorate there and specialized in applied mathematics, with specific research interest in special functions, integral transforms, and the application of computer algebra techniques.

Gregory S. Yablonsky is professor at St. Louis University. He is a representative of the Soviet-Russian catalytic school, being a postdoctoral student of Mikhail Slin'ko and having collaborated with Georgii Boreskov. He also was a member of the Siberian chemico-mathematical team together with Valerii Bykov, Alexander Gorban, and Vladimir Elokhin. He gained international experience working in many universities of the world, including St. Louis, United States; Ghent, Belgium; Singapore; Belfast, N. Ireland, United Kingdom.

Dagmar R. D'hooge is assistant professor in Polymer Reaction Engineering and Industrial Processing of Polymers at Ghent University. He is a member of the Laboratory for Chemical Technology and associate member of the Department of Textiles. He was a visiting researcher in the Matyjaszewski Polymer Group at Carnegie Mellon University (Pittsburgh, United States) and the Macromolecular Chemistry Group at Karlsruhe Institute of Technology (Germany). His research focuses on the development and application of multiscale modeling platforms for designing industrial-scale polymerization processes and polymer processing units for both conventional and high-tech polymeric materials.

Joris W. Thybaut is full professor in Catalytic Reaction Engineering at the Laboratory for Chemical Technology at Ghent University. After obtaining his PhD on single-event microkinetic modeling of hydrocracking and hydrogenation, he went to the Institut des Recherches sur la Catalyse (Lyon, France) for a postdoc on high throughput experimentation. Today, he investigates rational catalyst and reactor design for the processing of conventional and, more particularly, alternative feedstocks.

Guy B. Marin is senior full professor in Catalytic Reaction Engineering and is head of the Laboratory for Chemical Technology at Ghent University. He was educated in the tradition of the thermodynamic and kinetic school of the Low Countries as well as of the American school, with Michel Boudart as a postdoctoral adviser, and benefited from the Dutch school of catalysis. The investigation of chemical kinetics, aimed at the modeling and design of chemical processes and products all the way from molecule up to full scale, constitutes the core of his research.

The authors abbreviated the title of this book to ADAMICE, which can suggest several phonetic associations, the most primordial of which refers to Adam. It is quite symbolic that Adam's name in Hebrew shares an etymology with clay and blood, both of which are catalytic substrates in chemistry and biochemistry. Maybe we can think of Adam as the first ever model, and a very advanced model, too.

We gratefully acknowledge the outstanding skills, tireless activity, patience, and understanding of our editing aide, Mrs Annelies van Diepen.

## CHAPTER 1

Introduction

1.1 Chemistry and Mathematics: Why They Need Each Other

Chemistry needs mathematics. Mathematics needs chemistry. Since the Dark Ages, in chemistry, or alchemy as it was called then, the so-called cross rule was well known for answering questions regarding the preparation of solutions of a certain desired concentration. This rule is not needed to answer simple questions, like which ratio of 100% alcohol and pure water must be mixed to prepare a 50% alcohol solution: 50:50. For more complicated questions, however, such as how to prepare a 30% alcohol solution from two solutions with concentrations of 90% and 10% alcohol, the cross rule comes in handy (see Scheme 1.1A).

The rule works as follows. Write the desired concentration in the middle and the input concentrations on the left. Then calculate the difference between the number on the top left (90) and the one in the middle (30), and then write that number in the bottom right (60). Then write the difference between the number on the bottom left (10) and the one in the middle (30) in the top right (20). From this it follows that the solutions need to be mixed in a ratio of 20:60=1:3. Scheme 1.1B shows the cross rule in generalized form, with $\alpha > \gamma > \beta$.

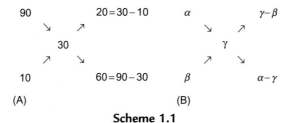

(A)   (B)

Scheme 1.1

Cross rule: (A) example and (B) generalized form.

Now, how can we prove this rule? Our procedure should meet the requirement:

$$a\alpha + b\beta = \gamma \qquad (1.1)$$

where α and β are known concentrations of solutions I and II and γ is the concentration of the desired solution III. The constants a and b are the parts of solutions I and II that are needed for preparing solution III. Obviously, the balance equation

$$a+b=1 \quad \text{or} \quad b=1-a \quad (\text{or } a+b=100\%) \qquad (1.2)$$

Advanced Data Analysis and Modelling in Chemical Engineering. http://dx.doi.org/10.1016/B978-0-444-59485-3.00001-1

must hold. Substituting Eq. (1.2) into Eq. (1.1) yields

$$a\alpha + (1-a)\beta = \gamma \tag{1.3}$$

from which follows that

$$a = \frac{\gamma - \beta}{\alpha - \beta} \tag{1.4}$$

Consequently,

$$b = 1 - a = 1 - \frac{\gamma - \beta}{\alpha - \beta} = \frac{\alpha - \gamma}{\alpha - \beta} \tag{1.5}$$

and the ratio is given by

$$a : b = \left(\frac{\gamma - \beta}{\alpha - \beta}\right) : \left(\frac{\alpha - \gamma}{\alpha - \beta}\right) = (\gamma - \beta) : (\alpha - \gamma) \tag{1.6}$$

which is the cross rule.

Thus, mathematics is needed for solving typical problems of chemistry. Such problems are solved in every chemical laboratory, every pharmacy, and even in most kitchens. To do so, in this case we are using what is known as linear algebra.

Mathematics needs chemistry too, not only for demonstrating its power and usefulness but also for formulating new mathematical problems.

1.2 Chemistry and Mathematics: Historical Aspects

A mutual attraction between mathematics and chemistry was growing gradually during the 18th, 19th, and 20th centuries. Antoine Lavoisier's law of the conservation of mass and John Dalton's law of multiple proportions could not have been discovered without precise weight measurements. Chemistry transformed into a quantitative science, and Lavoisier and Dalton became pioneers of chemical stoichiometry. From another side, Arthur Cayley, the creator of the theory of matrices, applied his theory to the study of chemical isomers in the 1860s and 1870s.

Since the mid-19th century, various new concepts of physical chemistry, in particular laws of chemical kinetics, have been developed. A few examples are

- the mass-action law by Guldberg and Waage.
- the "normal classification of reactions" by van't Hoff.
- the temperature dependence of the reaction rate by Arrhenius.
- electrochemical relationships by Nernst.
- quasi-steady-state principle by Chapman and Bodenstein.
- theory of chain reactions by Semenov and Hinshelwood.
- model for the growth of microorganisms by Monod.

All these principles have been proposed based on the solid mathematical foundation of algebraic and differential equations. Moreover, the greatest chemical achievement, Mendeleev's periodic table, was created using a detailed critical review of quantitative data, first of all, the atomic masses of known elements. A legend says that after the presentation of his table in 1869, Mendeleev was asked, "Why did you categorize the elements based on their mass and not on their first letter?" In fact, the mass of an element is not a chemical property; it is not its ability to react with other substances, nor its likeliness to explode, and it is not even a smell. So the number of an element in Mendeleev's periodic table remained very formal—just a number, the element's place in the table—until, in 1913, Moseley discovered a mathematical relationship between the wavelength of X-rays emitted by an element and its atom number. This discovery was a great step in the understanding that the number of the element in the periodic table is equal to the number of electrons (and the number of protons) belonging to a single atom of this element.

During the "Sturm-und-Drang" quantum period of the 1920s, Heisenberg's matrix mechanics and Schrödinger's equation have been established as "very mathematized" cornerstones of the rigorous theory of elementary chemical processes, which changed the field of chemistry dramatically.

Finally, in the 20th century, researchers discovered many critical chemical phenomena. In the first half of the century, these were mainly of a nonisothermal nature (Semenov, Bodenstein, Hinshelwood), while in the second half they concerned isothermal phenomena, in particular oscillating reactions (Belousov, Zhabotinsky, Prigogine, Ertl). In the 1950–60s special attention was paid to studying very fast reactions by the relaxation technique (Eigen). All reaction and reaction-diffusion nonsteady-state data have been interpreted based on the dynamic theories proposed by prominent mathematicians of this century, including Poincaré and Lyapunov, Andronov, Hopf, and Lorenz.

1.3 Chemistry and Mathematics: New Trends

A new period of the interaction between chemistry and mathematics was shaped out by the computer revolution of the 1950–60s. Making a general statement about the difference between the classical and contemporary scientific situations, one can carefully say that

- The classical situation was characterized by a search for general scientific laws. Albert Einstein described the discovery of these new laws as "a flight from the miracle": "The development of this world of thoughts is in a certain sense a continuous flight from the miracle."
- The focus of the contemporary scientific situation has changed. From the epoch of the great scientific revolutions we have moved to the epoch of the intellectual devices, mostly computers, from the Divine Plan to a collection of models.

This type of division had to be made carefully: never say "never" and a period of great scientific discoveries may still lie ahead. Nevertheless, chemistry and chemical engineering have been much influenced by the computer revolution, which is still ongoing. Apparently, this has been a strong additional factor for the shaping of the new discipline "mathematical chemistry." Although this term was used previously by Lomenosov in the 18th century, the systematic development of this area only started in the 1950s. At this time, Neal Amundson and Rutherford Aris published the first papers and books describing the corresponding mathematical models in much detail and creating the foundation of the discipline (Aris, 1965, 1969, 1975; Aris and Amundson, 1958a,b,c).

Presently, a whole battery of efficient computational methods has been developed for modeling chemical systems at different levels (nano-, micro-, meso-, and macrolevel). We are now able to cover a variety of chemical aspects using an enormous number of available methods, such as

- computational chemistry, in particular methods based on density functional theory.
- computational chemical kinetics, in particular CHEMKIN software and various "stiff" methods of solving sets of ordinary and partial differential equations.
- computational fluid dynamics.
- sophisticated statistical methods.

From the other side, during this period many rigorous theoretical results in mathematical chemistry have been obtained, especially in the area of chemical dynamics. Wei and Prater (1962) developed a concept of "lumping," presenting a general result for first-order mass action-law systems. Many results have been obtained at the boundary between chemical thermodynamics and chemical kinetics. In 1972, Horn and Jackson (1972) posed the problem of searching the relationships between the structure of the detailed mechanism and its kinetic behavior. This theory was developed further in the 1970s and 1980s by Yablonsky et al. (Gorban et al., 1986; Yablonskii et al., 1983, 1991) and independently by Feinberg (1987, 1989), Clarke (1974), and Ivanova (1979). Horn (1964) was the first to pose the problem of attainable regions for chemical processes. In the 1970s, Gorban constructed a theory of "thermodynamically unattainable" regions, that is, regions that are impossible to reach from certain initial conditions (Glansdorff and Prigogine, 1971; Gorban, 1984). For more details on this chenucogeometric theory based on thermodynamic Lyapunov functions (see Gorban, 1984; Yablonsky et al., 1991).

Theoretical breakthroughs in the understanding of chemical critical phenomena (multiplicity, oscillations, waves) were achieved by different groups of chemists. The Belgian school of irreversible thermodynamics (Prigogine and Nicolis) provided insight into the origin of these phenomena based on the concepts of the "thermodynamic stability criterion" and "dissipative structure" (Glansdorff and Prigogine, 1971; Nicolis and Prigogine, 1977; Prigogine, 1967). The group led by Aris (Minneapolis, Minnesota, USA) presented a uniquely scrupulous study of the dynamic properties of the nonisothermal continuous stirred-tank reactor (CSTR) based on bifurcation theory (Farr and Aris, 1986). The first part of this paper's title, "Yet who would have thought the old man to have had so much blood in him?" is a quotation from Shakespeare's *Macbeth*. Finally, Ertl and coworkers (Berlin, Germany) constructed a theoretical model of kinetic oscillations in the oxidation of carbon monoxide over platinum (Ertl et al., 1982; Ertl, 2007), which may be called the "Mona Lisa of heterogeneous catalysis" as it has been an attractive object of catalysis studies for a long time. The model developed by Ertl et al. (1982) describes many peculiarities of this reaction, including surface waves.

Two recent trends in mathematical chemistry are model reduction and chemical calculus. In model reduction, two approaches are popular: manifold analysis (Maas and Pope, 1992) and so-called asymptotology of chemical reaction networks (Gorban et al., 2010), which is a generalization of the concept of a rate-limiting step. The chemical calculus approach, which was created in collaboration between Washington University in St. Louis, Missouri (USA) and Ghent University (Belgium), is focused on precise catalyst characterization using pulse-response data generated in a temporal analysis of products (TAP) reactor (Constales et al., 2001; Gleaves et al., 1988, 1997; Yablonsky et al., 2007). With the TAP method, during an experiment the change of the catalyst composition as a result of a gas pulse is insignificant. At present, chemical calculus can be considered as a mathematical basis for precise nonsteady-state characterization of the chemical activity of solid materials. For a critical review of theoretical results, see Marin and Yablonsky (2011).

Despite evident progress in the development of both computational and theoretical areas, there still are only a few achievements that can be considered as fruits of synergy between the essential physicochemical understanding, a rigorous mathematical analysis and efficient computing. When we talk about "advanced data analysis," the adjective "advanced" means an attempt to achieve this unachievable ideal. From this point of view, our book is such an attempt. In every chapter, we try to present different constituents of the modeling of chemical systems, such as a physicochemical model framework, some aspects of data analysis including primary data analysis, and rigorous results obtained by mathematical analysis.

The term "advanced" may also relate to a new, high level of physicochemical understanding and it does not necessarily entail a more complex model. In many situations, "to understand" means "to simplify" and we are not afraid to present simple but essential knowledge. Finally, to be "advanced" means to be able to pose new practical and theoretical questions and answer them. Formulating new questions is extremely important for the progress of our science, which is still, as chemists say, "in statu nascendi."

1.4 Structure of This Book and Its Building Blocks

Now we are going to briefly discuss the contents of this book, explaining its logic and reasoning. Chapters 2–8 are organized as a sequence of building blocks: "chemical composition"—"complex chemical reaction"—"chemical reactor." Chapter 9 complements these chapters, providing information about model discrimination and parameter estimation methods. Chapter 10 deals with specific problems of modeling polymerization processes and Chapter 11 describes advanced theoretical analysis focusing on computer algebra. Next, a more detailed, chapter-by-chapter description of the book follows.

In Chapter 2, a new original approach is presented based on a new unified C-matrix, which generates three main matrices of chemical systems representing both chemical composition and complex chemical transformation. These three matrices are the molecular matrix, the stoichiometric matrix, and the matrix of Horiuti numbers, and the original algorithm is derived in this chapter.

In Chapter 3, the theoretical "minimum minimarum" is explained. This toolbox serves the primary analysis of data of chemical transformations and the construction of typical kinetic models, both steady-state and nonsteady-state. In the course of the primary analysis, the reaction rate is extracted from the observed net rate of production of a chemical component. Although the procedure is rather simple, its understanding is not trivial.

Using methods from graph theory, it is explained which part of the kinetic description is influenced by the complexity of the detailed reaction mechanism and which part is not. This is an important step in the so-called "gray-box" approach, which is widely applied in chemical engineering modeling.

As the title of Chapter 4 suggests, this chapter lays out physicochemical principles of simplifications of complex models, that is, assumptions on quasi-equilibrium and quasi-steady-state, assumptions on the abundance of some chemical components, and the assumption of a rate-limiting step. These principles are systematically used in the primary analysis of complex models.

Chapter 5 is devoted to the basic reactor models used in chemical engineering: the batch reactor, the CSTR, the plug-flow reactor, and the TAP reactor with its modification, the thin-zone TAP reactor. Special attention is paid to the reaction-diffusion reactor, the detailed analysis of which opens up wide perspectives for the understanding of different types of reactors, such as catalytic, membrane and biological reactors.

Chapter 6 reflects on new areas and problems at the boundary of chemical thermodynamics and chemical kinetics, that is, estimating features of kinetic behavior based on thermodynamic characteristics. New methods of chemico-geometric analysis are described. Recently found, original, equilibrium relationships between nonequilibrium data, which can be observed in special experiments with symmetrical initial conditions, are theoretically analyzed in detail for different types of chemical reactors. The known problem of kinetic control versus thermodynamic control is explained in a new way.

Chapter 7 is about stability and bifurcations. Typical catalytic mechanisms and corresponding models for the interpretation of complex dynamic behavior, in particular multiplicity of steady states and oscillations, are presented.

Chapter 8 provides original results for optimal configurations of multizone reaction-diffusion reactors, and compares these with results for a cascade of reactors known from the literature.

Chapter 9 offers a classical toolbox of model discrimination and parameter identification methods. Such methods are constantly modified and they are a "golden treasure" to every modeler.

Chapter 10 provides a formal apparatus, in particular integrodifferential equations, for describing the physicochemical properties of polymers and their complex transformations.

Finally, the focus of Chapter 11 is on advanced methods of theoretical analysis, especially computer algebra. Most of the results presented in this chapter, that is, critical simplification, the analytical criteria for distinguishing nonlinear from linear behavior, are original.

Summarizing, the building of mathematical chemistry has not been completed. This book covers only a number of selected rooms of this structure. It is not easy to live in a building that has been only partly completed, but there is no alternative, and generally in science this is a typical story. In addition, in many countries of the Near East where rainfall is scarce, it is customary for families to move into new houses long before all rooms have been prepared. So, we are moving...

References

Aris, R., 1965. Introduction to the Analysis of Chemical Reactors. Prentice-Hall, Englewood Cliffs, NJ.

Aris, R., 1969. Elementary Chemical Reactor Analysis. Prentice-Hall, Englewood Cliffs, NJ.

Aris, R., 1975. The Mathematical Theory of Diffusion and Reaction in Permeable Catalysts. Clarendon Press, Oxford.

Aris, R., Amundson, N.R., 1958a. An analysis of chemical reactor stability and control-I: the possibility of local control, with perfect or imperfect control mechanisms. Chem. Eng. Sci. 7, 121–131.

Aris, R., Amundson, N.R., 1958b. An analysis of chemical reactor stability and control-II: the evolution of proportional control. Chem. Eng. Sci. 7, 132–147.

Aris, R., Amundson, N.R., 1958c. An analysis of chemical reactor stability and control-III: the principles of programming reactor calculations. Some extensions. Chem. Eng. Sci. 7, 148–155.

Clarke, B.L., 1974. Stability analysis of a model reaction network using graph theory. J. Chem. Phys. 60, 1493–1501.

Constales, D., Yablonsky, G.S., Marin, G.B., Gleaves, J.T., 2001. Multi-zone TAP-reactors theory and application: I. The global transfer matrix equation. Chem. Eng. Sci. 56, 133–149.

Ertl, G., (2007). Reactions at surfaces: from atoms to complexity, Nobel Lecture, December 8.

Ertl, G., Norton, P.R., Rüstig, J., 1982. Kinetic oscillations in the platinum-catalyzed oxidation of CO. Phys. Rev. Lett. 49, 177–180.

Farr, W.W., Aris, R., 1986. Yet who would have thought the old man to have had so much blood in him? Reflections on the multiplicity of steady states of the stirred tank reactor. Chem. Eng. Sci. 41, 1385–1402.

Feinberg, M., 1987. Chemical reaction network structure and the stability of complex isothermal reactors—I. The deficiency zero and deficiency one theorems. Chem. Eng. Sci. 42, 2229–2268.

Feinberg, M., 1989. Necessary and sufficient conditions for detailed balancing in mass action systems of arbitrary complexity. Chem. Eng. Sci. 44, 1819–1827.

Glansdorff, P., Prigogine, I., 1971. Thermodynamic Theory of Structure, Stability, and Fluctuations. Wiley, London.

Gleaves, J.T., Ebner, J.R., Kuechler, T.C., 1988. Temporal analysis of products (TAP)—a unique catalyst evaluation system with submillisecond time resolution. Catal. Rev. Sci. Eng. 30, 49–116.

Gleaves, J.T., Yablonskii, G.S., Phanawadee, P., Schuurman, Y., 1997. TAP-2: an interrogative kinetics approach. Appl. Catal. A 160, 55–88.

Gorban, A.N., 1984. Equilibrium Encircling: Equations of Chemical Kinetics and Their Thermodynamic Analysis. Nauka, Novosibirsk.

Gorban, A.N., Bykov, V.I., Yablonskii, G.S., 1986. Thermodynamic function analogue for reactions proceeding without interaction of various substances. Chem. Eng. Sci. 41, 2739–2745.

Gorban, A.N., Radulescu, O., Zinovyev, A.Y., 2010. Asymptotology of chemical reaction networks. Chem. Eng. Sci. 65, 2310–2324.

Horn, F.J.M., 1964. Attainable and Non-Attainable Regions in Chemical Reaction Technique. In: Proceedings of the third European Symposium on Chemical Reaction Engineering. Pergamon Press, London, pp. 1–10.

Horn, F., Jackson, R., 1972. General mass action kinetics. Arch. Rat. Mech. Anal. 47, 81–116.

Ivanova, A.N., 1979. Conditions of uniqueness of steady state related to the structure of the reaction mechanism. Kinet. Katal. 20, 1019–1023 (in Russian).

Maas, U., Pope, S.B., 1992. Simplifying chemical kinetics: intrinsic low-dimensional manifolds in composition space. Combust. Flame 88, 239–264.

Marin, G.B., Yablonsky, G.S., 2011. Kinetics of Chemical Reactions—Decoding Complexity. Wiley-VCH, Weinheim.

Nicolis, G., Prigogine, I., 1977. Self-Organization in Non-Equilibrium Systems: From Dissipative Structures to Order Through Fluctuations. Wiley, New York, NY.

Prigogine, I., 1967. Introduction to Thermodynamics of Irreversible Processes, third ed. Wiley-Interscience, New York, NY.

Wei, J., Prater, C.D., 1962. The structure and analysis of complex reaction systems. In: Eley, D.D. (Ed.), Advances in Catalysis. Academic Press, New York, NY, pp. 203–392.

Yablonskii, G.S., Bykov, V.I., Gorban, A.N., 1983. Kinetic Models of Catalytic Reactions. Nauka, Novosibirsk.

Yablonskii, G.S., Bykov, V.I., Gorban, A.N., Elokhin, V.I., 1991. Kinetic Models of Catalytic Reactions. Comprehensive Chemical Kinetics, vol. 32. Elsevier, Amsterdam.

Yablonsky, G.S., Constales, D., Shekhtman, S.O., Gleaves, J.T., 2007. The Y-procedure: how to extract the chemical transformation rate from reaction-diffusion data with no assumption on the kinetic model. Chem. Eng. Sci. 62, 6754–6767.

## CHAPTER 2

Chemical Composition and Structure: Linear Algebra

2.1 Introduction

The following questions often arise in the practice of chemical engineering:

- Given a set of components of known atomic composition, establish which of them are key to determine, together with the element balances, the amounts of all others. Furthermore, find a way of generating all possible reactions involving these components. This will be addressed by the augmented molecular matrix, introduced in Section 2.2.
- Given a set of reactions, determine which of these are key to describing all other reactions as combinations of this set and how to do this. Also determine which additional balances appear and which thermodynamic characteristics (enthalpy change, equilibrium coefficient) are dependent. This will be addressed by the augmented stoichiometric matrix, introduced in Section 2.3.
- For a set of reactions involving intermediates, which typically cannot be measured, determine the overall reactions, that is, the ones not involving such intermediates, and find the numbers by which these can be written as combinations of the given reactions (the so-called Horiuti numbers). This will be addressed by augmenting a specially crafted stoichiometric matrix, as explained in Section 2.4.

What is a *matrix?* It is a square or rectangular array of numbers or expressions, placed between parentheses, for example:

$$\begin{bmatrix} 1 & 3 & -7 \\ 2 & -4 & 3 \end{bmatrix} \tag{2.1}$$

In this case there are two rows and three columns, and we call this a 2×3 matrix, mentioning first the number of rows and then the number of columns. There can be no empty spaces left in the matrix. Matrices are widely used in engineering to represent a variety of mathematical objects and relationships. We will gradually introduce them, in close connection with the requirements from chemistry and engineering that justify their use.

Let us start with chemical reactions. A statement such as

$$2H_2 + O_2 \rightarrow 2H_2O \tag{2.2}$$

Advanced Data Analysis and Modelling in Chemical Engineering. http://dx.doi.org/10.1016/B978-0-444-59485-3.00002-3

describes a chemical process, a reaction, that consumes two molecules of hydrogen gas, H_2, and one molecule of oxygen gas, O_2, to produce two molecules of water, H_2O. Such a reaction does not actually occur as the collision of the three gas molecules involved, but rather as a complicated, not yet fully elucidated set of reaction steps. Nevertheless, Eq. (2.2) describes the overall reaction and can readily be tested experimentally: two moles of hydrogen and one mole of oxygen are transformed into two moles of water. The statement in Eq. (2.2) implies that all H_2 and O_2 should be completely transformed into H_2O so that no H_2 or O_2 should be left. Thus, Eq. (2.2) is a falsifiable statement, and therefore scientific.

If we ignore the direction of the arrow in Eq. (2.2), it can be viewed as an equality of sorts:

$$2H_2 + O_2 \triangleq 2H_2O \tag{2.3}$$

where "$\triangleq$"denotes the equality of all atom counts on both sides. This is not an absolute equality: O_2 and $2H_2$ together are not the same as $2H_2O$, but they are amenable to transformation into each other. Eq. (2.3) can be rewritten as

$$-2H_2 - O_2 + 2H_2O \triangleq zero \tag{2.4}$$

By convention, the terms that originate from the left-hand side, the reactants, are assigned a negative sign, and the products are assigned a positive sign. Maybe we could associate this equation to a (1×3) matrix? The matrix

$$\begin{bmatrix} -2 & -1 & 2 \end{bmatrix} \tag{2.5}$$

would be the most natural choice, but since the meaning of the columns (H_2, O_2, and H_2O) would be lost, we introduce a typographical convention for indicating the meaning of the columns:

$$\begin{matrix} H_2 & O_2 & H_2O \\ \begin{bmatrix} -2 & -1 & 2 \end{bmatrix} \end{matrix} \tag{2.6}$$

Note that the chemical symbols in Eq. (2.6) are hints, *not* part of the matrix!

Another reaction might occur between the same reactants, namely the synthesis of hydrogen peroxide:

$$H_2 + O_2 \rightarrow H_2O_2 \tag{2.7}$$

This reaction can be represented by the matrix:

$$\begin{matrix} H_2 & O_2 & H_2O_2 \\ \begin{bmatrix} -1 & -1 & 1 \end{bmatrix} \end{matrix} \tag{2.8}$$

and it is logical to represent the system of both reactions, Eqs. (2.2), (2.7), by the 2×4 matrix:

$$
\begin{array}{cccc}
H_2 & O_2 & H_2O & H_2O_2 \\
\end{array}
$$
$$
\begin{bmatrix}
-2 & -1 & 2 & 0 \\
-1 & -1 & 0 & 1
\end{bmatrix}
\tag{2.9}
$$

Note that for this matrix the rows are not labeled. We could add more hints by adding the names of the reaction products to the left of the matrix, as in

$$
\begin{array}{cccc}
& H_2 & O_2 & H_2O & H_2O_2 \\
H_2O & \begin{bmatrix} -2 & -1 & 2 & 0 \\ H_2O_2 & -1 & -1 & 0 & 1 \end{bmatrix}
\end{array}
\tag{2.10}
$$

but initially we will consider the rows of this stoichiometric matrix as merely stoichiometric information: how many molecules of reactants (negative entries) produce how many molecules of products (positive entries)? Components that do not occur in a particular reaction have a zero entry in that row. Reactions are the building blocks of a chemical description, but before we can handle them more formally, we must also express how each component combines its constituent elements. This is done in the molecular matrix.

2.2 The Molecular Matrix and Augmented Molecular Matrix

2.2.1 The Molecular Matrix

For every component present in the reaction mixture under consideration, its elemental composition must be specified in order to verify that all elements are properly preserved. The most convenient and complete way of specifying the elemental composition is by a matrix in which the columns correspond to the elements, with possibly one column reserved for electric charge, and the rows signify the different molecules, ions, radicals, and so forth, that are being considered. This is called the *molecular matrix* **M**. For instance, for a mixture consisting of H_2, CH_4, C_2H_6, and C_2H_4, the molecular matrix is given by

$$
\mathbf{M} =
\begin{array}{c}
\\
H_2 \\
CH_4 \\
C_2H_6 \\
C_2H_4
\end{array}
\begin{array}{c}
C\ H \\
\begin{bmatrix}
0 & 2 \\
1 & 4 \\
2 & 6 \\
2 & 4
\end{bmatrix}
\end{array}
\tag{2.11}
$$

Based on this molecular matrix, the molar masses of the components can be determined from the atomic masses of the elements. For instance, the molar mass of C_2H_6 is two times the atomic mass of C plus six times that of H. To express such relationships in a very compact manner, mathematicians have devised a special way of multiplying matrices: the atomic masses of the elements can be grouped in a column vector $\mathbf{m_A}$ (in this case consisting of two rows),

$$\mathbf{m_A} = \begin{matrix} H \\ C \end{matrix} \begin{bmatrix} 1.008 \\ 12 \end{bmatrix} \tag{2.12}$$

and the molar masses of the components can go in a 4×1 column vector $\mathbf{m_M}$:

$$\mathbf{m_M} = \begin{matrix} H_2 \\ CH_4 \\ C_2H_6 \\ C_2H_4 \end{matrix} \begin{bmatrix} 2.016 \\ 16.032 \\ 30.048 \\ 28.032 \end{bmatrix} \tag{2.13}$$

Then in compact form, $\mathbf{m_M} = \mathbf{Mm_A}$. In general, a product of two matrices $\mathbf{B}$ and $\mathbf{C}$ is defined only if the number of columns of $\mathbf{B}$ equals the number of rows of $\mathbf{C}$, and the product $\mathbf{P} = \mathbf{BC}$ will have the number of rows of $\mathbf{B}$ and the number of columns of $\mathbf{C}$. A general entry P_{ij} of $\mathbf{P}$ is then given by definition as a sum of products of entries of row i of $\mathbf{B}$ and entries of column j of $\mathbf{C}$:

$$P_{ij} = \mathbf{B}_{i1}\mathbf{C}_{1j} + \mathbf{B}_{i2}\mathbf{C}_{2j} + \cdots + \mathbf{B}_{ip}\mathbf{C}_{pj} \tag{2.14}$$

where p is the number of columns of $\mathbf{B}$ (and rows of $\mathbf{C}$).

Furthermore, when a set of mathematical objects $\nu_1, \nu_2, \ldots, \nu_n$ can be multiplied by real numbers a_i and the sum of the products satisfies the equation $a_1\nu_1 + a_2\nu_2 + \cdots + a_n\nu_n = 0$, where not all real numbers $a_1, a_2, \ldots, a_n$ are zero, they are said to be *linearly dependent*, otherwise they are called *linearly independent*. For instance, the equations $x + y = 5$ and $x - y = 9$ are linearly independent, but if we add the equation $x = 7$, they become dependent, since $x + y = 5$ plus $x - y = 9$ minus twice $x = 7$ yields $0 = 0$. A set of linearly independent objects that is maximal, that is, to which no object can be added without losing linear independence, is called a *basis*. In the previous example, $x + y = 5$ by itself is not a basis since $x - y = 9$ can be added to it without losing linear independence, but the set of these two equations does form a basis since no third equation can be added to this set independently.

2.2.2 Application of the Molecular Matrix to Element Balances

In practice, it is often crucial to calculate the balance for each of the elements present in the components of a mixture. Now the questions arise how many of these balances are independent and which amounts can be deduced from other ones given the element balance values. To solve this problem in a systematic way, a certain ordering of the components must be chosen; typically, the best known or most easily measured components should be listed first. Then the molecular matrix can be augmented by adding a unit matrix, that is, a matrix with ones on the main diagonal and zeros elsewhere, to its right. For the molecular matrix represented by Eq. (2.11), we obtain

$$
\begin{array}{c}
\\
H_2 \\
CH_4 \\
C_2H_6 \\
C_2H_4
\end{array}
\begin{array}{cccccc}
C_t & H_t & H_2 & CH_4 & C_2H_6 & C_2H_4 \\
\left[\begin{array}{cccccc}
0 & 2 & 1 & 0 & 0 & 0 \\
1 & 4 & 0 & 1 & 0 & 0 \\
2 & 6 & 0 & 0 & 1 & 0 \\
2 & 4 & 0 & 0 & 0 & 1
\end{array}\right]
\end{array}
\tag{2.15}
$$

This matrix is best interpreted as a form of double bookkeeping as used in accounting: in the first two columns, the counts of the elements C and H, now called C_t and H_t, for total C and H, are tallied (debit); in the next four columns, the components are figured (credit).

An augmented molecular matrix can be transformed to a *Reduced Row Echelon Form* or *RREF*. This method is essential to all matrix transformations in this chapter. The idea behind the RREF is that we work from the first column all the way to the rightmost one. For each column we determine whether it is possible to eliminate it by finding a nonzero entry, or *pivot,* in a row that has not been considered before. If not, we skip to the next column. If a pivot is found, we use it to eliminate all other entries in that row. We also move the pivot row up as far as possible. We cannot tell in advance where all the pivots will be found; we must find them one by one since the elimination procedure can change zero entries into nonzero ones and vice versa. In general, we also do not know in advance how many pivots will be found. However, in the special case of a matrix augmented with a unit matrix, we do know that their number will be equal to the number of rows.

As an example, consider the augmented matrix for a mixture consisting of H_2, CH_4, C_2H_6, and C_2H_4 given by Eq. (2.15). In order to find the RREF, we start with the first column and find a nonzero entry in the second row. We move this second row up to the first position:

$$
\begin{array}{cccccc}
C_t & H_t & H_2 & CH_4 & C_2H_6 & C_2H_4 \\
\left[\begin{array}{cccccc}
1 & 4 & 0 & 1 & 0 & 0 \\
0 & 2 & 1 & 0 & 0 & 0 \\
2 & 6 & 0 & 0 & 1 & 0 \\
2 & 4 & 0 & 0 & 0 & 1
\end{array}\right]
\end{array}
\tag{2.16}
$$

The pivot is in bold. Note that the hints labeling the rows are absent, since their identity will not be preserved during the reduction of the matrix.

Next, we use the first row to eliminate all other entries of the first column. The second row requires no action since the value there is already zero. We subtract twice the first row from the third row, changing the matrix to

$$\begin{array}{cccccc} C_t & H_t & H_2 & CH_4 & C_2H_6 & C_2H_4 \end{array}$$
$$\begin{bmatrix} \mathbf{1} & 4 & 0 & 1 & 0 & 0 \\ 0 & 2 & 1 & 0 & 0 & 0 \\ 0 & -2 & 0 & -2 & 1 & 0 \\ 2 & 4 & 0 & 0 & 0 & 1 \end{bmatrix} \tag{2.17}$$

Similarly, we subtract twice the first row from the fourth row, resulting in

$$\begin{array}{cccccc} C_t & H_t & H_2 & CH_4 & C_2H_6 & C_2H_4 \end{array}$$
$$\begin{bmatrix} \mathbf{1} & 4 & 0 & 1 & 0 & 0 \\ 0 & 2 & 1 & 0 & 0 & 0 \\ 0 & -2 & 0 & -2 & 1 & 0 \\ 0 & -4 & 0 & -2 & 0 & 1 \end{bmatrix} \tag{2.18}$$

Now we tackle the second column. The first row has already been assigned a pivot, so we do not consider it for elimination purposes. In the second row, we find a nonzero value, 2, by which we divide the entries in this row:

$$\begin{array}{cccccc} C_t & H_t & H_2 & CH_4 & C_2H_6 & C_2H_4 \end{array}$$
$$\begin{bmatrix} \mathbf{1} & 4 & 0 & 1 & 0 & 0 \\ 0 & \mathbf{1} & 1/2 & 0 & 0 & 0 \\ 0 & -2 & 0 & -2 & 1 & 0 \\ 0 & -4 & 0 & -2 & 0 & 1 \end{bmatrix} \tag{2.19}$$

Now we use this second row for the elimination of the entries in the other rows of the second column, including the first row. So we subtract four times the second row from the first, add twice the second row to the third, and add four times the second row to the fourth. The result of these operations is

$$\begin{array}{cccccc} C_t & H_t & H_2 & CH_4 & C_2H_6 & C_2H_4 \end{array}$$
$$\begin{bmatrix} \mathbf{1} & 0 & -2 & 1 & 0 & 0 \\ 0 & \mathbf{1} & 1/2 & 0 & 0 & 0 \\ 0 & 0 & 1 & -2 & 1 & 0 \\ 0 & 0 & 2 & -2 & 0 & 1 \end{bmatrix} \tag{2.20}$$

When looking at the third column to find a pivot, we do not consider the first and second rows because they have already been assigned pivots. We do, however, find a suitable entry, 1, in the third row and use this as a pivot. We add twice the third row to the first, subtract half of it from the second, and subtract twice the third row from the fourth.

This results in

$$\begin{array}{cccccc} C_t & H_t & H_2 & CH_4 & C_2H_6 & C_2H_4 \end{array}$$
$$\begin{bmatrix} 1 & 0 & 0 & -3 & 2 & 0 \\ 0 & 1 & 0 & 1 & -1/2 & 0 \\ 0 & 0 & 1 & -2 & 1 & 0 \\ 0 & 0 & 0 & 2 & -2 & 1 \end{bmatrix} \qquad (2.21)$$

Finally, in the fourth column we do not consider the first three rows, because they already have pivots, but find a nonzero value in the fourth row, 2, and choose it as a pivot. First we divide the entries in the fourth row by this value,

$$\begin{array}{cccccc} C_t & H_t & H_2 & CH_4 & C_2H_6 & C_2H_4 \end{array}$$
$$\begin{bmatrix} 1 & 0 & 0 & -3 & 2 & 0 \\ 0 & 1 & 0 & 1 & -1/2 & 0 \\ 0 & 0 & 1 & -2 & 1 & 0 \\ 0 & 0 & 0 & 1 & -1 & 1/2 \end{bmatrix} \qquad (2.22)$$

and then add three times the fourth row to the first, subtract it once from the second, and add it twice to the third:

$$\begin{array}{cccccc} C_t & H_t & H_2 & CH_4 & C_2H_6 & C_2H_4 \end{array}$$
$$\begin{bmatrix} 1 & 0 & 0 & 0 & -1 & 3/2 \\ 0 & 1 & 0 & 0 & 1/2 & -1/2 \\ 0 & 0 & 1 & 0 & -1 & 1 \\ 0 & 0 & 0 & 1 & -1 & 1/2 \end{bmatrix} \qquad (2.23)$$

We have now obtained the RREF. Fig. 2.1 shows a map of this matrix indicating the subareas.

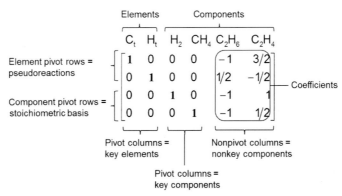

Fig. 2.1
Map of subareas in the RREF of Eq. (2.23).

Because the molecular matrix was augmented with a unit matrix, the pivot columns together form a unit submatrix indicating the *key elements* and *key components*. Key elements of a certain chemical mixture are those the amounts of which cannot be determined from the amounts of other elements; key components are those the amounts of which, together with the amounts of the key elements, uniquely determine the amounts of the other, nonkey components.

Every row has a pivot; the first two rows have pivots in element columns and are related to pseudoreactions in which precisely one atom of the element is "created" as the result of a transformation of components. Contrary to ordinary chemical reactions, there is no balance of the pivotal element in these pseudoreactions, but instead an excess of a single atom in the products. For instance, the first row in Eq. (2.23) represents the pseudoreaction $C_2H_6 \rightleftarrows 3/2\ C_2H_4$ with an excess of one C atom in the product: reactants and products contain two and three C atoms, respectively. Similarly, the pseudoreaction in the second row, $1/2\ C_2H_4 \rightleftarrows 1/2\ C_2H_6$, has an excess of one H atom (two H atoms in reactants, three in products).

The pivots of the third and fourth rows are not elements but the components H_2 and CH_4. They represent the reactions $C_2H_6 \rightleftarrows H_2 + C_2H_4$, in which a single unit of H_2 is formed, and $C_2H_6 \rightleftarrows CH_4 + 1/2\,C_2H_4$, in which a single unit of CH_4 is formed. These rows corresponding to reactions form a stoichiometric basis for all reactions involving the components considered.

The first four columns are pivot columns, in which only entries 1 occur as pivots and all other entries are zero. The fifth and sixth columns have no pivot, and correspond to components that are not key. Their amounts can be deduced from the element balances and the key components. The coefficients can be read off at once in the columns, which were obtained from the pseudoreactions and reactions. In the fifth column, the amount of C_2H_6 equals

$$n_{C_2H_6} = -n_{C_t} + 1/2 n_{H_t} - n_{H_2} - n_{CH_4} \tag{2.24}$$

and in the sixth, the amount of C_2H_4 equals

$$n_{C_2H_4} = 3/2 n_{C_t} - 1/2 n_{H_t} + n_{H_2} + 1/2 n_{CH_4} \tag{2.25}$$

In this example, if we assume that $n_{C_2H_4}$ can be written as a linear combination:

$$n_{C_2H_4} = \alpha n_{C_t} + \beta n_{H_t} + \gamma n_{H_2} + \delta n_{CH_4} \tag{2.26}$$

and consider the effect of the first pseudoreaction, $C_2H_6 \rightleftarrows 3/2\,C_2H_4$, in which C is "created" and H_2 and CH_4 do not participate, n_{C_t} increases by 1, $n_{C_2H_4}$ by 3/2, and n_{H_t}, n_{H_2}, and n_{CH_4} are not affected.

Consequently, α must equal 3/2. In the same way, β, γ, and δ are determined uniquely by the second, third, and fourth pseudoreactions and reactions.

Summarizing, to analyze the element balances

(1)   list the elements and components present in the order of decreasing ease of measurement.
(2)   form the molecular matrix by entering the number of atoms of an element for each component in its own row.
(3)   augment this matrix by adding a unit matrix to its right.
(4)   perform the operations resulting in the RREF.

Then, in the result:

(1)   every row is still stoichiometrically valid.
(2)   every pivot column corresponds to a value that must be known, either that of an element balance or of a key component.
(3)   every nonpivot column corresponds to an unknown value that can be deduced from the pivots using the matrix coefficients in that column.
(4)   the rows for which all elemental entries are zero form the basis of stoichiometrically possible reactions.

Because the matrix is augmented with a unit matrix, the total number of pivot columns always equals the number of rows, which is also the number of components. It also equals the number of key elements plus the number of key components. The number of key elements can be less than the total number of elements, when some element columns have no pivot.

Example 2.1 Fixed Ratio of Elements

Consider a mixture consisting of the isomers $c\text{-}C_4H_8$, $t\text{-}C_4H_8$, and the dimer C_8H_{16}. In order to determine which are the key elements, and the key and nonkey components, we first construct the molecular matrix as

$$
\begin{array}{c}
 \\
c\text{-}C_4H_8 \\
t\text{-}C_4H_8 \\
C_8H_{16}
\end{array}
\begin{array}{cc}
C & H \\
\left[\begin{array}{cc} 4 & 8 \\ 4 & 8 \\ 8 & 16 \end{array}\right]
\end{array}
\tag{2.27}
$$

We augment it to the matrix:

$$
\begin{array}{ccccc}
C_t & H_t & c\text{-}C_4H_8 & t\text{-}C_4H_8 & C_8H_{16}
\end{array}
$$
$$
\left[\begin{array}{ccccc}
4 & 8 & 1 & 0 & 0 \\
4 & 8 & 0 & 1 & 0 \\
8 & 16 & 0 & 0 & 1
\end{array}\right]
\tag{2.28}
$$

with RREF

$$\begin{array}{ccccc} C_t & H_t & c\text{-}C_4H_8 & t\text{-}C_4H_8 & C_8H_{16} \\ \begin{bmatrix} \mathbf{1} & 2 & 0 & 0 & 1/8 \\ 0 & 0 & \mathbf{1} & 0 & -1/2 \\ 0 & 0 & 0 & \mathbf{1} & -1/2 \end{bmatrix} \end{array} \tag{2.29}$$

An unusual event occurs here: the second column, even though elemental, has no pivot. It therefore contains a nonkey element, and we can read from it that H_t is twice C_t in all components of the mixture. Next, $c\text{-}C_4H_8$ and $t\text{-}C_4H_8$ are the two key components, and the amount of C_8H_{16} is always given by

$$n_{C_8H_{16}} = 1/8 n_{C_t} - 1/2 n_{c\text{-}C_4H_8} - 1/2 n_{t\text{-}C_4H_8} \tag{2.30}$$

The general result that can be derived from this example is the fundamental equation:

Number of key components = number of components − number of key elements   (2.31)

The number of key components is thus a property of the reaction mixture, but the key components actually chosen depend on the ordering: the components listed first will be chosen to be key components. This is why it is useful to order the components by decreasing ease of measurement, so that the components whose values can be measured more easily will be selected as key components.

The RREF method is completely mechanical, so it requires no insight and when performed correctly will always give the same result: even though different nonzero entries might be chosen as pivots in its course, the final result will always be the same. Conceptually, the RREF means that given an ordering of variables, we try and eliminate all possible (pivot) variables when they can be expressed in terms of other variables that come later in the ordering.

Verifying that the element balances hold is an essential way of making sure that no components were overlooked. Then the determination of the key components is not arbitrary, since it only depends on knowing the composition of the mixture of all participating components.

Example 2.2 Different Order of Components

Rework the example with the molecular matrix given by Eq. (2.11), but now with the reverse order of components.

Solution:

The new molecular matrix is

$$\begin{array}{cc} & C\ H \\ \begin{array}{c} C_2H_4 \\ C_2H_6 \\ CH_4 \\ H_2 \end{array} & \begin{bmatrix} 2 & 4 \\ 2 & 6 \\ 1 & 4 \\ 0 & 2 \end{bmatrix} \end{array} \tag{2.32}$$

After augmentation with a unit matrix, this matrix becomes

$$
\begin{array}{cccccc}
C_t & H_t & C_2H_4 & C_2H_6 & CH_4 & H_2
\end{array}
$$
$$
\begin{bmatrix}
2 & 4 & 1 & 0 & 0 & 0 \\
2 & 6 & 0 & 1 & 0 & 0 \\
1 & 4 & 0 & 0 & 1 & 0 \\
0 & 2 & 0 & 0 & 0 & 1
\end{bmatrix}
\tag{2.33}
$$

Its RREF is readily computed to be

$$
\begin{array}{cccccc}
C_t & H_t & C_2H_4 & C_2H_6 & CH_4 & H_2
\end{array}
$$
$$
\begin{bmatrix}
1 & 0 & 0 & 0 & 1 & -2 \\
0 & 1 & 0 & 0 & 0 & 1/2 \\
0 & 0 & 1 & 0 & -2 & 2 \\
0 & 0 & 0 & 1 & -2 & 1
\end{bmatrix}
\tag{2.34}
$$

so that now the key components are C_2H_4 and C_2H_6 (because they come earlier in the ordering). The basic reactions are $C_2H_4 + 2\,CH_2 \rightleftarrows 2\,CH_4$ and $C_2H_6 + H_2 \rightleftarrows 2\,CH_4$. The amounts of the nonkey components work out to

$$
n_{CH_4} = n_{C_t} - 2n_{C_2H_4} - 2n_{C_2H_6}
\tag{2.35}
$$

and

$$
n_{H_2} = -2n_{C_t} + 1/2\,n_{H_t} + 2n_{C_2H_4} + n_{C_2H_6}
\tag{2.36}
$$

Example 2.3 Mixture Containing a One-Atom Molecule

Calculate the balances for a mixture containing the molecules C (atomic carbon), CO, CO_2, and O_2.

Solution:

The fact that atomic carbon also occurs among the components does not require any special treatment; only the usual care must be taken to distinguish the C_t elemental column from the C atomic carbon column. The molecular matrix is

$$
\begin{array}{c}
 \\
C \\
CO \\
CO_2 \\
O_2
\end{array}
\begin{array}{c}
C \quad O \\
\begin{bmatrix}
1 & 0 \\
1 & 1 \\
1 & 2 \\
0 & 2
\end{bmatrix}
\end{array}
\tag{2.37}
$$

and after augmentation,

$$
\begin{array}{cccccc}
C_t & O_t & C & CO & CO_2 & O_2
\end{array}
$$
$$
\begin{bmatrix}
1 & 0 & 1 & 0 & 0 & 0 \\
1 & 1 & 0 & 1 & 0 & 0 \\
1 & 2 & 0 & 0 & 1 & 0 \\
0 & 2 & 0 & 0 & 0 & 1
\end{bmatrix}
\tag{2.38}
$$

Its RREF is then

$$
\begin{array}{cccccc}
C_t & O_t & C & CO & CO_2 & O_2
\end{array}
$$
$$
\begin{bmatrix}
1 & 0 & 0 & 0 & 1 & -1 \\
0 & 1 & 0 & 0 & 0 & 1/2 \\
0 & 0 & 1 & 0 & -1 & 1 \\
0 & 0 & 0 & 1 & -1 & 1/2
\end{bmatrix}
\tag{2.39}
$$

The key components are C and CO (as dictated by the order of components) and the amount of the nonkey components can be calculated from

$$
n_{CO_2} = n_{C_t} - n_C - n_{CO} \tag{2.40}
$$

and

$$
n_{O_2} = -n_{C_t} + 1/2 n_{O_t} + n_C + 1/2 n_{CO} \tag{2.41}
$$

Example 2.4 Combination of Two Groups of Elements

Consider the components acetylene (C_2H_2), hydrochloric acid (HCl), and vinyl chloride (C_2H_3Cl). Determine the key components and balances.

Solution:

The augmented molecular matrix is

$$
\begin{array}{cccccc}
C_t & H_t & Cl_t & C_2H_2 & HCl & C_2H_3Cl
\end{array}
$$
$$
\begin{bmatrix}
2 & 2 & 0 & 1 & 0 & 0 \\
0 & 1 & 1 & 0 & 1 & 0 \\
2 & 3 & 1 & 0 & 0 & 1
\end{bmatrix}
\tag{2.42}
$$

with RREF

$$
\begin{array}{cccccc}
C_t & H_t & Cl_t & C_2H_2 & HCl & C_2H_3Cl
\end{array}
$$
$$
\begin{bmatrix}
1 & 0 & -1 & 0 & -3/2 & 1/2 \\
0 & 1 & 1 & 0 & 1 & 0 \\
0 & 0 & 0 & 1 & 1 & -1
\end{bmatrix}
\tag{2.43}
$$

so that C and H are the independent elements and Cl is dependent according to

$$
n_{Cl_t} = -n_{C_t} + n_{H_t} \tag{2.44}
$$

In this case, C_2H_2 is the only key component because C_2H_3Cl is a combination of the two groups of elements C_2H_2 and HCl.

The amounts of the nonkey components are

$$n_{HCl} = -3/2 n_{C_t} + n_{H_t} + n_{C_2H_2} \qquad (2.45)$$

and

$$n_{C_2H_3Cl} = 1/2 n_{C_t} - n_{C_2H_2} \qquad (2.46)$$

2.2.3 Remarks

The method indicated in this section works equally well for ions and radicals, ionic or not. In the case of negative ions, the excess electrons must be specified as well, since electrons are distinct from all elements; similarly, in the case of positive ions, the deficit in electrons must be indicated. This will ensure that charge is always properly preserved. When different isotopes of an element occur, they must be counted separately along with the different components they occur in. Sometimes unconventional "elements" such as active catalyst sites are used. Sometimes subcomponents can be used as pseudoelements (water in hydrates, building blocks of polymers, amino acids).

2.3 The Stoichiometric Matrix

In general, a *stoichiometric matrix* represents a set of reactions involving given components. The columns of this matrix correspond to the different components, the rows correspond to the different reactions, and the entries are stoichiometric coefficients, which by convention are negative for the reactants and positive for the products.

In Section 2.2, we have shown how the molecular matrix, when augmented with a unit matrix and converted to the corresponding RREF, yields a basis for all stoichiometrically acceptable reactions. In reality, however, many of these reactions may be chemically impossible. Therefore, a special stoichiometric matrix can be considered, in which only selected reactions occur.

2.3.1 Application of the Stoichiometric Matrix to Reactions

Consider the set of reactions

$$2H_2 + O_2 \rightleftharpoons 2H_2O \qquad (2.47)$$

$$H_2 + O_2 \rightleftharpoons H_2O_2 \qquad (2.48)$$

which is represented by the stoichiometric matrix $\mathbf{S}$

$$
\begin{array}{cccc}
\text{H}_2 & \text{O}_2 & \text{H}_2\text{O} & \text{H}_2\text{O}_2
\end{array}
$$
$$
\begin{bmatrix}
-2 & -1 & 2 & 0 \\
-1 & -1 & 0 & 1
\end{bmatrix}
\tag{2.49}
$$

while the molecular matrix $\mathbf{M}$ of the components that participate in these reactions is

$$
\begin{array}{c}
 \\
\text{H}_2 \\
\text{O}_2 \\
\text{H}_2\text{O} \\
\text{H}_2\text{O}_2
\end{array}
\begin{array}{c}
\text{H\ O} \\
\begin{bmatrix}
2 & 0 \\
0 & 2 \\
2 & 1 \\
2 & 2
\end{bmatrix}
\end{array}
\tag{2.50}
$$

If we wish to verify, for instance, that hydrogen atoms are conserved by the first reaction, Eq. (2.47), we must count them in every term on either side, and then subtract these values: there are four hydrogen atoms on both sides and subtraction yields zero. An equivalent but mathematically more elegant method is to use the entries of the stoichiometric matrix and the molecular matrix as follows: we read -2 for the coefficient of H_2 in the first row of the stoichiometric matrix and multiply by 2, the number of H atoms in H_2 as indicated by the molecular matrix, and so on. The result is then $(-2)(2)+(-1)(0)+(2)(2)+(0)(2)=0$, which by definition is precisely the entry $(1,1)$ of the product of the stoichiometric and molecular matrices $\mathbf{SM}$. Similarly, the entry (ij) of that product expresses the difference in atom counts of element j in reaction i. Consequently, we can express the balance of all reactions for all elements compactly as the zero matrix with all entries zero:

$$
\mathbf{SM} = 0
\tag{2.51}
$$

It is useful to compute the RREF of a stoichiometric matrix because the pivots will reveal the key components, and the nonpivot columns will reveal the balances obeyed by the nonkey components. In this example, the RREF equals

$$
\begin{array}{cccc}
\text{H}_2 & \text{O}_2 & \text{H}_2\text{O} & \text{H}_2\text{O}_2
\end{array}
$$
$$
\begin{bmatrix}
1 & 0 & -2 & 1 \\
0 & 1 & 2 & -2
\end{bmatrix}
\tag{2.52}
$$

so that the key components are H_2 and O_2, and the relationships are read off in the nonpivot columns, 3 and 4, as

$$
\Delta n_{\text{H}_2\text{O}} = -2\Delta n_{\text{H}_2} + 2\Delta n_{\text{O}_2}
\tag{2.53}
$$

and

$$
\Delta n_{\text{H}_2\text{O}_2} = \Delta n_{\text{H}_2} - 2\Delta n_{\text{O}_2}
\tag{2.54}
$$

where Δ indicates that we are expressing the changes in amounts that are caused by these reactions.

Example 2.5 Effect of the Reactions Chosen

Returning to the molecules $c\text{-}C_4H_8$, $t\text{-}C_4H_8$, and the dimer C_8H_{16} of Example 2.2, we can propose the reactions

$$c\text{-}C_4H_8 + t\text{-}C_4H_8 \rightleftarrows C_8H_{16} \tag{2.55}$$

and

$$c\text{-}C_4H_8 \rightleftarrows t\text{-}C_4H_8 \tag{2.56}$$

which are represented by the stoichiometric matrix

$$\begin{array}{ccc} c\text{-}C_4H_8 & t\text{-}C_4H_8 & C_8H_{16} \end{array} \\ \begin{bmatrix} -1 & -1 & 1 \\ -1 & 1 & 0 \end{bmatrix} \tag{2.57}$$

with RREF

$$\begin{array}{ccc} c\text{-}C_4H_8 & t\text{-}C_4H_8 & C_8H_{16} \end{array} \\ \begin{bmatrix} 1 & 0 & -1/2 \\ 0 & 1 & -1/2 \end{bmatrix} \tag{2.58}$$

so that $c\text{-}C_4H_8$ and $t\text{-}C_4H_8$ are key components, and the balance in the nonpivot column reads

$$\Delta n_{C_8H_{16}} = -1/2\Delta n_{c\text{-}C_4H_8} - 1/2\Delta n_{t\text{-}C_4H_8} \tag{2.59}$$

If, however, we retain only the first reaction, overlooking the second, isomerization reaction, the stoichiometric matrix is

$$\begin{array}{ccc} c\text{-}C_4H_8 & t\text{-}C_4H_8 & C_8H_{16} \end{array} \\ \begin{bmatrix} -1 & -1 & 1 \end{bmatrix} \tag{2.60}$$

with RREF

$$\begin{array}{ccc} c\text{-}C_4H_8 & t\text{-}C_4H_8 & C_8H_{16} \end{array} \\ \begin{bmatrix} 1 & 1 & -1 \end{bmatrix} \tag{2.61}$$

Then there is only one key component, $c\text{-}C_4H_8$, and the nonpivot columns indicate the balances

$$\Delta n_{t\text{-}C_4H_8} = \Delta n_{c\text{-}C_4H_8} \tag{2.62}$$

and

$$\Delta n_{C_8H_{16}} = -\Delta n_{c\text{-}C_4H_8} \tag{2.63}$$

Comparing this use of the stoichiometric matrix to that of the molecular matrix (Section 2.2), we observe the following:

- When considering the molecular matrix, the element balances (C_t, H_t, etc.) provide us with absolute values. In the current setting of reactions, we can state facts only about changes in the number of molecules of the components, not about their total amounts.

- Because the stoichiometric matrix may contain fewer reactions than the ones generated by the molecular matrix, there may be more balances and fewer key components in the stoichiometric matrix than in the molecular matrix.
- The determination of key components from the stoichiometric matrix depends on more choices than in the case of the molecular matrix, since it involves selecting specific reactions. Therefore, the result of this procedure may be different.
- Again, since the components listed later tend to be expressed in terms of the ones listed earlier, the order of listing should be according to decreasing ease of measurement.

In this application, the stoichiometric matrix was not augmented, and we cannot tell by which linear combination of the original reactions the final result was obtained. Sometimes it is useful to keep track of the original reactions, for example when thermodynamic data such as enthalpy changes are required or for the calculation of equilibrium coefficients.

2.3.2 The Augmented Stoichiometric Matrix

Adding the hypothetical reaction

$$2H_2O + O_2 \rightleftarrows 2H_2O_2 \tag{2.64}$$

to the previous system, Eqs. (2.47), (2.48), results in the stoichiometric matrix

$$
\mathbf{S} = \begin{array}{cccc} H_2 & O_2 & H_2O & H_2O_2 \\ \begin{bmatrix} -2 & -1 & 2 & 0 \\ -1 & -1 & 0 & 1 \\ 0 & -1 & -2 & 2 \end{bmatrix} \end{array} \tag{2.65}
$$

Augmenting this matrix with a unit matrix

$$
\begin{array}{ccccccc} H_2 & O_2 & H_2O & H_2O_2 & R_1 & R_2 & R_3 \\ \begin{bmatrix} -2 & -1 & 2 & 0 & 1 & 0 & 0 \\ -1 & -1 & 0 & 1 & 0 & 1 & 0 \\ 0 & -1 & -2 & 2 & 0 & 0 & 1 \end{bmatrix} \end{array} \tag{2.66}
$$

eventually results in the following RREF:

$$
\begin{array}{c} R'_1 \\ R'_2 \\ R'_3 \end{array} \begin{array}{ccccccc} H_2 & O_2 & H_2O & H_2O_2 & R_1 & R_2 & R_3 \\ \begin{bmatrix} 1 & 0 & -2 & 1 & 0 & -1 & 1 \\ 0 & 1 & 2 & -2 & 0 & 0 & -1 \\ 0 & 0 & 0 & 0 & 1 & -2 & 1 \end{bmatrix} \end{array} \tag{2.67}
$$

Fig. 2.2 shows a map of this matrix indicating the subareas. Every row represents a reaction, possibly a zero reaction, in which there are no reactants or products. Here this is the case for R'_3, for which the coefficients are read off the third row in the columns representing the reactions: $(1)R_1 + (-2)R_2 + (1)R_3$, where R_1, R_2, and R_3 represent the reactions in, respectively, Eqs. (2.47), (2.48), (2.64). This produces the zero reaction R'_3. The first row in the

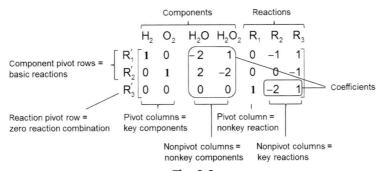

Fig. 2.2

Map of subareas in the RREF of Eq. (2.67).

reaction columns of the RREF corresponds to the reaction R_1', $2H_2O \rightleftarrows H_2 + H_2O_2$, obtained by the combination $(0)R_1 + (-1)R_2 + (1)R_3$. The second row corresponds to the reaction R_2', $2H_2O_2 \rightleftarrows O_2 + 2H_2O$, obtained by the combination $(0)R_1 + (0)R_2 + (-1)R_3$.

The reaction columns in the RREF can be interpreted as follows. If we assign a rate to each of the original reactions, r_1 for R_1, etc., and similarly rates r_1' for R_1', etc., for the new set of reactions, the problem solved by these reaction columns is how to express the original rates r_i in terms of the new ones, r_i'; for instance, the column R_2 indicates that $r_2 = (-1)r_1' + (0)r_2' + (-2)r_3'$.

2.3.2.1 How to find the key and nonkey reactions

The RREF of the augmented stoichiometric matrix indicates that every reaction with a pivot in its column can be written as a linear combination of nonpivot reactions. All entries to the left of the pivot are necessarily zero, so that the row containing the pivot must represent a zero reaction, that is, the pivot reaction added to the reactions to its right multiplied by their entries, yields zero. Hence, the pivot reaction itself is minus the sum of these multiplied reactions, and dependent on them. In this example, since R_3' is the zero reaction and R_1 is the corresponding pivot column, from $(1)R_1 + (-2)R_2 + (1)R_3 = R_3' = 0$, it follows that $R_1 = (2)R_2 + (-1)R_3$.

Contrary to the key components, which occur in pivot columns, the reactions in pivot columns are the nonkey ones. In the present example, the second and third reactions are key (nonpivot) and the first is nonkey (pivot). As before, the ordering of the reactions will influence the choice of key and nonkey reactions, and the earlier a reaction occurs in the ordering, the more likely it is not to be a key reaction. Again, we can count the number of pivot columns in two ways: on the one hand, because the matrix was augmented with a unit matrix, it equals the number of reactions. On the other hand, each pivot column either corresponds to a key component or to a nonkey reaction. Consequently,

$$\text{Number of key components} + \text{number of nonkey reactions} = \text{number of reactions} \qquad (2.68)$$

Subtracting the number of nonkey reactions from both sides yields the fundamental equality

$$\text{Number of key components} = \text{number of key reactions} \tag{2.69}$$

In this example, these numbers are both equal to two; H_2 and O_2 are the key components, and R_2 and R_3 are the key reactions.

The number of balances equals the number of nonkey components as indicated before. They are caused by the reactions, not by element balances. The coefficients by which to multiply the change of the amounts of the key components are read off on the column as before.

2.3.2.2 Enthalpy change and equilibrium coefficients

The RREF can also be used to find the enthalpy change of the reactions given in the rows: if $\Delta_r H_1$ is the enthalpy change of reaction R_1, etc., then the enthalpy change of the reaction in the first row of the RREF, $2H_2O \rightleftharpoons H_2 + 2H_2O_2$, is $-\Delta_r H_2 + \Delta_r H_3$ and that of the reaction in the second row, $2H_2O_2 \rightleftharpoons O_2 + 2H_2O$, is $-\Delta_r H_3$. The last row, which represents a zero reaction, must have an enthalpy change of zero, so that $0 = \Delta_r H_1 - 2\Delta_r H_2 + \Delta_r H_3$. The latter is an example of how the value of the enthalpy change of a reaction in a pivot column can be written in terms of the values of subsequent columns, in this case $\Delta_r H_1 = 2\Delta_r H_2 - \Delta_r H_3$. This example also illustrates that it is useful to list the reactions with lesser-known thermodynamic properties first.

A similar reasoning can be applied to the logarithms of the equilibrium coefficients $K_{eq,1}$, etc. The equilibrium coefficient of the first reduced reaction must equal $K_{eq,3}/K_{eq,2}$, that of the second $1/K_{eq,3}$, and that of the third, zero reaction leads to $1 = K_{eq,1}K_{eq,3}/K_{eq,2}^2$, or in solved-for form, $K_{eq,1} = K_{eq,2}^2/K_{eq,3}$. In general, the entries in a nonpivot reaction row are (with changed sign) the exponents in the product expression for the corresponding equilibrium coefficient.

2.4 Horiuti Numbers

During modeling of catalytic reactions, often unmeasured quantities enter the equations, for example when surface intermediates are considered. In such cases, one would like to know which are the overall reactions and how many of these reactions are independent.

The overall reaction is obtained by multiplying the reactions with certain coefficients, the so-called Horiuti numbers σ, and then adding the results.

Consider the following set of reactions:

$$Z + H_2O \rightleftharpoons ZO + H_2 \tag{2.70}$$

$$ZO + CO \rightleftarrows Z + CO_2 \tag{2.71}$$

where Z is a symbolic element denoting an active catalyst site. What is/are the overall reaction(s)? Again, this problem can be readily solved by using the RREF. Until now, we have stressed that in choosing the order of components occurring in the molecular matrix and in the stoichiometric matrix, it is advisable to list them in order of decreasing ease of measure, so that the components that are difficult to measure can then be expressed in terms of those that are easier to measure.

However, in the Horiuti setting, the purpose is not to express the short-lived intermediates (in this case ZO and Z), which are difficult or even impossible to measure, in terms of the long-lived components (H_2, H_2O, CO, and CO_2), but to find relations between the long-lived components that do not involve the intermediates at all. That is why, in this setting, in order to eliminate the intermediates using an RREF of the stoichiometric matrix, the intermediates must be listed *first*, not last:

$$\begin{array}{cccccc} Z & ZO & H_2 & H_2O & CO & CO_2 \\ \begin{bmatrix} -1 & 1 & 1 & -1 & 0 & 0 \\ 1 & -1 & 0 & 0 & -1 & 1 \end{bmatrix} \end{array} \tag{2.72}$$

Again we augment this matrix with a unit matrix to keep track of the original reactions:

$$\begin{array}{cccccccc} Z & ZO & H_2 & H_2O & CO & CO_2 & R_1 & R_2 \\ \begin{bmatrix} -1 & 1 & 1 & -1 & 0 & 0 & 1 & 0 \\ 1 & -1 & 0 & 0 & -1 & 1 & 0 & 1 \end{bmatrix} \end{array} \tag{2.73}$$

The RREF of this matrix is

$$\begin{array}{cccccccc} Z & ZO & H_2 & H_2O & CO & CO_2 & R_1 & R_2 \\ \begin{bmatrix} \mathbf{1} & -1 & 0 & 0 & -1 & 1 & 0 & 1 \\ 0 & 0 & \mathbf{1} & -1 & -1 & 1 & 1 & 1 \end{bmatrix} \end{array} \tag{2.74}$$

The rows in which all intermediates have zero entries provide a basis of the overall reactions. Note that if no such rows occur, there is no overall reaction because intermediates are always involved. There can be several different overall reactions; and the same overall reaction may be found by two different sets of Horiuti numbers, but in that case their difference will produce a zero overall reaction $0 \rightleftarrows 0$, which will show up as such in the RREF.

In the present case, we ignore the first row but consider the second: $H_2O + CO \rightleftarrows H_2 + CO_2$, which is the overall reaction. its Horiuti numbers are the coefficients affecting the original reactions R_1 and R_2 in that row, that is, 1 and 1. If we bundle these numbers in a so-called Horiuti matrix, $\boldsymbol{\sigma}$, and consider only the intermediate part $\mathbf{S}_{int}$ of the stoichiometric matrix $\mathbf{S}$,

$$\sigma = \begin{matrix} R_1 \\ R_2 \end{matrix} \begin{bmatrix} 1 \\ 1 \end{bmatrix}; \quad S_{int} = \begin{matrix} & Z & ZO \\ R_1 & \begin{bmatrix} -1 & 1 \\ R_2 & 1 & -1 \end{bmatrix} \end{matrix} \tag{2.75}$$

then we have the remarkable matrix product property that $\sigma^T S_{int} = 0$. Here σ^T denotes the transpose of σ, that is, the matrix obtained by interchanging the rows and columns of σ, so that here

$$\sigma^T = \begin{matrix} R_1 \ R_2 \\ [1 \quad 1] \end{matrix} \tag{2.76}$$

Example 2.6 Reaction With Three Intermediates

Consider the set of reactions:

$$2Z + O_2 \rightleftarrows 2ZO \tag{2.77}$$

$$Z + CO \rightleftarrows ZCO \tag{2.78}$$

$$ZO + CO \rightleftarrows Z + CO_2 \tag{2.79}$$

$$ZO + ZCO \rightleftarrows 2Z + CO_2 \tag{2.80}$$

Determine the overall reaction(s) and Horiuti numbers.

Solution:

The augmented matrix is

$$\begin{matrix} & Z & ZO & ZCO & O_2 & CO & CO_2 & R_1 & R_2 & R_3 & R_4 \\ & \begin{bmatrix} -2 & 2 & 0 & -1 & 0 & 0 & 1 & 0 & 0 & 0 \\ -1 & 0 & 1 & 0 & -1 & 0 & 0 & 1 & 0 & 0 \\ 1 & -1 & 0 & 0 & -1 & 1 & 0 & 0 & 1 & 0 \\ 2 & -1 & -1 & 0 & 0 & 1 & 0 & 0 & 0 & 1 \end{bmatrix} \end{matrix} \tag{2.81}$$

with RREF

$$\begin{matrix} & Z & ZO & ZCO & O_2 & CO & CO_2 & R_1 & R_2 & R_3 & R_4 \\ & \begin{bmatrix} 1 & 0 & -1 & 0 & 1 & 0 & 0 & 0 & -1 & 1 \\ 0 & 1 & -1 & 0 & 2 & -1 & 0 & 0 & -2 & 1 \\ 0 & 0 & 0 & 1 & 2 & -2 & -1 & 0 & -2 & 0 \\ 0 & 0 & 0 & 0 & 0 & 0 & 0 & 1 & -1 & 1 \end{bmatrix} \end{matrix} \tag{2.82}$$

The first two rows are discarded, since the intermediates occur in these rows, but the third and fourth are kept. The overall reactions are $2CO_2 \rightleftarrows 2CO + O_2$ and the zero reaction; the respective Horiuti numbers are $-1, 0, -2, 0$ and $0, 1, -1, 1$, and the Horiuti matrix is given by

$$\sigma = \begin{matrix} R_1 \\ R_2 \\ R_3 \\ R_4 \end{matrix} \begin{bmatrix} -1 & 0 \\ 0 & 1 \\ -2 & -1 \\ 0 & 1 \end{bmatrix} \tag{2.83}$$

Fig. 2.3 shows a map of the RREF represented by Eq. (2.82) indicating the subareas.

In practice, the Horiuti numbers of the zero overall reaction are not considered as such, but indicate the freedom that remains in choosing a set of Horiuti numbers: any linear combination of zero-reaction Horiuti numbers can be added to or subtracted from those of any overall reaction. For instance, subtracting them from the overall ones in this case yields $-1, -1, -1, -1$, or, equivalently, 1, 1, 1, 1 for the reverse overall reaction. Such a set of values is visually more pleasing but cannot be expected to be favored by a mechanical procedure such as the RREF procedure, which instead tries to eliminate as many entries as possible (in all pivot columns corresponding to reactions).

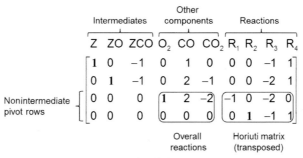

Fig. 2.3

Map of subareas in the RREF of Eq. (2.83).

Example 2.7 Two Types of Active Sites

Consider the set of reactions:

$$A + Z_1 \rightleftarrows AZ_1 \tag{2.84}$$

$$B + Z_2 \rightleftarrows BZ_2 \tag{2.85}$$

$$AZ_1 + BZ_2 \rightleftarrows AB + Z_1 + Z_2 \tag{2.86}$$

Determine the overall reaction(s) and Horiuti numbers.

Solution: The augmented matrix is

$$
\begin{array}{ccccccccccc}
Z_1 & Z_2 & AZ_1 & BZ_2 & A & B & AB & R_1 & R_2 & R_3 \\
\end{array}
$$

$$
\begin{bmatrix}
-1 & 0 & 1 & 0 & -1 & 0 & 0 & 1 & 0 & 0 \\
0 & -1 & 0 & 1 & 0 & -1 & 0 & 0 & 1 & 0 \\
1 & 1 & -1 & -1 & 0 & 0 & 1 & 0 & 0 & 1
\end{bmatrix} \tag{2.87}
$$

The RREF of this matrix is

$$
\begin{array}{ccccccccccc}
Z_1 & Z_2 & AZ_1 & BZ_2 & A & B & AB & R_1 & R_2 & R_3 \\
\end{array}
$$

$$
\begin{bmatrix}
1 & 0 & -1 & 0 & 0 & -1 & 1 & 0 & 1 & 1 \\
0 & 1 & 0 & -1 & 0 & 1 & 0 & 0 & -1 & 0 \\
0 & 0 & 0 & 0 & 1 & 1 & -1 & -1 & -1 & -1
\end{bmatrix} \tag{2.88}
$$

We consider only the third row, and read out the overall reaction as AB⇌A + B with Horiuti numbers −1, −1, −1. In practice, these numbers would be more likely reported equivalently as 1, 1, 1 for the reverse reaction, A + B⇌AB. This is also the form that would have been obtained if the component AB had been listed before A and B instead of after them.

2.5 Summary

Tables 2.1 and 2.2 summarize the characteristic features of the RREFs of the augmented molecular matrix and the augmented stoichiometric matrix.

In any case, the number of key components equals the number of key reactions. If fewer reactions are retained, there will be fewer key components due to surplus balances. The following relations hold

Number of key components of molecular matrix
= number of components − number of key elements

Number of key components of stoichiometric matrix
≤ number of key components of molecular matrix

Number of key components of stoichiometric matrix
= number of key reactions of stoichiometric matrix

Table 2.1 Characteristics of the RREF of the augmented molecular matrix

Column/Row	Entry	Comment
Column		
Pivot element	Key (or independent) element	
Nonpivot element	Nonkey (or dependent) element	Read column for coefficients
Pivot component	Key component	
Nonpivot component	Nonkey component	Read column for coefficients expressing amounts
Row		
With element pivot	Pseudoreaction using nonkey components to create a single atom of that element	
With component pivot	Pseudoreaction producing that component and nonkey components from nonkey components	By retaining only the rows with component pivots and discarding the element columns, the most general stoichiometric matrix for the given set of components is obtained

Table 2.2 Characteristics of the RREF of the augmented stoichiometric matrix

Column/Row	Entry	Comment
Column		
Pivot component Nonpivot component Pivot reaction	Key component Nonkey component Nonkey reaction	Read column for coefficients Read rest of row for coefficients, changing the sign
Nonpivot reaction	Key reaction	
Row		
With component pivot	Basic reaction to create one unit of that component without participation of any other component	
With reaction pivot	Zero reaction combination of the reactions given	

The construction of the augmented stoichiometric matrix to determine the Horiuti numbers has the same general properties in the RREF. Furthermore, the rows not involving any intermediates are of special interest, since they correspond to the overall reactions and exhibit the Horiuti numbers in the augmented columns. In addition, a Horiuti matrix can be constructed as part of the general analysis of the augmented stoichiometric matrix. The equilibrium coefficient of each overall reaction is then given by the product of the equilibrium coefficients of the elementary reactions raised to powers given by the Horiuti numbers.

Appendix. The RREF in Python

Software for computing the RREF of a matrix is readily available in computer algebra packages. For completeness, here we list a script that offers the same functionality, but works in the popular open-source package Python:

#!/usr/bin/python

 # -- coding: utf-8 -*-*

 # These ^^first two lines are comments geared towards Unix implementations

 # We need the fractions package for its fast greatest common divisor implementation

import fractions

 # rref(a) where a is a matrix represented as a Python list of lists of integers

 # will return the RREF of a

def rref(a):

```
# The number of rows m is the number of lists in the list

   m = len(a)

# The number of columns n is obtained from the first list; we assume but do not check
# that all lists in a have this same length

   n = len(a[1])

# Initially, no rows have been completed in the sense of having a pivot

   rowsdone = 0

# And no columns have been done either; but we traverse the n columns to work on them
# in this for loop

   for columnsdone in range(n):

# Ignoring the rows that were done, we look for a nonzero entry in column i

      i = rowsdone
      while i < m and a[i][columnsdone] == 0:
         i = i + 1

# Did we find one?

      if i < m:

# Yes, we found a nonzero entry; if it is not yet in uppermost position...

         if i ! = rowsdone:

# ... we move it up to there

            for j in range(n):
               t = a[i][j]
               a[i][j] = a[rowsdone][j]
               a[rowsdone][j] = t

# Now the nonzero entry is in pivot position we use it to zero all other entries in its column:

            for i in range(m):

# ... we do not zero the pivot itself

               if i ! = rowsdone:

# ... but all other entries in its column by multiplying with these coefficients f and g
```

$$f = a[rowsdone][columnsdone]$$

$$g = a[i][columnsdone]$$

... *this loop handles the linear combination*

```
for j in range(n):

    a[i][j] = a[i][j] * f - a[rowsdone][j] * g
```

... *now we normalize the results by calculating their greatest common divisor, h*

```
h = 0

for j in range(n):

    h = fractions.gcd(h, a[i][j])
```

... *and if it is nonzero, divide by it*

```
if h ! = 0:

    for j in range(n):

        a[i][j] = a[i][j] / h
```

... *this completes the work and we register the new row as being completed*

```
rowsdone = rowsdone + 1
```

Summarizing: the matrix is assumed to be given as a list of lists of integers. Its dimensions m and n are obtained from this representation, and then the code successively inspects each of the columns to see if it contains a pivot. Because all arithmetic is exact using the built-in unlimited precision integers of Python, the greatest common divisor of every new row is eliminated along the way. Whenever a pivot is found in a column, a new pivot row has been found and the number of rows done is increased accordingly. Finally, the matrix a has been modified to be in RREF form.

Nomenclature

Symbols

a_i real number

$\Delta_r H_i$ enthalpy change of reaction i (J mol^{-1})

$K_{eq,i}$ equilibrium coefficient of reaction i

M   molecular matrix

m$_A$ column vector of atomic masses (kg mol^{-1})

m$_M$ column of molar masses (kg mol^{-1})

n_i amount of component i (mol)

$P_{i,j}$ entry of matrix **P**
R_i referring to reaction i
R_i' referring to "new" reaction i
r_i rate of reaction i (depends)
r_i' rate of "new" reaction i (depends)
S   stoichiometric matrix
$\mathbf{S}_{int}$ stoichiometric matrix of intermediates
x variable
y variable

Greek symbols

Δ change
ν_i mathematical object
$\boldsymbol{\sigma}$ Horiuti matrix
σ Horiuti number

Subscripts

t   total

Superscripts

T   transpose

Further Reading

Aris, R., 1965. Introduction to the Analysis of Chemical Reactors. Prentice-Hall, Englewood Cliffs, NJ, 317 p.

Aris, R., 1969. Elementary Chemical Reactor Analysis. Prentice-Hall, Englewood Cliffs, NJ, 352 p.

Erdi, P., Toth, J., 1989. Mathematical Models of Chemical Reactions: Theory and Application of Deterministic and Stochastic Models. Manchester University Press, Manchester, UK, 267 p.

Smith, W.R., Missen, R.W., 1997. Using Mathematica and Maple to obtain chemical equations. J. Chem. Educ. 74, 1369–1371.

Complex Reactions: Kinetics and Mechanisms – Ordinary Differential Equations – Graph Theory

3.1 Primary Analysis of Kinetic Data

3.1.1 Introduction

Kinetic experiments are experiments in which the change of chemical composition of the reacting mixture is monitored in time and/or as a function of the various parameters of the system, in particular the inlet or initial chemical composition, temperature, catalyst state, and so on. The main paradigm of contemporary chemical kinetics is the following: a chemical reaction is complex and consists of a set of elementary reactions for which the kinetic laws are assumed to be known. Kinetic analysis focuses on revealing this set of reactions and developing the corresponding mathematical model, that, is the kinetic model.

The goals of the primary analysis of kinetic data are

- extracting the rates (rates of change of reacting components, rates of reactions, etc.) based on the data obtained from kinetic experiments, especially concentration measurements.
- determining the coherency or incoherency between the observed kinetic dependences, and categorizing the dependences of the "rate parameters".
- proposing primary assumptions on the kinetic model and its mathematical description.

The procedures for pursuing these goals depend on the type of kinetic experiments.

3.1.2 Types of Reactors for Kinetic Experiments

Kinetic experiments are performed in various types of chemical reactors, which can be classified as either *open* or *closed*, depending on whether or not there is exchange of matter with the surroundings. This classification was taken from classical thermodynamics. Closed reactors can exchange energy with the surroundings, but they cannot exchange matter, while open reactors can exchange either matter and energy or only matter. In chemical kinetics and engineering, a closed reactor is better known as a *batch reactor*; an open reactor as a

Advanced Data Analysis and Modelling in Chemical Engineering. http://dx.doi.org/10.1016/B978-0-444-59485-3.00003-5

continuous-flow reactor. There also exist *semiopen (or semiclosed) reactors*, commonly known as *semibatch reactors.* In these reactors only some type of material is exchanged with the surroundings. Another type of reactor is the *pulse reactor.* In this type of reactor a small quantity of a chemical substance is injected into the reactor.

3.1.3 Requirements of the Experimental Information

The primary information about chemical transformations is obtained by the measurement of the chemical composition, that is, concentration measurements. The chemical processes occurring are complex and do not consist only of chemical reactions but also of physical phenomena, such as mass and heat transport. The major goal of chemical kinetic studies is to extract intrinsic kinetic information related to complex chemical reactions. Therefore, the transport regime in the reactor has to be well defined and its mathematical description has to be reliable. A typical strategy in kinetic experiments is the minimization of the effects of mass and heat transport on the rate of change of the chemical composition. In accordance with this, a kinetic experiment ideally has to fulfill two main requirements: isothermicity of the active zone and uniformity of the chemical composition, which can be accomplished by, for example, perfect mixing within the reaction zone.

A kinetic experiment should usually be performed under (near) isothermal conditions. However, the temperature may be changed between two experiments. Temperature gradients across the reactor can be minimized in various ways, for example by intensive heat exchange between the reactor and the surroundings, by dilution of the reactive medium, or by its rapid recirculation. Uniformity of the chemical composition at reactor scale is achieved by intensive mixing using special mixing devices. Internal mixing can be achieved by using impellers, external mixing by the use of recirculation pumps. Both isothermicity of the active zone and uniformity of the chemical composition can also be attained in reactors in which the reaction zone is very small, for example in differential plug-flow reactors (PFRs), shallow beds, and so on.

3.2 Material Balances: Extracting the Net Rate of Production

For any chemical component in any chemical reactor, the material balance can be presented qualitatively as follows:

$$\text{temporal change of amount of component} = \text{transport change} + \text{change due to reaction} \quad (3.1)$$

in which the temporal change of the amount of a component, often termed accumulation, is its change with respect to time at a fixed position; the transport change is the change caused by flow through the reactor; and the reaction change is the change caused by chemical reaction. Eq. (3.1) can be used as a basis for the classification and qualitative description of

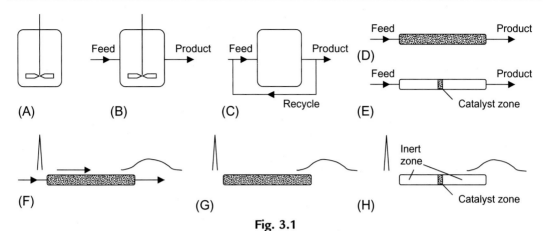

Fig. 3.1
Reactors for kinetic experiments: (A) batch reactor; (B) continuous stirred-tank reactor;
(C) continuous-flow reactor with recirculation; (D) plug-flow reactor; (E) differential plug-flow
reactor; (F) convectional pulse reactor; (G) diffusional pulse reactor or TAP reactor; and
(H) thin-zone TAP reactor.

different types of reactors for kinetic studies. Fig. 3.1 shows schematic representations of a
number of reactor types.

In reactors operating at steady state, the temporal change of the concentration of component
i is zero, $dc_i/dt = 0$, while in non-steady-state operation, $dc_i/dt \neq 0$. Equations with a zero
derivative dc_i/dt are algebraic models and equations with a nonzero derivative are differential
models of chemical processes.

3.2.1 Transport in Reactors

In laboratory reactors, a well-defined transport regime can be used as a "measuring stick"
for extracting the intrinsic kinetic dependences. Examples of well-defined transport regimes
are pure convection and pure diffusion. For convection, the molar flow rate F_i of component
i is determined as the product of its concentration c_i and the total volumetric flow rate q_V:

$$F_i = q_V c_i \tag{3.2}$$

For diffusion, in the simplest case, the molar flow rate is determined in accordance with Fick's
first law, which in one spatial dimension is:

$$F_i = -D_i A \frac{\partial c_i}{\partial z} \tag{3.3}$$

where D_i is the diffusion coefficient, A is the cross-sectional area of the reactor available for
fluid flow, and z is the axial reactor coordinate.

Quite often the importance of transport phenomena has to be assessed at different scales, that of the reactor being the largest. Characteristic times for transport by convection and diffusion are given by Eqs. (3.4), (3.5).

$$\tau_{\text{conv.}} = \frac{V}{q_V} = \frac{L}{u} \tag{3.4}$$

$$\tau_{\text{diff.}} = \frac{A}{D_i} \tag{3.5}$$

In Eq. (3.4), V is the reaction volume, L is the reactor length, and u is the superficial fluid velocity. The so-called first Damköhler number, Da_I, is defined as the ratio of the chemical reaction rate to the rate of convective mass transport (both with unit s^{-1}) or equivalently as the ratio of the characteristic time for convective mass transport to that for chemical reaction (both with unit s).

$$Da_I = \frac{\tau_{\text{conv.}}}{\tau_{\text{reaction}}} \tag{3.6}$$

For a first-order reaction, with reaction rate coefficient k,

$$Da_I = \frac{\tau_{\text{conv.}}}{1/k} = k\tau_{\text{conv.}} \tag{3.7}$$

The second Damköhler number, Da_{II}, is defined as the ratio of the characteristic time for diffusive mass transport to that for chemical reaction.

In solid-catalyzed reactions, the scale of the catalyst pellets also has to be considered. The influence of inter- and intraparticle transport on the reaction rate has to be eliminated experimentally and/or estimated quantitatively prior to the kinetic experiments.

In perfectly mixed convectional reactors, the "transport change" can be represented as the difference of convectional molar flow rates, $q_{V0}c_{i0} - q_V c_i$, or as $q_{V0}(c_{i0} - c_i)$ if $q_V = q_{V0}$.

In purely diffusional reactors, the "transport change" in the simplest case can be represented as the difference between diffusional flow rates in and out, F_{i0} and F_i. Both flow rates are written in accordance with Fick's first law:

$$F_{i0} = -D_i A \frac{\partial c_i}{\partial z}\bigg|_z \; ; \; F_i = -D_i A \frac{\partial c_i}{\partial z}\bigg|_{z+\Delta z} \tag{3.8}$$

Then,

$$F_{i0} - F_i = \left(-D_i A \frac{\partial c_i}{\partial z}\bigg|_z \right) - \left(-D_i A \frac{\partial c_i}{\partial z}\bigg|_{z+\Delta z} \right) = D_i A \frac{\partial^2 c_i}{\partial z^2} \Delta z \tag{3.9}$$

Non-steady-state models in which only ordinary derivatives of the type dc/dt and dc/dz occur, that is, models consisting of ordinary differential equations, are termed *lumped models*.

Non-steady-state models in which partial derivatives such as $\partial c/\partial t$, $\partial c/\partial z$, $\partial^2 c/\partial z^2$, etc., occur, that is, models containing partial differential equations, are termed *distributed models*.

3.2.2 Change Due to the Chemical Reaction—Conversion

In chemical kinetics and chemical engineering, the concept of fractional conversion, or simply conversion, is widely used. The conversion X_i of component i is defined as

$$X_i = \frac{F_{i0} - F_i}{F_{i0}}; \quad F_i = F_{i0}(1 - X_i) \tag{3.10}$$

or, when the reaction volume is constant, as

$$X_i = \frac{c_{i0} - c_i}{c_{i0}}; \quad c_i = c_{i0}(1 - X_i) \tag{3.11}$$

X_i is dimensionless and varies between 0 and 1.

For a first-order reaction, the reaction rate can be expressed as

$$r = kc_i = kc_{i0}(1 - X_i) \tag{3.12}$$

with k the reaction rate coefficient (s^{-1}), which depends on the temperature (see Section 3.3.2).

3.2.3 Ideal Reactors (Homogeneous)

In chemical kinetics, four basic reactor concepts are used:

- batch reactor (BR)
- continuous stirred-tank reactor (CSTR)
- plug-flow reactor (PFR)
- temporal analysis of products (TAP) reactor

Reactors can be considered ideal if the conditions of flow and mixing are perfect, that is, not hampered by phenomena such as dispersion, short-circuiting, or dead spaces. Well-defined convectional transport is used in the BR, the CSTR, and the PFR. These reactors can be used for both homogeneous and heterogeneous reactions. The PFR is usually used for solid-catalyzed, gas-phase reactions, while in BRs and CSTRs generally at least one liquid phase is present.

Well-defined diffusional transport is used in the TAP reactor for studying heterogeneous catalytic reactions. This reactor is discussed in Section 3.2.4.

First, let us analyze ideal reactors of constant reaction volume in which a stoichiometrically single reaction takes place without explicitly taking into account the presence of a solid catalyst, that is, we are assuming the reaction is not catalyzed or is homogeneously catalyzed.

Reaction rates r are all expressed in moles per unit of reaction volume per second. If solid catalysts are involved, it is more convenient to express reaction rates per unit mass (r_W) or unit surface area (r_S) of catalyst.

3.2.3.1 The batch reactor

In an ideal BR, that is, a non-steady-state closed reactor with perfect mixing, the transport term is absent and Eq. (3.1) becomes

$$\text{temporal change of amount of component} = \text{change due to reaction} \qquad (3.13)$$

By definition, the BR is a non-steady-state reactor. The simplest mathematical model for the temporal change of the concentration of component i in a BR of constant reaction volume is

$$\frac{dc_i}{dt} = R_i = \nu_i r \qquad (3.14)$$

where R_i is the net rate of production of component i per unit reaction volume, ν_i is the stoichiometric coefficient, and r is the reaction rate. The convention is to assign negative stoichiometric coefficients to reactants and positive coefficients to products. Thus, R_i is also negative for reactants and positive for products.

BRs can be operated in two modes:

- The reaction is performed in the gas phase. The volume of the reaction mixture is considered to be constant and equal to the reactor volume. The pressure of the reaction mixture may change as a function of reactant conversion.
- The reaction is performed in the liquid phase. The volume of the reaction mixture may change as a function of reactant conversion. The pressure is typically constant.

3.2.3.2 The continuous stirred-tank reactor

A CSTR is an open reactor with perfect mixing (a gradientless reactor) and convective flow only. Mixing can be achieved by both internal and external recirculation. The material balance for any component in a non-steady-state CSTR can be written as

$$\frac{dc_i}{dt} = R_i + \frac{F_{i0} - F_i}{V} \qquad (3.15)$$

Obviously, a BR is a particular case of a CSTR, in which there is no exchange of matter between the reactor and the surroundings, so $F_i = F_{i0} = 0$.

Typically, CSTR experiments are performed in the steady-state regime, that is, $dc_i/dt = 0$. At steady state, the net rate of production of component i can be determined from

$$R_i = -\frac{F_{i0} - F_i}{V} = -\frac{q_{V0}c_{i0} - q_V c_i}{V} \tag{3.16}$$

If $q_V = q_{V0}$, Eq. (3.16) can be expressed as

$$R_i = -\frac{q_{V0}(c_{i0} - c_i)}{V} = -\frac{c_{i0} - c_i}{\tau} \tag{3.17}$$

where $\tau = V/q_{V0}$ is the *space time*. It is called space time because its definition involves a spatial variable, V, which distinguishes it from the *astronomic time*. Space time corresponds to the average residence time in an isothermal CSTR. In the steady-state CSTR, the net rate of production is equal to the difference between the outlet and inlet concentrations divided by the space time. In terms of conversion, see Eqs. (3.10), (3.11), for a first-order reaction, Eq. (3.17) can be written as

$$k(1 - X_i) = \frac{X_i}{\tau} \tag{3.18}$$

or

$$X_i = \frac{k\tau}{1 + k\tau} \tag{3.19}$$

The term $k\tau$, also known as the first Damköhler number, Da_I, that is, the ratio of the time scale for transport from inlet to outlet of the reactor to the time scale of the reaction (see also Section 3.2.1), is the main characteristic of the CSTR. For large $k\tau$ ($\gg 1$) conversion is complete, $X_i = 1$. If $k\tau$ is small ($\ll 1$), $X_i \approx k\tau$. Knowing the conversion, an apparent first-order rate coefficient can be determined:

$$k = \frac{X_i}{1 - X_i}\frac{1}{\tau} \tag{3.20}$$

3.2.3.3 The plug-flow reactor

In an ideal PFR, axial diffusion effects are neglected. It is also assumed that perfect uniformity is achieved in the radial direction, which is the direction perpendicular to that of the flow. This is relatively easy to achieve in tubular reactors with a high large length-to-diameter ratio, the so-called aspect ratio. The composition of the fluid phase varies along the reactor, so the material balance for any component of the reaction mixture must be established for a differential element:

$$dV\frac{dc_i}{dt} = R_i dV - q_V dc_i \tag{3.21}$$

More rigorously, Eq. (3.21) can be written as a partial differential equation:

$$dV\frac{\partial c_i}{\partial t} = R_i dV - q_V \frac{\partial c_i}{\partial z}dz \tag{3.22}$$

Using $q_V = uA$ and $dV = Adz$, Eq. (3.22) can be written as

$$\frac{\partial c_i}{\partial t} = R_i - u\frac{\partial c_i}{\partial z} \qquad (3.23)$$

or

$$\frac{\partial c_i}{\partial t} + \frac{\partial c_i}{\partial \tau} = R_i \qquad (3.24)$$

with $\tau = z/u$.

At steady state, when $\partial c_i/\partial t = 0$, the model equation for an ideal PFR can be expressed by the ordinary differential equation

$$\frac{dc_i}{d\tau} = R_i \qquad (3.25)$$

which remarkably is almost identical to the expression for a BR, Eq. (3.14). The only difference is the meaning of the term "time" used. In the model for the BR, the time is the time of the experimental observation or "astronomic time," whereas the time in the model for the PFR is the space time, τ.

To obtain the net rate of production, we have to find the derivative in Eq. (3.25) based on experimental data. For a steady-state PFR, the longitudinal concentration profile is usually difficult to measure, and the measured characteristic is the gas-phase concentration at the reactor exit. For any given reactor length, the space time will change by systematically increasing or decreasing the flow rate. This is the so-called differential method of analyzing PFR data. In practice, however, the experimental error is rather large because net rates of production of components cannot be measured directly, in contrast with the CSTR data analysis.

In some cases, a so-called differential PFR model is used. The differential PFR can be considered as a hybrid between a CSTR and an ordinary PFR, with conversions sufficiently small not to affect the reaction rate. This model can be applied to reactors with a thin active zone or a high feed flow rate or both. In a differential PFR, the reaction zone can be assumed to be perfectly mixed. The concentrations inside the reaction zone can be taken as the inlet or outlet concentration or as the average of these two concentrations. For a differential PFR, Eq. (3.25) is reduced to Eq. (3.17).

3.2.4 Ideal Reactors With Solid Catalyst

3.2.4.1 The batch reactor

The equivalent of Eq. (3.14) for a BR containing a solid catalyst is

$$\frac{1}{\rho_{cat}}\frac{\varepsilon_b}{1-\varepsilon_b}\frac{dc_i}{dt} = R_{W,i} = \nu_i r_W \qquad (3.26)$$

where $R_{W,i}$ is the net rate of production of component i per unit mass of catalyst, r_W is the reaction rate per unit mass of catalyst, ρ_{cat} is the density of the catalyst pellet, and ε_b is the void fraction of the catalyst bed. The void fraction is the ratio of the fluid volume to that of the total reactor volume, $V_f/V = V_f/(V_f + V_{cat})$. Typical values are between 0.4 and 0.46 (Ergun, 1952).

3.2.4.2 The continuous stirred-tank reactor

In the case of a CSTR with a solid catalyst, the material balance for component i, Eq. (3.15), can be written as

$$\frac{1}{\rho_{cat}}\frac{\varepsilon_b}{1-\varepsilon_b}\frac{dc_i}{dt} = R_{W,i} + \frac{F_{i0} - F_i}{W_{cat}} \tag{3.27}$$

with W_{cat} the mass of catalyst in the reactor. The net rate of production of component i at steady state, that is, $dc_i/dt = 0$, can be determined from

$$R_{W,i} = -\frac{F_{i0} - F_i}{W_{cat}} \tag{3.28}$$

With

$$X_i = \frac{F_{i0} - F_i}{F_{i0}} \tag{3.29}$$

we obtain

$$R_{W,i} = -\frac{F_{i0}X_i}{W_{cat}} \tag{3.30}$$

where W_{cat}/F_{i0} is also often referred to as the space time.

3.2.4.3 The plug-flow reactor

For the PFR with a solid catalyst, the equivalent of Eq. (3.21) is

$$\frac{1}{\rho_{cat}}\frac{\varepsilon_b}{1-\varepsilon_b}dW_{cat}\frac{dc_i}{dt} = R_{W,i}dW_{cat} - dF_i \tag{3.31}$$

At steady state, that is, $dc_i/dt = 0$,

$$R_{W,i} = \frac{dF_i}{dW_{cat}} = q_V\frac{dc_i}{dW_{cat}} \tag{3.32}$$

or

$$R_{W,i} = -\frac{F_{i0}dX_i}{dW_{cat}} = -\frac{dX_i}{d(W_{cat}/F_{i0})} \tag{3.33}$$

3.2.4.4 The temporal analysis of products reactor

The TAP reactor, created by John Gleaves in 1988, typically contains a fixed catalyst bed. A small amount of a chemical substance is injected into the reactor during a small time interval. In a conventional pulse reactor, the substance is pulsed into an inert steady carrier-gas stream. The relaxation of the outlet composition following the perturbation by this pulse provides information about the reaction kinetics. In the TAP reactor, no carrier-gas stream is used and the substance is pulsed directly into the reactor. Transport occurs by diffusion only, in particular by well-defined Knudsen diffusion. The Knudsen diffusion coefficient does not depend on the composition of the reacting gas mixture. In a thin-zone TAP reactor (TZTR), the catalyst is located within a narrow zone only, similar to the differential PFR.

The net rate of production of component i in the catalyst zone of the TZTR is the difference between two diffusional flow rates at the boundaries of the thin active zone divided by the mass of catalyst in the reactor:

$$R_{W,i} = -\frac{F_{i0}(t) - F_i(t)}{W_{\text{cat}}} \tag{3.34}$$

To some extent this is analogous to the case of the steady-state differential PFR and the steady-state CSTR, but in these reactors the net rate of production is given by the difference between convectional flow rates.

In industry, pulse regimes are used in a number of adsorption/desorption processes, in particular in swing adsorption processes in which gaseous components are separated from a mixture of gases under pressure based on a difference in affinity for the adsorbent material.

3.2.5 Summary

The conceptual material balance equation, Eq. (3.1), is often written as

temporal change of amount of component = flow in − flow out + reaction change $\qquad$ (3.35)

In a BR, the "flow in" and "flow out" terms are absent, while in a CSTR, both flow terms are present. In a PFR, both flow terms are present too, but "flow in − flow out" is presented in differential form. In pulse reactors, initially there is only the "flow in" term, while later there is only the "flow out" term.

The conceptual difference between the different methods of measuring the net rate of production is presented in Table 3.1. Experimentally, these three types of rate measurement differ regarding both the reactor construction and the extraction of the value of R_i. Technically, the PFR is the simplest reactor for measuring steady-state data, but the differential method of analyzing PFR data is a quite complicated experimental procedure. The CSTR has an advantage over the BR and PFR regarding the accuracy of the estimated rate: the rate in a CSTR

Table 3.1 Determining the net rate of production in ideal reactors

Reactor	Homogeneous	Heterogeneous
BR (non-steady-state)	$R_i = \dfrac{dc_i}{dt}$	$R_{W,i} \propto \dfrac{dc_i}{dt}$
CSTR (steady state)	$R_i = -\dfrac{F_{i0} - F_i}{V} = -\dfrac{q_{v0}c_{i0} - q_v c_i}{V}$ At constant q_V: $R_i = -\dfrac{c_{i0} - c_i}{\tau}; \tau = \dfrac{V}{q_{v0}}$	$R_{W,i} = -\dfrac{F_{i0} - F_i}{W_{cat}}$
PFR (steady state)	$R_i = \dfrac{dc_i}{d\tau}; \tau = \dfrac{z}{u}$	$R_{W,i} = \dfrac{dF_i}{dW_{cat}}$
TZTR	Not applicable	$R_{W,i} = -\dfrac{F_{i0}(t) - F_i(t)}{W_{cat}}$

can be determined directly from the difference between the inlet and outlet concentrations, see Eq. (3.16), whereas the rate in a BR or PFR must be determined from the concentration derivative, leading to a larger experimental error. The TZTR has great potential for extracting detailed non-steady-state kinetic information. Statistical problems in kinetic analysis will be discussed in Chapter 9.

3.3 Stoichiometry: Extracting the Reaction Rate From the Net Rate of Production

3.3.1 Complexity and Mechanism: Definition of Elementary Reactions

Most chemical reactions are complex in nature. They proceed through a so-called reaction mechanism or, equivalently, detailed mechanism, or just mechanism that consists of a number of steps, referred to as *elementary steps*. Each elementary step comprises a forward and a reverse *elementary reaction*. Rigorously, every step and every overall reaction is reversible but in reality many steps and overall reactions can be considered to be irreversible. An elementary reaction takes place exactly as written and elementary reaction is characterized by one energetic barrier. The rate of an elementary reaction can be defined as the number of elementary acts of chemical transformation per unit volume of the reaction mixture (or per unit catalyst surface area, etc.) per unit time.

If an elementary reaction involves one reactant molecule ($A \rightarrow B$), it is classified as a unimolecular reaction or a first-order reaction. If two molecules take part in the reaction (eg, $2A \rightarrow B$ or $A + B \rightarrow C$), the reaction is called bimolecular or second order. With the participation of three molecules ($3A \rightarrow B$, or $2A + B \rightarrow C$), the reaction is said to be termolecular or third order. The simultaneous interaction of more than three reactant molecules in one elementary reaction is believed to be highly improbable and even termolecular reactions are very rare.

Examples of monomolecular or first-order reactions are

$$OH\cdot \rightarrow O\cdot + H\cdot \qquad (3.36)$$

and

$$HO_2\cdot \rightarrow OH\cdot + O\cdot \qquad (3.37)$$

The rates of these reactions are $r = kc_{OH\cdot}$ and $r = kc_{HO_2\cdot}$, respectively.

The reactions

$$2OH\cdot \rightarrow H_2O_2 \qquad (3.38)$$

and

$$H_2 + O_2 \rightarrow 2OH\cdot \qquad (3.39)$$

are bimolecular or second order with rates $r = kc_{OH\cdot}^2$ and $r = kc_{H_2}c_{O_2}$, respectively.

The reaction

$$2NO + O_2 \rightarrow 2NO_2 \qquad (3.40)$$

is one of the few examples of a termolecular or third order reaction, with $r = kc_{NO}^2 c_{O_2}$.

As a rule, in elementary reactions, reactants and products are different. An example of an exception to this rule is one of the steps in the thermal dissociation of hydrogen:

$$H\cdot + H_2 \rightarrow 3H\cdot \qquad (3.41)$$

In the forward reaction of this step, one of the reactants (radical $H\cdot$) is also the product of the reaction. This is an example of a so-called autocatalytic reaction.

3.3.2 Homogeneous Reactions: Mass-Action Law

The rate of an elementary step is determined by the difference between the rates of the forward and the reverse reactions:

$$r = r^+ - r^- \qquad (3.42)$$

where r, r^+, and r^- are the rate of the step, the rate of the forward reaction and the rate of the reverse reaction, respectively.

An elementary step can be described by

$$\sum \alpha_i A_i \underset{k^-}{\overset{k^+}{\rightleftharpoons}} \sum \beta_i B_i \qquad (3.43)$$

where A_i and B_i are reactants and products with α_i and β_i the absolute values of their stoichiometric coefficients, and k^+ and k^- are the rate coefficients for the forward and reverse reactions. In addition to the mentioned limitations on the values of α_i and β_i (for elementary reactions these coefficients equal one, or two, or, rarely three), the sums of the coefficients α_i and β_i must also not be larger than three. The dependence of the rates of the forward and reverse reactions on the concentrations of reactants is expressed in terms of the *mass-action law* as

$$r^+ = k^+ c_{A_1}^{\alpha_1} c_{A_2}^{\alpha_2} \cdots = k^+ \Pi c_{A_i}^{\alpha_i} \tag{3.44}$$

$$r^- = k^- c_{B_1}^{\beta_1} c_{B_2}^{\beta_2} \cdots = k^- \Pi c_{B_i}^{\beta_i} \tag{3.45}$$

where c_{Ai} and c_{Bi} are the concentrations of reactants and products. Mass-action-law equations for elementary reactions can be considered as particular cases of *power-law* relationships. These semiempirical equations characterize actual experimental kinetic dependences for many reactions. Unlike for elementary reactions, the coefficients α_i and β_i in power-law equations may have values other than one, two, or three, including zero and fractional values, and may be positive or negative. The rate coefficients determine the reaction rates of the forward and the reverse reaction at unitary values of reactant concentrations. They are governed by the Arrhenius dependences and increase exponentially with inverse absolute temperature:

$$k^+ = k_0^+ \exp\left(-\frac{E_a^+}{R_g T}\right) \tag{3.46}$$

$$k^- = k_0^- \exp\left(-\frac{E_a^-}{R_g T}\right) \tag{3.47}$$

Here k_0^+ and k_0^- are the pre-exponential factors, E_a^+ and E_a^- are the activation energies, R_g is the universal gas constant, and T is the absolute temperature. The ratio of the rate coefficients of the forward and reverse reaction determines the equilibrium coefficient K_{eq}:

$$\frac{k^+}{k^-} = K_{eq} \tag{3.48}$$

At constant pressure, the difference between the activation energies for the forward and reverse reaction determines the reaction enthalpy:

$$\Delta_r H = E_a^+ - E_a^- \tag{3.49}$$

For an exothermic reaction, that is, a reaction in which heat is released, $\Delta_r H < 0$. For an endothermic reaction, in which heat is consumed, $\Delta_r H > 0$.

Table 3.2 shows orders of magnitude of the kinetic parameters for first-order reactions. The pre-exponential factor for a monomolecular reaction is about 10^{13} s^{-1}.

3.3.3 Heterogeneous Reactions

In the case of solid-catalyzed gas-phase reactions, reactants in elementary steps can be gaseous components or surface intermediates. Any chemical step that involves a catalyst can be written as

Table 3.2 Orders of magnitude of the kinetic parameters for first-order homogeneous reactions

Reaction Type	Rate Coefficient k (s^{-1})	Reaction Type	Activation Energy E_a (kJ mol^{-1})
Slow	$<10^{-3}-10^{-2}$	Low	<20
Moderate	$10^{-2}-10$	Moderate	$20-130$
Fast	>10	High	>130

$$\sum \alpha_i A_i + \sum \alpha_j X_j \underset{k^-}{\overset{k^+}{\rightleftharpoons}} \sum \beta_i B_i + \sum \beta_j Y_j \qquad (3.50)$$

where A_i and B_i are reactants and products in the gas phase with α_i and β_i the absolute values of their stoichiometric coefficients and X_j and Y_j are surface intermediates with α_j and β_j the absolute values of their stoichiometric coefficients. Typically, Eq. (3.50) is of the form

$$\alpha A + \sum \alpha_j X_j \underset{k^-}{\overset{k^+}{\rightleftharpoons}} \beta B + \sum \beta_j Y_j \qquad (3.51)$$

Moreover, α and β are either one or zero, because in an elementary catalytic reaction only one molecule from the gas phase reacts (eg, $CH_4 + Z \rightarrow CH_2Z + H_2$) or none at all (eg, $CHOHZ \rightarrow COZ + H_2$). The rates of the forward and reverse reactions can be written analogously to Eqs. (3.44), (3.45):

$$r_S^+ = k^+ \Pi c_{A_i}^{\alpha_i} \Pi \theta_{X_j}^{\alpha_j} \qquad (3.52)$$

$$r_S^- = k^- \Pi c_{B_i}^{\beta_i} \Pi \theta_{Y_j}^{\beta_j} \qquad (3.53)$$

in which α_i and β_i are either zero or one and $\alpha_j \leq 2$ if $\alpha_i = 1$, $\alpha_j \leq 3$ if $\alpha_i = 0$, $\beta_j \leq 2$ if $\beta_i = 1$ and $\beta_j \leq 3$ if $\beta_i = 0$. r_S^+ and r_S^- are the forward and reverse reaction rates per unit catalyst surface area. The θ_{X_j} and θ_{Y_j} are the normalized surface concentrations or fractional surface coverages of surface intermediates X_j and Y_j:

$$\theta_{X_j} = \frac{\Gamma_{X_j}}{\Gamma_t}; \quad \theta_{Y_j} = \frac{\Gamma_{Y_j}}{\Gamma_t} \qquad (3.54)$$

with Γ_{X_j} and Γ_{Y_j} the surface concentrations of intermediates X_j and Y_j and Γ_t the total concentration of surface intermediates, including free active sites. The total concentration of active sites can be determined in separate experiments such as adsorption experiments or multipulse response experiments under high-vacuum conditions. In the case of solid-catalyzed liquid-phase reactions, reactants in elementary steps can be bulk liquid components or intermediates on the catalyst surface. The reaction may occur within a thin liquid film on the catalyst surface.

3.3.4 Rate Expressions for Single Reactions: Stoichiometry

For a homogeneous reaction system without exchange of matter with the surrounding medium, the rate of a stoichiometrically single reaction, either elementary or complex, can be expressed as

$$r = -\frac{1}{\alpha_i V}\frac{dn_{A_i}}{dt} = \frac{1}{\beta_i V}\frac{dn_{B_i}}{dt} \tag{3.55}$$

Here n_{A_i} and n_{B_i} are the number of moles of reactants and products. For a heterogeneously catalyzed reaction in a closed system, the reaction rate can be expressed as, for instance,

$$r_S = -\frac{1}{\alpha_i S_{cat}}\frac{dn_{A_i}}{dt} = \frac{1}{\beta_i S_{cat}}\frac{dn_{B_i}}{dt} \tag{3.56}$$

Here S_{cat} is the catalyst surface area and the rate r_S is the rate per unit catalyst surface area. The reaction rate can also be expressed per unit volume of catalyst, r_V, or per unit mass of catalyst r_W. These rates can be transformed easily into each other:

$$r_S = \frac{V_{cat}}{S_{cat}}r_V = \frac{V_{cat}\rho_{cat}}{S_{cat}}r_W = \frac{W_{cat}}{S_{cat}}r_W \tag{3.57}$$

For chemical processes without a change in the number of moles during the course of the reaction, the reaction rate takes the traditional form:

$$r = -\frac{1}{\alpha_i}\frac{dc_{A_i}}{dt} = \frac{1}{\beta_i}\frac{dc_{B_i}}{dt} \tag{3.58}$$

Similarly, the rate of a solid-catalyzed reaction can then be written as

$$r_S = -\frac{V_f}{\alpha_i S_{cat}}\frac{dc_{A_i}}{dt} = \frac{V_f}{\beta_i S_{cat}}\frac{dc_{B_i}}{dt} \tag{3.59}$$

3.3.5 Relationships Between Reaction Rate and Net Rate of Production

When interpreting information from kinetic measurements one should take into account that there is a difference between the measured value of the *net rate of production* of component i, R_i, and the *reaction rate*, r. For a single stoichiometric reaction step, the relationship between r and R_i can be expressed as follows:

$$r = \frac{R_i}{\nu_i} \quad \text{or} \quad R_i = \nu_i r \tag{3.60}$$

where ν_i is the stoichiometric coefficient of component i.

For example, the reaction rate for the elementary step $A \rightleftarrows B$ is given by

$$r = r^+ - r^- = k^+ c_A - k^- c_B \tag{3.61}$$

Now,

$$r = \frac{R_A}{\nu_A} = \frac{R_B}{\nu_B} = \frac{R_A}{-1} = \frac{R_B}{1} \Rightarrow r = -R_A = R_B \tag{3.62}$$

For the elementary step $2A + B \rightleftarrows 3C$,

$$r = r^+ - r^- = k^+ c_A^2 c_B - k^- c_C^3 \tag{3.63}$$

and

$$r = -\frac{R_A}{2} = -R_B = \frac{R_C}{3} \quad \text{or} \quad R_A = -2r, R_B = -r, R_C = 3r \tag{3.64}$$

In the case that a component is participating in multiple reactions, R_i is a linear combination of the rates in which this component is consumed or formed in the steps taking place, r_s. The coefficients in this linear combination are the stoichiometric coefficients ν_{is} of the component in each of the steps:

$$R_i = \sum_s r_{is} = \sum_s \nu_{is} r_s \tag{3.65}$$

Both the net rate of production and the reaction rate are used in many further data processing procedures, such as the determination of rate coefficients k, pre-exponential factors k_0, activation energies E_a, kinetic orders, and so on. The definitions of these rates have to be carefully distinguished. The net rate of production of a component is an experimentally observed characteristic. It is the change of the number of moles of a component per unit volume of reactor (or catalyst surface, volume, or mass) per unit time. The reaction rate r can be introduced only after a reaction equation has been assumed with the corresponding stoichiometric coefficients. Then, the value of the reaction rate can be calculated based on the *assumed* stoichiometric reaction equation. This is an important conceptual difference between the *experimentally observed* net rate of production and the *calculated* reaction rate, which is a result of our interpretation. The main methodological lesson is: Do not mix experimental measurements and their interpretation.

3.4 Distinguishing Kinetic Dependences Based on Patterns and Fingerprints

3.4.1 Single Reactions

3.4.1.1 Irreversible reaction

Consider a BR in which a single irreversible reaction $A \rightarrow B$ takes place. The mathematical model for this reaction is the ordinary differential equation

$$\frac{dc_A}{dt} = R_A = -r \tag{3.66}$$

An obvious fingerprint of the irreversible reaction is that given sufficient time the final conversion of the reactants is complete.

The kinetic expression can be written as

$$r = kc_A^n = kc_{A0}^n(1 - X_A)^n \tag{3.67}$$

where n is an apparent reaction order. One can distinguish the apparent order n based on the change of concentration in time. A useful concept here is the "half-life," the time during which the reactant concentration is reduced to half its initial value.

(1) First-order reaction, $r = kc_A$.

Solving the differential equation (3.66) with the first-order rate expression leads to

$$\ln c_A = \ln c_{A0} - kt \quad \text{or} \quad \ln\left(\frac{c_{A0}}{c_A}\right) = kt \tag{3.68}$$

The result is a linear dependence "$\ln c_A$ versus t." A plot of $\ln c_A$ as a function of time t gives as straight line with slope $-k$. The half-life, $t_{1/2}$, at which $c_A = c_{A0}/2$, is then given by

$$t_{1/2} = \frac{\ln 2}{k} \approx \frac{0.693}{k} \tag{3.69}$$

The half-life for this reaction does not depend on the initial concentration of A.

(2) Second-order reaction, $r = kc_A^2$.

Solving Eq. (3.66) gives

$$\frac{1}{c_A} = \frac{1}{c_{A0}} + kt \tag{3.70}$$

A linear dependence "$(1/c_A)$ versus t" is obtained. The slope of the straight line obtained is k. In this case, the half-life is inversely proportional to the initial concentration:

$$t_{1/2} = \frac{1}{kc_{A0}} \tag{3.71}$$

(3) Zero-order reaction, $r = k$.

A linear dependence "c_A versus t" is observed:

$$c_A = c_{A0} - kt \tag{3.72}$$

The slope of the resulting straight line is $-k$. In this case, the half-life is proportional to the initial concentration:

$$t_{1/2} = \frac{c_{A0}}{2k} \qquad (3.73)$$

(4) nth order reaction, $r = kc_A^n, 0 < n < 1$

A linear dependence "$(1/c_A)^{(n-1)}$ versus t" is obtained:

$$\left(\frac{1}{c_A}\right)^{(n-1)} = \left(\frac{1}{c_{A0}}\right)^{(n-1)} + (n-1)kt \qquad (3.74)$$

Fig. 3.2 illustrates the dependence of the reaction rate on the reactant concentration for zero-, first-, and second-order reactions.

3.4.1.2 Reversible reaction

For a first-order reversible reaction $A \rightleftarrows B$, $r = r^+ - r^- = k^+ c_A - k^- c_B$, where k^+ and k^- are the rate coefficients of the forward and reverse reaction. In accordance with the law of mass conservation, assuming the reactor initially only contains A, $c_A + c_B = c_{A0}$.

At equilibrium, the solution of Eq. (3.66) is

$$c_A - c_{A,eq} = \left(c_{A0} - c_{A,eq}\right) \exp\left(-(k^+ + k^-)t\right) \qquad (3.75)$$

where the equilibrium concentration of A is given by

$$c_{A,eq} = \frac{k^-}{k^+ + k^-} c_{A0} = \frac{1}{K_{eq} + 1} c_{A0} \qquad (3.76)$$

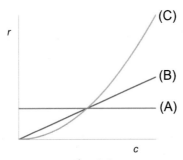

Fig. 3.2

Typical dependences of the reaction rate on the reactant concentration: (A) zero order; (B) first order; and (C) second order. In all cases the rate coefficient equals one.

with equilibrium coefficient $K_{eq} = k^+/k^-$. At $t_{1/2}$, substitution of the first part of Eq. (3.76) in Eq. (3.75) with $c_A = c_{A0}/2$ leads to

$$\frac{k^+ - k^-}{2(k^+ + k^-)} = \frac{k^+}{k^+ + k^-} \exp\left(-(k^+ + k^-)t_{1/2}\right) \tag{3.77}$$

or, after rearrangement,

$$\frac{2k^+}{k^+ - k^-} = \exp\left((k^+ + k^-)t_{1/2}\right) \tag{3.78}$$

from which the half-life can be determined to be

$$t_{1/2} = \frac{1}{k^+ + k^-} \ln\frac{2K_{eq}}{K_{eq} - 1} \tag{3.79}$$

An obvious fingerprint of a reversible reaction is that the final conversion of the reactants is not complete, $c_{A,final} = c_{A,eq} \neq 0$.

It is interesting to note that for both the irreversible reaction $A \rightarrow B$ and the reversible reaction $A \rightleftarrows B$, at the point of intersection of the concentrations of reactant A and product B as a function of time, $c_A = c_B = c_{A0}/2$. This is the simple consequence of the material balance $c_A + c_B = c_{A0}$. The intersection point is at the half-time point $t_{1/2}$.

3.4.1.3 Reversible reaction with Langmuir-Hinshelwood-Hougen-Watson kinetics

In many cases, kinetic data are characterized by a well-distinguished plateau, see Fig. 3.3.

Such data are described by an equation of the Langmuir-Hinshelwood-Hougen-Watson (LHHW) type:

$$r = \frac{k^+ f^+ (c_r) - k^- f^- (c_p)}{\sum_l k_l \prod_i c_i^{p_{li}}} = \frac{k^+ \left(f^+ (c_r) - \dfrac{f^- (c_p)}{K_{eq}}\right)}{\sum_l k_l \prod_i c_i^{p_{li}}} \tag{3.80}$$

or,

Fig. 3.3
Typical dependences of the reaction rate on the reactant concentration for LHHW kinetics.

$$r = \frac{f^+(\mathbf{c_r}) - \dfrac{f^-(\mathbf{c_p})}{K_{eq}}}{\sum_l \bar{k}_l \prod_i c_i^{p_{li}}} \tag{3.81}$$

Here, $\mathbf{c_r}$ and $\mathbf{c_p}$ are sets of reactant and product concentrations; k_l are apparent kinetic parameters of the steady-state kinetic description, $\bar{k}_l$ ($= k_l/k^+$) are modified kinetic parameters, and p_{li} is a positive integer. The numerator in these equations is often referred to as the driving force, while the denominator may be termed the "kinetic resistance."

For an irreversible overall reaction, Eq. (3.80) becomes:

$$r = \frac{f^+(\mathbf{c_r})}{\sum_l \bar{k}_l \prod_i c_i^{p_{li}}} \tag{3.82}$$

This equation, the Langmuir-Hinshelwood equation, was first proposed by Langmuir and Hinshelwood in the 1920–30s for solid-catalyzed gas-phase reactions under the assumption that adsorption and desorption rates are high compared with rates of other chemical transformations on the catalyst surface. In this model, adsorption-desorption steps are considered to be at equilibrium. Later, Hougen, and Watson proposed a similar equation, the Hougen-Watson equation, for a reversible catalytic reaction, again under the assumption that the adsorption-desorption steps are at equilibrium.

In the 1970–80s, Yablonsky et al. demonstrated that the LHHW equation can be used far beyond the equilibrium assumption. For details see Yablonskii et al. (1991) and Marin and Yablonsky (2011). It was shown that the typical kinetic descriptions of Eqs. (3.80), (3.81) have interesting properties related to the level of the detailed mechanism:

- The numerator can be constructed based on the overall reaction only. This numerator does not depend on the complexity of the reaction mechanism.
- The denominator, that is, the kinetic resistance, reflects the complexity of the reaction mechanism.
- In the denominator, every term is a monomial of concentrations. Every such term corresponds to some path for producing a particular intermediate. The reactions of this path involve gaseous reactants and products, the concentrations of which are incorporated in this term.
- The kinetic parameters of the kinetic description, k_l and $\bar{k}_l$, are exponential functions of the temperature or a sum of such functions.

It was also shown (Lazman and Yablonsky, 2008) that the LHHW equation can even be applied to problems "far from equilibrium" as a "thermodynamic branch" of a more complicated description. In fact, the LHHW equation is a particular case of the so-called kinetic polynomial, which is an implicit representation, $f(r,c,T)=0$, and not the traditional explicit one, $r=f(c,T)$ (see Chapter 9 in Marin and Yablonsky, 2011).

In some situations, the kinetic descriptions of Eqs. (3.80), (3.81) can be even more simplified. Considering that the reaction rate can be determined from experimental data and that for an irreversible reaction the value of $f^+(c_r)$ is known and that for a reversible reaction the value of $f^+(c_r)-f^-(c_p)/K_{eq}$, is known, the kinetic resistance can be determined. Using this resistance in a description of steady-state kinetic data can be advantageous because in contrast with the original LHHW equation, this equation is linear regarding the estimated coefficients $\bar{k}_l$.

3.4.2 Distinguishing Between Single and Multiple Reactions

In a single reaction (elementary or complex), the relation between the net rates of production of the reactants and products is given by the stoichiometric coefficients of the reaction, which are integers. Stoichiometric equations for "real" complex reactions are free of this limitation.

For example, for the oxidation reaction of ethylene oxide, the stoichiometric equation is

$$2C_2H_4O+5O_2 \rightarrow 4CO_2+4H_2O \tag{3.83}$$

and the ratios of the net rates of production of C_2H_4O, O_2, CO_2, and H_2O are $-2:-5:4:4$. This reaction is a single one but not an elementary one. If measured net rates of production do not correspond to a single stoichiometric relationship, there is not a single reaction, but multiple reactions occur. For example, the oxidation of ethylene to ethylene oxide involves three reactions:

$$2C_2H_4+O_2 \rightarrow 2C_2H_4O \tag{3.84}$$

$$C_2H_4+3O_2 \rightarrow 2CO_2+2H_2O \tag{3.85}$$

and oxidation of the formed ethylene oxide, reaction (3.83). Proportional relationships between the net rates of production of all participating reactants and products, that is, C_2H_4, O_2, C_2H_4O, CO_2, and H_2O, are not observed.

3.4.3 Distinguishing Between Elementary and Complex Reactions

There are two fingerprints of complex reactions in comparison with elementary ones:

- Integer stoichiometric coefficients
  As mentioned above, for complex reactions the stoichiometric coefficients may be larger than three, in contrast with those for elementary reactions.

The limitation that the sum of these coefficients in each reaction, forward or reverse, must not be larger than three, is also absent for complex reactions.

- Kinetic orders

  For complex reactions, the reaction rate cannot be described by *mass-action law* dependences, see Eqs. (3.44), (3.45), with stoichiometric coefficients α_i and β_i and their sum ≤ 3. Furthermore, if the rate coefficients of Eqs. (3.44), (3.45) are not governed by Arrhenius exponential dependences on the temperature, Eqs. (3.46), (3.47), this can also be considered as a fingerprint of a complex reaction.

A reaction is not necessarily elementary if only one of the limitations mentioned above is true. For example, many reactions in which one, two, or three molecules are participating are not elementary. Furthermore, in some cases the kinetic law of a complex reaction may be approximated by the kinetic mass-action law of an elementary reaction. *Kinetic orders of single reactions*, apparent or true, can be found based on patterns presented by Eqs. (3.68), (3.70), (3.72).

3.4.4 Summary of Strategy

A typical strategy of distinguishing the kinetic dependences based on kinetic fingerprints and patterns is presented in Figs. 3.4–3.6.

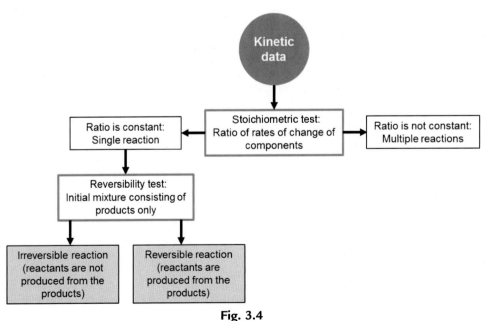

Fig. 3.4

Distinguishing between single and multiple reactions and between irreversible and reversible reactions.

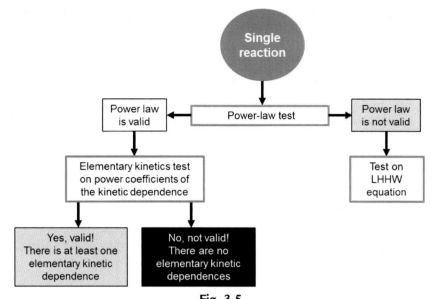

Fig. 3.5
Power-law test for single reactions.

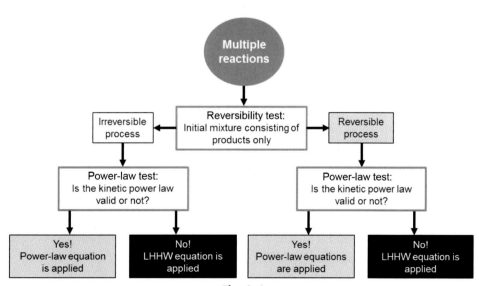

Fig. 3.6
Distinguishing between irreversible and reversible multiple reactions and power-law test.

3.5 Ordinary Differential Equations

The non-steady-state (dynamic) behavior of chemical reactions is described in terms of differential equations of the type

$$\frac{d\mathbf{c}}{dt}=f(\mathbf{c},\mathbf{k}) \tag{3.86}$$

in which $\mathbf{c}$ is a vector of concentrations and $\mathbf{k}$ is a vector of kinetic parameters. The space of vectors $\mathbf{c}$ is the phase space of Eq. (3.86). Its points are specified by coordinates $c_1,c_2,\ldots,c_{N_c}$. The set of points in the phase space is the set of all possible states of the chemical reaction mixture. The phase space can be the complete vector space but also only a certain part. In chemical kinetics, variables are either concentrations or quantities of chemical components in the mixture. Their values cannot be negative. Eq. (3.86) describes the temporal evolution of a chemical reaction mixture. Typically, this evolution occurs from an initial state to a final state, which may be stable or unstable. In physical chemistry, such a state is termed the *equilibrium state* or just *equilibrium* for a closed chemical system, that is, a system that does not exchange matter with its surroundings. For an open chemical system, which is characterized by the exchange of matter with the surroundings, this state is termed the *steady state*. Since a closed system is a particular case of an open system, the equilibrium state is a particular case of the steady state.

Usually the right-hand side of Eq. (3.89) does not explicitly contain the time variable t. In that case, we refer to the chemical reaction mixture as autonomous.

Three methods exist for studying non-steady-state behavior:

- change in time t
- change of parameters $\mathbf{k}$
- change of a concentration with respect to others

These methods correspond to changes of position in the dynamic space $(\mathbf{c}, t)$, the parametric space $(\mathbf{c}, \mathbf{k})$, and the phase space, respectively.

We will now analyze the dynamics of a catalytic isomerization reaction, $A \rightleftarrows B$, for which the detailed mechanism is the following:

$$\begin{aligned}(1)\ &A+Z\rightleftarrows AZ\\(2)\ &AZ\rightleftarrows BZ\\(3)\ &BZ\rightleftarrows B+Z\end{aligned} \tag{3.87}$$

For this reaction mechanism, Eq. (3.86) can be written out for the reactant, product, and intermediates:

$$\frac{dc_A}{dt}=\frac{S_{cat}}{V_f}(-1\cdot r_1-1\cdot r_1 c_A+0\cdot r_2+0\cdot r_3)+\frac{q_{v0}(c_{A0}-c_A)}{V_f} \tag{3.88}$$

$$\frac{dc_B}{dt} = \frac{S_{cat}}{V_f}(0 \cdot r_1 + 0 \cdot r_2 + 1 \cdot r_3 + 1 \cdot r_3 c_B) + \frac{q_{v0}(c_{B0} - c_B)}{V_f} \tag{3.89}$$

where r_1, r_2, and r_3 are the rates of steps 1, 2, and 3.

$$\frac{d\theta_Z}{dt} = -1 \cdot r_1 + 0 \cdot r_2 + 1 \cdot r_3 = -r_1 + r_3 \tag{3.90}$$

$$\frac{d\theta_{BZ}}{dt} = 0 \cdot r_1 + 1 \cdot r_2 - 1 \cdot r_3 = r_2 - r_3 \tag{3.91}$$

Also, $\sum_i c_i = c_A + c_B = c_t = \text{constant}$ and $\sum_j \theta_j = \theta_Z + \theta_{AZ} + \theta_{BZ} = 1$.

The rates of the steps can be expressed as

$$r_1 = r_1^+ - r_1^- = \left(k_1^+ c_A\right)\theta_Z - k_1^- \theta_{AZ} \tag{3.92}$$

$$r_2 = r_2^+ - r_2^- = k_2^+ \theta_{AZ} - k_2^- \theta_{BZ} \tag{3.93}$$

$$r_3 = r_3^+ - r_3^- = k_3^+ \theta_{BZ} - \left(k_3^- c_B\right)\theta_Z \tag{3.94}$$

where r_1^+, r_2^+, and r_3^+ are the reaction rates of the forward reactions with rate coefficients k_1^+, k_2^+, and k_3^+ and r_1^-, r_2^-, and r_3^- are the reaction rates of the reverse reactions with rate coefficients k_1^-, k_2^-, and k_3^-.

Assuming that the concentrations of the gaseous components are constant, the non-steady-state model of surface transformations can be written as

$$\frac{d\theta_Z}{dt} = -\left(k_1^+ c_A\right)\theta_Z + k_1^- \theta_{AZ} + k_3^+ \theta_{BZ} - \left(k_3^- c_B\right)\theta_Z \tag{3.95}$$

$$\frac{d\theta_{AZ}}{dt} = \left(k_1^+ c_A\right)\theta_Z - k_1^- \theta_{AZ} - k_2^+ \theta_{AZ} + k_2^- \theta_{BZ} \tag{3.96}$$

$$\frac{d\theta_{BZ}}{dt} = k_2^+ \theta_{AZ} - k_2^- \theta_{BZ} - k_3^+ \theta_{BZ} + \left(k_3^- c_B\right)\theta_Z \tag{3.97}$$

The solution to this model is of the exponential form:

$$\theta_Z = \alpha e^{\lambda t} \tag{3.98}$$

$$\theta_{AZ} = \beta e^{\lambda t} \tag{3.99}$$

$$\theta_{BZ} = \gamma e^{\lambda t} \tag{3.100}$$

where λ is a set of characteristic roots, or eigenvalues.

We then obtain:

$$\lambda \alpha e^{\lambda t} = -\left(k_1^+ c_A\right)\alpha e^{\lambda t} + k_1^- \beta e^{\lambda t} + k_3^+ \gamma e^{\lambda t} - \left(k_3^- c_B\right)\alpha e^{\lambda t} \tag{3.101}$$

$$\lambda \beta e^{\lambda t} = \left(k_1^+ c_A\right)\alpha e^{\lambda t} - k_1^- \beta e^{\lambda t} - k_2^+ \beta e^{\lambda t} + k_2^- \gamma e^{\lambda t} \tag{3.102}$$

$$\lambda \gamma e^{\lambda t} = k_2^+ \beta e^{\lambda t} - k_2^- \gamma e^{\lambda t} - k_3^+ \gamma e^{\lambda t} + \left(k_3^- c_B\right)\alpha e^{\lambda t} \tag{3.103}$$

Dividing by $e^{\lambda t}$ and rearranging, yields

$$-\alpha\left[\left(k_1^+ c_A\right) + \left(k_3^- c_B\right) + \lambda\right] + \beta k_1^- + \gamma k_3^+ = 0 \tag{3.104}$$

$$\alpha\left(k_1^+ c_A\right) - \beta\left(k_1^- + k_2^+ + \lambda\right) + \gamma k_2^- = 0 \tag{3.105}$$

$$\alpha\left(k_3^- c_B\right) + \beta k_2^+ - \gamma\left(k_3^+ + k_2^- + \lambda\right) = 0 \tag{3.106}$$

The set of Eqs. (3.104)–(3.106) can be represented as a matrix. Elements of its columns have factors α, β, and γ, respectively. Dividing the elements by these factors we obtain

$$\begin{bmatrix} -\left(k_1^+ c_A + k_3^- c_B + \lambda\right) & k_1^- & k_3^+ \\ k_1^+ c_A & -\left(k_1^- + k_2^+ + \lambda\right) & k_2^- \\ k_3^- c_B & k_2^+ & -\left(k_3^+ + k_2^- + \lambda\right) \end{bmatrix} = 0 \tag{3.107}$$

This is the characteristic equation for this isomerization mechanism and its corresponding model. The determinant of the matrix of Eq. (3.107) is

$$-\left(k_1^+ c_A + k_3^- c_B + \lambda\right)\begin{vmatrix} -\left(k_1^- + k_2^+ + \lambda\right) & k_2^- \\ k_2^+ & -\left(k_3^+ + k_2^- + \lambda\right) \end{vmatrix}$$
$$-k_1^- \begin{vmatrix} k_1^+ c_A & k_2^- \\ k_3^- c_B & -\left(k_3^+ + k_2^- + \lambda\right) \end{vmatrix} + k_3^+ \begin{vmatrix} k_1^+ c_A & -\left(k_1^- + k_2^+ + \lambda\right) \\ k_3^- c_B & k_2^+ \end{vmatrix} = 0 \tag{3.108}$$

from which it follows that

$$0 = \lambda^2 + \lambda\left(k_1^+ c_A + k_2^+ + k_3^+ + k_1^- + k_2^- + k_3^- c_B\right) + \left\{\left(k_1^+ k_2^+ + k_1^+ k_3^+ + k_1^+ k_2^-\right)c_A\right.$$
$$\left. + \left(k_2^+ k_3^- + k_1^- k_3^- + k_2^- k_3^-\right)c_B + k_2^+ k_3^+ + k_3^+ k_1^- + k_1^- k_2^-\right\} \tag{3.109}$$

From Eq. (3.109) we obtain three characteristic roots, λ_0, λ_1, and λ_2:

$$\begin{cases} \lambda_0 = 0 \\ \lambda_1 + \lambda_2 = -\left(k_1^+ c_A + k_2^+ + k_3^+ + k_1^- + k_2^- + k_3^- c_B\right) \\ \lambda_1 \lambda_2 = \left(k_1^+ k_2^+ + k_1^+ k_3^+ + k_1^+ k_2^-\right)c_A + \left(k_2^+ k_3^- + k_1^- k_3^- + k_2^- k_3^-\right)c_B + k_2^+ k_3^+ + k_3^+ k_1^- + k_1^- k_2^- \end{cases} \tag{3.110}$$

Since one of the characteristic roots is zero, a solution of the set of equations (3.95)–(3.97) can be written as

$$\theta_Z(t) = \theta_{Z,ss} + \alpha_1 e^{\lambda_1 t} + \alpha_2 e^{\lambda_2 t} \tag{3.111}$$

$$\theta_{AZ}(t) = \theta_{AZ,ss} + \beta_1 e^{\lambda_1 t} + \beta_2 e^{\lambda_2 t} \tag{3.112}$$

$$\theta_{BZ}(t) = \theta_{BZ,ss} + \gamma_1 e^{\lambda_1 t} + \gamma_2 e^{\lambda_2 t} \tag{3.113}$$

where $\theta_{Z,ss}$, $\theta_{AZ,ss}$, and $\theta_{BZ,ss}$ are the steady-state normalized concentrations of free active sites and surface intermediates. At $t \to \infty$, the concentrations of the surface intermediates reach these steady-state values.

If all steps of the mechanism would be irreversible, this would yield the following characteristic roots:

$$\begin{cases} \lambda_1 + \lambda_2 = -\left(k_1^+ c_A + k_2^+ + k_3^+\right) \\ \lambda_1 \lambda_2 = k_1^+ c_A \left(k_2^+ + k_3^+\right) + k_2^+ k_3^+ \end{cases} \tag{3.114}$$

It has to be stressed that the characteristic roots determining the relaxation to the steady state are not rate coefficients but rather complicated functions of rate coefficients.

The linear model has a unique positive solution. It is possible to show (see, eg, Yablonskii et al., 1991, p. 126), that the roots have negative real values or complex values with a negative real part ($\lambda < 0$). Complex roots are related to damped oscillations. Therefore, the relaxation process may lead to a single steady state with possibly damped oscillations. However, in reality the influence of the imaginary part of the roots will be insignificant and the damped oscillations will not be observable.

3.6 Graph Theory in Chemical Kinetics and Chemical Engineering

3.6.1 Introduction

Methods of mathematical graph theory have found wide applications in different areas of chemistry and chemical engineering. A graph is a set of points, *nodes*, connected by lines, *edges*. A graph can include *cycles*, which are finite sequences of graph edges with the same beginning and end node. It can correspond to, for instance, an electrical diagram, a representation of a railway network, a sequence of events, or, most relevant here, the mechanism of a complex chemical reaction.

Since the 1950s, graph methods have been widely used for representing complex networks of chemical reactions and deriving rate equations of complex reactions. King and Altman (1956) were the first to use these methods for representing enzyme-catalyzed reactions. In such reactions, typically all elementary reactions, forward and reverse, involve only one molecule of any intermediate. Therefore, mechanisms for enzyme-catalyzed reactions belong to the class of *linear reaction mechanisms*. These mechanisms generate a set of linear differential equations as a non-steady-state model, and a steady-state solution of this model is obtained by the solution of the corresponding system of linear algebraic equations.

In the methodology used by King and Altman (1956), the intermediates in the reaction mechanism are the nodes of the graph and the reactions are the edges. The direction of the reactions is indicated by arrows on the edges. Such graphs depict only a sequence of intermediate transformations from node to node. Nonintermediates, that is, reactants and products, participating in the elementary reactions may be indicated by placing their symbol at an arrow arriving at or leaving from an edge. The amounts of these nonintermediates are considered to be constant during the reaction.

3.6.2 Derivation of a Steady-State Equation for a Complex Reaction Using Graph Theory

Let us explain one of the most efficient applications of graph theory in chemistry. A steady-state kinetic model for intermediate j can be represented by the following equation:

$$-f_c\left(\theta_1, \theta_2, \ldots, \theta_j, \ldots, \theta_{N_{int}}\right) + f_g\left(\theta_1, \theta_2, \ldots, \theta_k, \ldots, \theta_{N_{int}}\right) = 0 \qquad (3.115)$$

where f_c and f_g are the kinetic dependences characterizing the consumption and generation of intermediate j, and θ_j is the normalized concentration of intermediate j. Typically, we do not consider an intermediate j to be involved in its production, that is, no autocatalysis occurs. Therefore, $j \neq k$ in f_g and the generation term does not depend on the concentration of intermediate j. In the general case, the model of Eq. (3.115) is nonlinear in the concentrations of the intermediates, while for linear mechanisms, that is, mechanisms in which all elementary reactions involve only one molecule of an intermediate, Eq. (3.115) is obviously linear.

3.6.2.1 Example: Two-step Temkin-Boudart mechanism

In heterogeneous catalysis, the simplest mechanism of a catalytic cycle is the two-step Temkin-Boudart mechanism. A particular case of this mechanism is the two-step mechanism of the water-gas shift (WGS) reaction (see Table 3.3). Table 3.4 shows a general representation of the

Table 3.3 Representation of the water-gas shift reaction

	σ
(1) $H_2O + Z \rightleftarrows H_2 + OZ$	1
(2) $OZ + CO \rightleftarrows Z + CO_2$	1
$H_2O + CO \rightleftarrows H_2 + CO_2$	

Table 3.4 General representation of the Temkin-Boudart mechanism

	σ
(1) $A + Z \rightleftarrows AZ$	1
(2) $AZ + B \rightleftarrows AB + Z$	1
$A + B \rightleftarrows AB$	

Temkin-Boudart mechanism. Catalytic intermediate Z is consumed in the first reaction and produced in the second one, while intermediate AZ (or OZ in the WGS mechanism) is produced in the first reaction and consumed in the second. Multiplying these reactions by the corresponding Horiuti numbers σ, which here equal one, and adding them, yields the overall reaction.

The steady-state kinetic model of the general Temkin-Boudart mechanism is written in accordance with the mass-action law as

$$Z: -\left(k_1^+ c_A\right)\theta_Z + \left(k_1^-\right)\theta_{AZ} + \left(k_2^+ c_B\right)\theta_{AZ} - \left(k_2^- c_{AB}\right)\theta_Z = 0 \tag{3.116}$$

$$AZ: \left(k_1^+ c_A\right)\theta_Z - \left(k_1^-\right)\theta_{AZ} - \left(k_2^+ c_B\right)\theta_{AZ} + \left(k_2^- c_{AB}\right)\theta_Z = 0 \tag{3.117}$$

The total normalized concentration of active sites must obey the active site balance:

$$\theta_Z + \theta_{AZ} = 1 \tag{3.118}$$

In Eqs. (3.116), (3.117), the coefficients in parentheses are "apparent" rate coefficients (Balandin, 1964) or "frequencies" (Schwab, 1982). These coefficients contain the rate coefficients of the reactions and the concentrations of reactants or products, which are considered to be constant. In graph theory, these coefficients are termed the weights of the edges or reaction weights w. We distinguish between the weights of the forward and reverse reactions, w_s^+ and w_s^-. The reaction weight is equal to the reaction rate at unitary concentrations of the reacting intermediate, that is, the rate divided by the concentration of the intermediate. The reaction weight equals a rate coefficient multiplied by the concentration of a gaseous component if it this component participates in the reaction. In all our considerations, it is assumed that no or only one gas molecule participates in an elementary reaction. The set of Eqs. (3.116)–(3.118) can easily be solved for θ_Z and θ_{AZ}. Substituting the obtained expressions for θ_Z and θ_{AZ} in the expressions for the rates of the two steps of the overall reaction,

$$r_1 = r_1^+ - r_1^- = \left(k_1^+ c_A\right)\theta_Z - k_1^- \theta_{AZ} = w_1^+ \theta_Z - w_1^- \theta_{AZ} \tag{3.119}$$

and

$$r_2 = r_2^+ - r_2^- = \left(k_2^+ c_A\right)\theta_{AZ} - \left(k_2^- c_{AB}\right)\theta_Z = w_2^+ \theta_{AZ} - w_2^- \theta_Z \tag{3.120}$$

we obtain

$$r_1 = r_2 = r = \frac{w_1^+ w_2^+ - w_1^- w_2^-}{w_1^+ + w_2^+ + w_1^- + w_2^-} \tag{3.121}$$

The outcome $r_1 = r_2 = \cdots = r$ is true for any reaction that proceeds via a single-route mechanism. At steady state, the rates of all elementary steps are equal and also equal to the rate of the overall reaction. There is a simple analogy from hydrodynamics illustrating this: if the flow through a closed pipeline system consisting of a set of different tubes is at steady state, the flow rate in any section of the pipeline is the same.

3.6.2.2 *General steady-state kinetic equations*

The general form of a set of steady-state kinetic equations for a linear mechanism can be represented by

$$\mathbf{W(c)}\boldsymbol{\theta}=0 \tag{3.122}$$

where $\mathbf{c}$ is the column vector of the concentrations of reactants and products participating in the overall reaction, $\boldsymbol{\theta}$ is the column vector of the normalized concentrations of the intermediates, and $\mathbf{W(c)}$ is the matrix of reaction weights:

$$\mathbf{W(c)} = \begin{bmatrix} -w_{11} & w_{12} & \cdots & w_{1n} \\ w_{21} & -w_{22} & \cdots & w_{2n} \\ \cdots & \cdots & \cdots & \cdots \\ w_{n1} & w_{n2} & \cdots & -w_{nn} \end{bmatrix} \tag{3.123}$$

In addition, the law of active site conservation must be fulfilled, in general dimensionless form:

$$\sum_{j=1}^{N_{\text{int}}} \boldsymbol{\theta}_j = 1 \tag{3.124}$$

The linear set of equations, Eq. (3.122), is typically solved using the well-known Cramer's rule. For a set of equations represented as

$$\mathbf{A}\boldsymbol{\theta} = \mathbf{b} \tag{3.125}$$

Cramer's rule states that

$$\boldsymbol{\theta}_j = \frac{\det\left(\mathbf{A}_j\right)}{\det\left(\mathbf{A}\right)} \tag{3.126}$$

where $\mathbf{A}_j$ is the matrix formed by replacing the jth column of $\mathbf{A}$ by the column vector $\mathbf{b}$. In our case, Eq. (3.122) has to be analyzed in combination with the linear balance of surface intermediates, Eq. (3.124). King and Altman (1956) used Cramer's rule as the basis for their pioneering graph approach. They modified it and found a relationship for the steady-state concentrations of intermediates. Later, Volkenstein and Goldstein (1966a,b) found a qualitative analogy between the King-Altman relationship and Mason's rule known in electrical engineering (Mason, 1953, 1956). They described the use of this rule for many biochemical reactions. A rigorous derivation of Mason's rule for kinetic equations is described in Yevstignejev and Yablonskii (1979). Details can be found in Yablonskii et al. (1983, 1991).

As already mentioned, the nodes of a graph are the catalytic reaction intermediates and the edges are the elementary reactions. The reaction weights are defined as

$$w_s^+ = \frac{r_s^+}{\theta_j^+}; \quad w_s^- = \frac{r_s^-}{\theta_j^-} \tag{3.127}$$

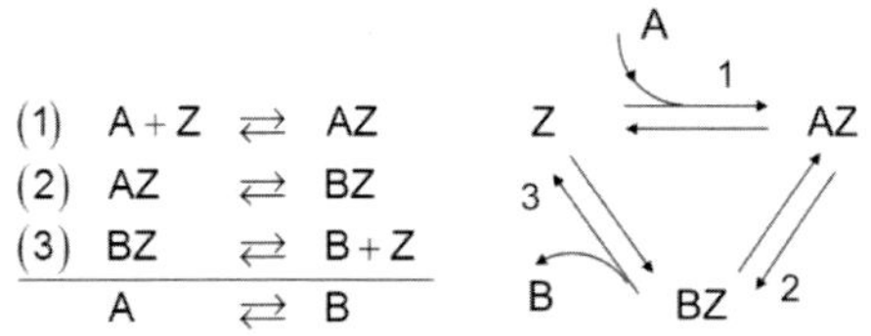

Fig. 3.7

Mechanism *(left)* and King-Altman graph *(right)* of the model isomerization reaction.

where θ_j^+ and θ_j^- are the normalized concentrations of the intermediates reacting in respectively the forward and reverse reaction. As an example, in the model isomerization reaction shown in Fig. 3.7, these weights are given by

$$w_1^+ = \frac{r_1^+}{\theta_Z} = \frac{k_1^+ c_A \theta_Z}{\theta_Z} = k_1^+ c_A; \quad w_1^- = \frac{r_1^-}{\theta_{AZ}} = \frac{k_1^- \theta_{AZ}}{\theta_{AZ}} = k_1^- \tag{3.128}$$

$$w_2^+ = \frac{r_2^+}{\theta_{AZ}} = \frac{k_2^+ \theta_{AZ}}{\theta_{AZ}} = k_2^+; \quad w_2^- = \frac{r_2^-}{\theta_{BZ}} = \frac{k_2^- \theta_{BZ}}{\theta_{BZ}} = k_2^- \tag{3.129}$$

$$w_3^+ = \frac{r_3^+}{\theta_{BZ}} = \frac{k_3^+ \theta_{BZ}}{\theta_{BZ}} = k_3^+; \quad w_3^- = \frac{r_3^-}{\theta_Z} = \frac{k_3^- c_B \theta_Z}{\theta_Z} = k_3^- c_B \tag{3.130}$$

3.6.2.3 Trees in graph theory

Now it is time to introduce a more complicated concept of graph theory, that of trees. A *tree* is any sequence of graph edges containing no cycles. In fact, a tree can relate to any combination of reactions. A *spanning tree*, or maximum tree, is a sequence of graph edges containing no cycles and joining all nodes of the graph. In order to create cycles, it is sufficient to add more edges. In accordance with this definition, a spanning tree cannot contain (1) two reactions starting from the same intermediate (eg, -1 and $+2$, or $+1$ and -3) or (2) two reactions of the same step (eg, $+1$ and -1). The physicochemical meaning of the spanning tree is as follows: it is the sequence of transformations through which a certain intermediate is formed from all other intermediates. Such a tree is called a node spanning tree. Fig. 3.8 shows all spanning trees for the graph of the isomerization reaction of Fig. 3.7. It can be easily shown that every node is characterized by three spanning trees and that their total number is $3^2 = 9$.

Three types of spanning trees can be distinguished in Fig. 3.8: forward spanning trees generated by a sequence of forward reactions, Fig. 3.8A; reverse spanning trees generated by a sequence of reverse reactions, Fig. 3.8B; and combined spanning trees generated by a sequence of both forward and reverse reactions, Fig. 3.8C. For a single-route reaction with N_{int} intermediates (nodes), there are N_s steps (edges), each consisting of a forward and a reverse reaction, and $N_{int} = N_s = N$. The total number of spanning trees is N^2. There are N forward spanning trees and N reverse spanning trees, while the number of combined spanning trees is $N(N-2) = N^2 - 2N$. The *spanning tree weight* is equal to the product of the weights of the

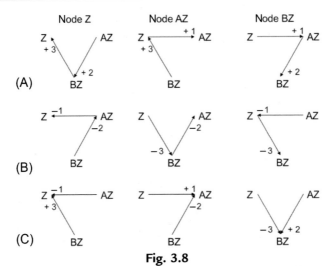

Fig. 3.8

Spanning trees of the model isomerization reaction of Fig. 3.7: (A) forward; (B) reverse; and (C) combined.

edges being part of the tree. For example, the weight of the first tree in Fig. 3.8, $W_{Z,1}$, equals the product $w_2^+ w_3^+$. Then the total weights of the node spanning trees are given by

$$W_Z = W_{Z,1} + W_{Z,2} + W_{Z,3} = w_2^+ w_3^+ + w_1^- w_2^- + w_3^+ w_1^- \tag{3.131}$$

$$W_{AZ} = W_{AZ,1} + W_{AZ,2} + W_{AZ,3} = w_1^+ w_3^+ + w_2^- w_3^- + w_1^+ w_2^- \tag{3.132}$$

$$W_{BZ} = W_{BZ,1} + W_{BZ,2} + W_{BZ,3} = w_1^+ w_2^+ + w_1^- w_3^- + w_2^+ w_3^- \tag{3.133}$$

The total weight of all spanning trees of the graph in Fig. 3.7 is defined by

$$W = W_Z + W_{AZ} + W_{BZ} \tag{3.134}$$

There is a simple analogy between Cramer's rule and the relationship based on the graph-theory approach:

$$\det(\mathbf{A}) = W \tag{3.135}$$

$$\det(\mathbf{A}_j) = W_j \tag{3.136}$$

Here W_j is the total weight of spanning trees for node j and W is the total weight of all spanning trees of the graph:

$$W = \sum_{j=1}^{N_{int}} W_j \tag{3.137}$$

Thus, in view of Eq. (3.126):

$$\theta_j = \frac{W_j}{W} \tag{3.138}$$

Note that this equation is valid for any dependence of the weights on the concentrations of the gaseous components. From the known expressions for the concentrations of the intermediates, we can easily determine the rate of any reaction. This is straightforward for linear mechanisms, but problems arise for nonlinear mechanisms. For our isomerization mechanism, the steady-state rates of all steps are equal and equal to the rate of consumption of A and to the rate of production of B. For example,

$$r = k_2^+ \theta_{AZ} - k_2^- \theta_{BZ} = w_2^+ \theta_{AZ} - w_2^- \theta_{BZ} \tag{3.139}$$

With Eq. (3.138), this yields

$$r = w_2^+ \frac{W_{AZ}}{W} - w_2^- \frac{W_{BZ}}{W} = \frac{w_2^+ W_{AZ} - w_2^- W_{BZ}}{W} \tag{3.140}$$

Substitution of Eqs. (3.131)–(133) after rearrangement leads to:

$$r = \frac{w_1^+ w_2^+ w_3^+ - w_1^- w_2^- w_3^-}{w_2^+ w_3^+ + w_1^- w_2^- + w_3^- w_1^- + w_1^+ w_3^+ + w_2^- w_3^- + w_1^+ w_2^- + w_1^+ w_2^+ + w_1^- w_3^- + w_2^+ w_3^-} \tag{3.141}$$

Then, after substituting Eqs. (3.128)–(130) and rearranging, we obtain:

$$r = \frac{k_1^+ k_2^+ k_3^+ c_A - k_1^- k_2^- k_3^- c_B}{k_1^+ c_A \left(k_2^+ + k_3^+ + k_2^-\right) + k_3^- c_B \left(k_2^+ + k_1^- + k_2^-\right) + k_2^+ k_3^+ + k_1^- k_2^- + k_3^+ k_1^-} \tag{3.142}$$

This equation shows the first important lesson on how and to what extent the structure of the steady-state overall equation reflects the details of the mechanism. The numerator can be written as $k^+ c_A - k^- c_B$ with $k^+ = k_1^+ k_2^+ k_3^+$ and $k^- = k_1^- k_2^- k_3^-$. In this form it relates to the overall reaction $A \rightleftarrows B$ obtained by adding the reactions of the detailed mechanism multiplied by the Horiuti numbers, in this case all equal to one. It is interesting that this numerator is absolutely independent of the mechanistic details. Irrespective of the number of assumed intermediates and reactions in our single-route linear mechanism, the numerator of the steady-state kinetic equation always corresponds to the rate law of the overall reaction, as if this were an elementary reaction obeying the mass-action law.

A simple "graph recipe" for deriving the steady-state rate equation of a single-route catalytic reaction is the following:

- construct a graph of the complex reaction.
- write the weight W_j of each reaction (edge), that is, the rate of the reaction divided by the normalized concentration of the intermediate participating in the reaction.
- for every intermediate (node) find the spanning trees by which the intermediate is produced from all others, the so-called node spanning trees.
- find the weight of every node spanning tree, which is the product of the weights of the reactions that are included in this tree.
- find the sum of weights of all node spanning trees W_j.

- find the total weight of all spanning trees of the graph W, Eq. (3.137);
- find the steady-state normalized concentrations of the free active sites and intermediates in accordance with Cramer's rule, Eq. (3.138);
- Find the steady-state reaction rate of any step;
- write the steady-state reaction rate in one of the following forms:

$$r = \frac{k^+ f^+ (\mathbf{c_r}) - k^- f^- (\mathbf{c_p})}{W} \tag{3.143}$$

$$r = \frac{k^+ \left(f^+ (\mathbf{c_r}) - \dfrac{f^- (\mathbf{c_p})}{K_{eq}} \right)}{\sum_l k_l \prod_i c_i^{p_{li}}} \tag{3.144}$$

$$r = \frac{f^+ (\mathbf{c_r}) - \dfrac{f^- (\mathbf{c_p})}{K_{eq}}}{\sum_l \bar{k}_l \prod_i c_i^{p_{li}}} \tag{3.145}$$

Since for a single-route mechanism the numerator of the rate does not depend on the details of the mechanism, it can be written directly based on the overall reaction according to the mass-action law. In such a "fast" derivation of the steady-state rate equation, steps (7) and (8) are omitted.

3.6.3 Derivation of Steady-State Kinetic Equations for Multiroute Mechanisms: Kinetic Coupling

In accordance with the theory developed by Yablonskii et al. (1991) for the analysis of multiroute linear mechanisms, the steady-state rate of step s is written as follows:

$$r_s = \frac{\sum_i C_{c,si} \kappa_i}{W} \tag{3.146}$$

where $C_{c,si}$ is the cycle characteristic and κ_i is the coupling parameter of the ith cycle that includes step s. This coupling parameter can be presented as follows. Every selected cycle that includes step s has common nodes with other cycles. Every cycle is characterized by the spanning trees of other cycles, that is, acyclic sequences of transformations through which the common intermediates are formed from other intermediates of other cycles. Then, the weight of a spanning tree is equal to the product of the weights of the edges being part of this tree. Also, every common node is characterized by its coupling parameter: the total weight of all spanning trees for this node, which is the sum of all corresponding spanning trees.

The coupling parameter of the *i*th cycle is the product of the coupling parameters of different common nodes.

Similar to a single-route reaction, in the general case the net rate of production of a given component equals the sum of the rates of all steps in which this component is involved.

Equation (3.146) reflects two nontrivial kinetic features of cycle coupling that are characteristic for multiroute mechanisms:

- All cycles including a certain step are taken into account, not just one cycle, as expressed in the numerator.
- the influence of other cycles on the one considered is reflected by the coupling parameter κ_i. For single-route mechanisms this coupling parameter is absent.

Two types of coupling exist: (i) cycles having a common intermediate but not a common step and (ii) cycles having one or more common steps. See Marin and Yablonsky (2011) and Yablonskii et al. (1991) for further details.

3.6.3.1 Cycles having a common intermediate

An example of cycles having a common intermediate is shown in Fig. 3.9.

In this case, every step is part of only one cycle; steps (1) and (2) are part of cycle I and steps (3) and (4) are part of cycle II. The cycle characteristics related to the two cycles are

$$C_{c,I} = k_1^+ k_2^+ c_A - k_1^- k_2^- c_B \qquad (3.147)$$

$$C_{c,II} = k_3^+ k_4^+ c_A - k_3^- k_4^- c_C \qquad (3.148)$$

For cycle I, the coupling parameter is equal to the total weight of the spanning trees of cycle II leading to the common intermediate Z:

$$\kappa_I = k_3^- + k_4^+ \qquad (3.149)$$

Similarly, for cycle II, the coupling parameter is equal to the total weight of the spanning trees of cycle I leading to the common intermediate Z:

$$\kappa_{II} = k_1^- + k_2^+ \qquad (3.150)$$

(1)	A + Z	$\rightleftarrows$	BZ
(2)	BZ	$\rightleftarrows$	B + Z
(3)	A + Z	$\rightleftarrows$	CZ
(4)	CZ	$\rightleftarrows$	C + Z
I	A	$\rightleftarrows$	B
II	A	$\rightleftarrows$	C

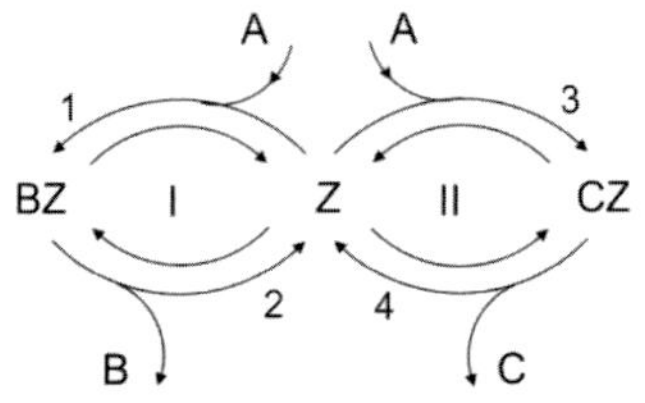

Fig. 3.9

Possible mechanism *(left)* and graph *(right)* of a two-route mechanism with a common intermediate.

The total weight of all spanning trees is

$$W = \left(k_1^+ c_A + k_1^- + k_2^+ + k_2^- c_B\right)\left(k_3^- + k_4^+\right) + \left(k_3^+ c_A + k_3^- + k_4^- + k_4^- c_C\right)\left(k_1^- + k_2^+\right) \tag{3.151}$$

The reaction rates for the two cycles can be obtained by substitution of Eqs. (3.147)–(3.150) into Eq. (3.146):

$$r_I = r_1 = r_2 = \frac{k_1^+ k_2^+ c_A - k_1^- k_2^- c_B}{W}\left(k_3^- + k_4^+\right) \tag{3.152}$$

$$r_{II} = r_3 = r_4 = \frac{k_3^+ k_4^+ c_A - k_3^- k_4^- c_C}{W}\left(k_1^- + k_2^+\right) \tag{3.153}$$

The ratio of the net rates of production of B and C can be expressed as

$$\frac{R_B}{R_C} = \frac{r_I}{r_{II}} = \frac{k_1^+ k_2^+ c_A - k_1^- k_2^- c_B}{k_3^+ k_4^+ c_A - k_3^- k_4^- c_C} \cdot \frac{\kappa_I}{\kappa_{II}} = \frac{\left(c_A - \dfrac{c_B}{K_{eq,I}}\right)}{\left(c_A - \dfrac{c_C}{K_{eq,II}}\right)} \cdot \overline{\kappa}(T) \tag{3.154}$$

in which $K_{eq,I} = K_{eq,1}K_{eq,2}$, $K_{eq,II} = K_{eq,3}K_{eq,4}$, and $\overline{\kappa}(T) = \dfrac{k_1^+ k_2^+}{k_3^+ k_4^+} \cdot \dfrac{k_3^- + k_4^+}{k_1^- + k_2^+}$. In this case, the

parameter $\overline{\kappa}$ depends only on the temperature, but it may also depend on concentrations, $\overline{\kappa}(c, T)$ for more complex mechanisms.

The net rate of production of A is

$$R_A = -(R_B + R_C) = -\frac{k_1^+ k_2^+ \left(c_A - \dfrac{c_B}{K_{eq,I}}\right)\left(k_3^- + k_4^+\right) + k_3^+ k_4^+ \left(c_A - \dfrac{c_C}{K_{eq,II}}\right)\left(k_1^- + k_2^+\right)}{W} \tag{3.155}$$

The selectivities to B and C are given by

$$S_B = \frac{R_B}{R_B + R_C} = \frac{k_1^+ k_2^+ \left(c_A - \dfrac{c_B}{K_{eq,I}}\right)\left(k_3^- + k_4^+\right)}{k_1^+ k_2^+ \left(c_A - \dfrac{c_B}{K_{eq,I}}\right)\left(k_3^- + k_4^+\right) + k_3^+ k_4^+ \left(c_A - \dfrac{c_C}{K_{eq,II}}\right)\left(k_1^- + k_2^+\right)} \tag{3.156}$$

and

$$S_C = \frac{R_C}{R_B + R_C} = 1 - S_B \tag{3.157}$$

3.6.3.2 Cycles having a common step

As an example of the coupling of two cycles with a common step, we consider the reaction mechanism shown in Fig. 3.10. The main difference with a cycle with only a common

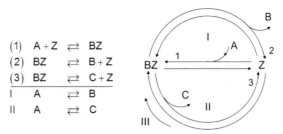

(1)	A + Z	$\rightleftarrows$	BZ
(2)	BZ	$\rightleftarrows$	B + Z
(3)	BZ	$\rightleftarrows$	C + Z
I	A	$\rightleftarrows$	B
II	A	$\rightleftarrows$	C

Fig. 3.10

Possible mechanism *(left)* and graph *(right)* of a two-route mechanism with a common intermediate and a common step.

intermediate and not a common step is that a step may be part of many cycles, at least two. In this example, every step is part of two cycles. Step (1) is part of cycles I and II, step (2) is part of cycles I and III, and step (3) is part of cycles II and III. The third cycle relates to the overall reaction $C \rightleftarrows B$. Only two of the three cycles are stoichiometrically independent.

The three cycle characteristics for this mechanism are

$$C_{c,I} = k_1^+ k_2^+ c_A - k_1^- k_2^- c_B \tag{3.158}$$

$$C_{c,II} = k_1^+ k_3^+ c_A - k_1^- k_3^- c_C \tag{3.159}$$

$$C_{c,III} = k_2^+ k_3^- c_C - k_3^+ k_2^- c_B \tag{3.160}$$

In this case, coupling parameters are absent, because there is no specific intermediate belonging to only one of the two cycles; both intermediates are common for all three cycles. Therefore, the coupling parameters are equal to one.

The net rates of production of B and C are given by

$$R_B = \frac{\left(k_1^+ k_2^+ c_A - k_1^- k_2^- c_B\right) + \left(k_2^+ k_3^- c_C - k_3^+ k_2^- c_B\right)}{W} \tag{3.161}$$

$$R_C = \frac{\left(k_1^+ k_3^+ c_A - k_1^- k_3^- c_C\right) - \left(k_2^+ k_3^- c_C - k_3^+ k_2^- c_B\right)}{W} \tag{3.162}$$

or

$$R_B = \frac{k_1^+ k_2^+ \left(c_A - \dfrac{c_B}{K_{eq,I}}\right) + k_2^+ k_3^- \left(c_C - \dfrac{c_B}{K_{eq,III}}\right)}{W} \tag{3.163}$$

$$R_C = \frac{k_1^+ k_3^+ \left(c_A - \dfrac{c_C}{K_{eq,II}}\right) - k_2^+ k_3^- \left(c_C - \dfrac{c_B}{K_{eq,III}}\right)}{W} \tag{3.164}$$

in which $K_{eq,I} = K_{eq,1} K_{eq,2}$, $K_{eq,II} = K_{eq,1} K_{eq,3}$ and $K_{eq,III} = K_{eq,2}/K_{eq,3}$ and

$$W = k_1^+ c_A + k_1^- + k_2^+ + k_2^- c_B + k_3^+ + k_3^- c_C \tag{3.165}$$

The net rate of production of A is

$$R_A = -(R_B + R_C) = -k_1^+ \frac{k_2^+ \left(c_A - \dfrac{c_B}{K_{eq,I}} \right) + k_3^+ \left(c_A - \dfrac{c_C}{K_{eq,II}} \right)}{W} \tag{3.166}$$

The expression for the ratio of the net rates of production of B and C is a little more complicated than in the case of just one common intermediate (Eq. 3.158), even though the number of intermediates is smaller:

$$\frac{R_B}{R_C} = \frac{k_2^+ \left(c_A - \dfrac{c_B}{K_{eq,I}} \right) + \left(c_C - \dfrac{c_B}{K_{eq,III}} \right)}{k_3^+ \left(c_A - \dfrac{c_C}{K_{eq,II}} \right) - \left(c_C - \dfrac{c_B}{K_{eq,III}} \right)} \tag{3.167}$$

For cycles having one common intermediate, this ratio is proportional to the ratio of driving forces, $(c_A - c_B/K_{eq,I})/(c_A - c_C/K_{eq,II})$, whereas for cycles with one common step, the equation contains a cross-term that reflects the participation of products in an additional cycle, the "global cycle." This is a specific feature of such cyclic mechanisms.

3.6.2.3 Cycles having two common steps

An example of the coupling of two cycles with two common steps is presented in Fig. 3.11.

The cycle characteristics and coupling parameters for this two-route mechanism with two common steps are

$$C_{c,I} = k_1^+ k_2^+ k_3^+ k_4^+ c_A - k_1^- k_2^- k_3^- k_4^- c_B \tag{3.168}$$

$$C_{c,II} = k_1^+ k_2^+ k_5^+ k_6^+ c_A - k_1^- k_2^- k_5^- k_6^- c_C \tag{3.169}$$

$$C_{c,III} = k_3^+ k_4^+ k_5^- k_6^- c_C - k_3^- k_4^- k_5^+ k_6^+ c_B \tag{3.170}$$

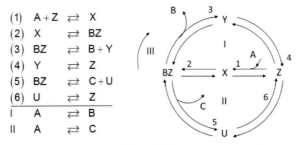

(1)	A + Z	$\rightleftarrows$	X
(2)	X	$\rightleftarrows$	BZ
(3)	BZ	$\rightleftarrows$	B + Y
(4)	Y	$\rightleftarrows$	Z
(5)	BZ	$\rightleftarrows$	C + U
(6)	U	$\rightleftarrows$	Z
I	A	$\rightleftarrows$	B
II	A	$\rightleftarrows$	C

Fig. 3.11
Possible mechanism *(left)* and graph *(right)* of a two-route mechanism with a common intermediate and two common steps.

For constructing the coupling parameter for cycle I, intermediates of cycle II that are not part of cycle I must be connected with the common intermediates Z and BZ. There is only one intermediate, U, that is part of only cycle II, and the goal is to connect it to cycle I. Thus, the coupling parameter of cycle I is equal to the total weight of the spanning trees

$$\kappa_I = k_5^- c_C + k_6^+ \tag{3.171}$$

Similarly, the coupling parameter of cycle II is given by

$$\kappa_{II} = k_3^- c_B + k_4^+ \tag{3.172}$$

and the coupling parameter of cycle III is given by

$$\kappa_{III} = k_2^+ + k_1^- \tag{3.173}$$

3.6.2.4 Different types of coupling between cycles

Comparing different types of multiroute mechanisms, it must be emphasized that there is a big difference between (i) cycles having common intermediates and (ii) cycles having common steps. Every step of cycles of type (i) is only part of one cycle, and the reaction rate of this step is described by a kinetic equation that is similar to the rate equation of a single-route mechanism. Just as the single-route reaction rate, this rate can be expressed as the difference between the forward and reverse reaction rates. What is important regarding cycles of type (i) is that the presence of other cycles quantitatively influences the reaction rate of the selected cycle, the reaction route rate. However, at a given temperature, other cycles cannot change the direction of the overall reaction corresponding to the selected cycle.

In cycles of type (ii), a step of one cycle is also part of other cycles and the reaction rate of this step is a linear combination of multiple reaction route rates. As a result, this reaction rate cannot be represented by the difference between a forward and reverse reaction rate. Other cycles influence the reaction rate of the selected step not only quantitatively but qualitatively as well. Indeed, other cycles can change the direction of the overall reaction corresponding to the selected cycle. This is the main difference between multiroute mechanisms of types (i) and (ii). Many examples of deriving such equations using graph theory can be found in the books by Yablonskii et al. (1991) and Marin and Yablonsky (2011).

3.6.4 Bipartite Graphs of Complex Reactions

Bipartite graphs for presenting complex mechanisms of chemical reactions have been proposed by Vol'pert (1972) and Hudyaev and Vol'pert (1985). These graphs contain nodes of two types: type X nodes corresponding to components X_i ($i = 1, 2, ..., N$) and type R nodes ascribed to the elementary forward reaction r_s^+ and reverse reaction r_s^- ($s = 1, 2, \ldots, N_s$) belonging to step s. Therefore, reversible steps are part of two edges. Edges connect reaction nodes and nodes of components taking part in the reaction.

A reaction step can be represented by the following general equation:

$$\sum_i \alpha_i X_i \rightleftarrows \sum_i \beta_i X_i \qquad (3.174)$$

An edge is oriented from a component X_i to a reaction $r_s^{+/-}$ if X_i is converted ($\alpha_i \neq 0$) and from a reaction $r_s^{+/-}$ to X_i if X_i is produced ($\beta_i \neq 0$). The stoichiometric coefficient α_i is the number of edges from component X_i to reaction $r_s^{+/-}$ and β_i is the number of edges from reaction $r_s^{+/-}$ to X_i. Figs. 3.12 and 3.13 show some of the simplest examples of bipartite graphs.

A more complicated example is the oxidation of CO over Pt. Two typical mechanisms exist for this reaction: (i) the Eley-Rideal or impact mechanism and (ii) the Langmuir-Hinshelwood or adsorption mechanism. The Eley-Rideal mechanism does not involve any interaction between catalytic intermediates; one component from the gas phase, in this case oxygen, adsorbs on the catalyst surface forming a surface intermediate, and another component from the gas phase, in this case carbon monoxide, reacts with this surface intermediate:

$$(1)\, O_2 + 2Pt \rightleftarrows 2PtO \qquad (3.175)$$

$$(2)\, CO + PtO \rightarrow CO_2 + Pt \qquad (3.176)$$

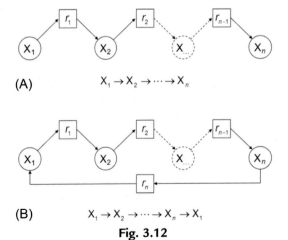

(A) $X_1 \rightarrow X_2 \rightarrow \cdots \rightarrow X_n$

(B) $X_1 \rightarrow X_2 \rightarrow \cdots \rightarrow X_n \rightarrow X_1$

Fig. 3.12

Examples of simple bipartite graphs for irreversible reactions: (A) acyclic mechanism and (B) cyclic mechanism.

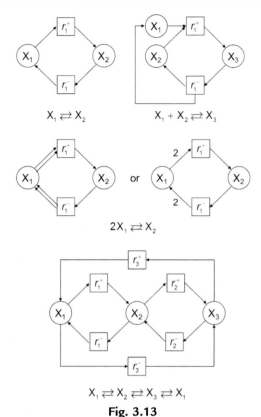

Fig. 3.13
Examples of simple bipartite graphs for reversible reactions.

In the Langmuir-Hinshelwood mechanism, two gaseous components, in this case oxygen and carbon monoxide, adsorb on the catalyst, and one step (step (3)) involves the interaction between the two different intermediates:

$$(1)\, O_2 + 2Pt \rightleftarrows 2PtO \tag{3.177}$$

$$(2)\, CO + Pt \rightleftarrows PtCO \tag{3.178}$$

$$(3)\, PtO + PtCO \rightarrow 2Pt + CO_2 \tag{3.179}$$

Assuming the concentration of gaseous components is constant, which is a typical assumption in studying kinetic models of non-steady-state solid-catalyzed reactions, the mechanisms can be presented as sequences of transformations of the surface intermediates. For the Eley-Rideal mechanism, these are the following:

$$(1)\, 2Pt \rightleftarrows 2PtO \tag{3.180}$$

$$(2)\, PtO \rightarrow Pt \tag{3.181}$$

The transformations of intermediates for the Langmuir-Hinshelwood mechanism are

$$(1)\, 2Pt \rightleftarrows 2PtO \tag{3.182}$$

$$(2)\, Pt \rightleftarrows PtCO \tag{3.183}$$

$$(3)\, PtO + PtCO \rightarrow 2Pt \tag{3.184}$$

Fig. 3.14 shows the corresponding graphs for both mechanisms.

In the analysis of bipartite graphs, the concept of cycles is crucial. The simplest class of reaction mechanisms is that with bipartite graphs that do not contain cycles (see Fig. 3.12A). These reaction mechanisms are called acyclic mechanisms and can be represented in general form as:

$$X_1 \rightarrow X_2 \rightarrow \cdots \rightarrow X_n \tag{3.185}$$

Obviously, this mechanism does not represent a catalytic reaction. Mechanisms for catalytic reactions always contain cycles. These cycles are oriented, that is, the direction of all edges in a cycle is the same and the end of the ith edge is the beginning of the $(i+1)$th edge.

Some graph cycles may be nonoriented. Fig. 3.15 shows the bipartite graph for the two-step reaction mechanism

$$(1)\, X_1 \rightarrow X_2 \tag{3.186}$$

$$(2)\, X_1 + X_2 \rightarrow X_3 \tag{3.187}$$

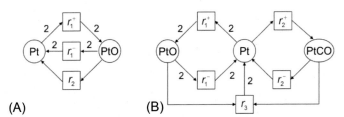

(A) (B)

Fig. 3.14

Bipartite graphs for the oxidation of CO on Pt: (A) Eley-Rideal mechanism and (B) Langmuir-Hinshelwood mechanism.

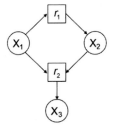

Fig. 3.15

Bipartite graph for the reaction mechanism $X_1 \rightarrow X_2$; $X_1 + X_2 \rightarrow X_3$.

This graph contains a nonoriented cycle: edges starting from X_1 have opposite directions and edges ending at r_2 are directed toward each other.

Bipartite graphs for the Eley-Rideal and Langmuir-Hinshelwood mechanisms contain both oriented and nonoriented cycles (see Figs. 3.16 and 3.17).

The structure of graphs of complex reactions is very important for the analysis of chemical dynamics. The absence of oriented cycles indicates relatively simple dynamic behavior. Clarke (1980) and Ivanova (Ivanova, 1979; Ivanova and Tarnopolskii, 1979) used bipartite graphs for

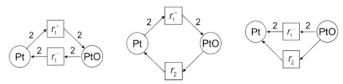

Fig. 3.16
Cycles for the oxidation of CO on Pt via the Eley-Rideal mechanism.

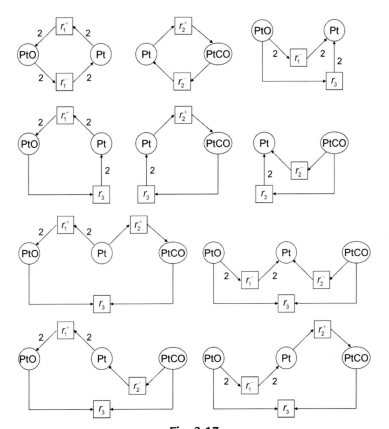

Fig. 3.17
Cycles for the oxidation of CO on Pt via the Langmuir-Hinshelwood mechanism.

the stability analysis of complex catalytic reactions, in particular to verify whether some critical phenomena, such as kinetic multiplicity of steady states and rate oscillations, can be explained within a given kinetic model. Using cycles, the characteristic equation for nonlinear detailed mechanisms can be constructed. Then, the eigenvalues of this equation, especially their signs, are analyzed. Based on this analysis, conclusions regarding the stability of the dynamic behavior can be drawn.

3.6.5 Graphs for Analyzing Relaxation: General Form of the Characteristic Polynomial

A non-steady-state kinetic model of a complex catalytic reaction with a linear mechanism can be formulated as

$$\frac{d\boldsymbol{\theta}}{dt} = \mathbf{W}(\mathbf{c})\boldsymbol{\theta} \tag{3.188}$$

where $\mathbf{c}$ is a set of measured (observed) concentrations of gaseous components, $\boldsymbol{\theta}$ is a set of nonmeasured (unobserved) normalized concentrations of intermediates, and $\mathbf{W}(\mathbf{c})$ is the matrix of the reaction weights. In addition, the law of active site conservation, Eq. (3.124), must be fulfilled. Eq. (3.188) is the non-steady-state kinetic model of the transformation of catalytic intermediates assuming that the concentrations of the observed components are constant. The solution of Eq. (3.188) is of the form

$$\theta_j(t) = \sum_{i=0}^{N_{\text{int}}-1} C_{ji}\exp(\lambda_i t) \quad j=1,\ldots,N_{\text{int}} \tag{3.189}$$

where λ_i are the roots (or eigenvalues) of the characteristic polynomial $P(\lambda)$, and C_{ji} are constants.

The characteristic polynomial of a square matrix $\mathbf{A}=|a_{ij}|$ of order n is called the determinant for the set of linear equations

$$\sum_{k=1}^{n}(a_{ik}-\delta_{ik}\lambda)x_k=0 \quad i=1,2,\ldots,n \tag{3.190}$$

where λ is a scalar and δ_{ik} is the Kronecker delta function,

$$\delta_{ik}=\begin{cases}1 & \text{if }i=k\\ 0 & \text{if }i\neq k\end{cases} \tag{3.191}$$

The characteristic polynomial $P(\lambda)$ can be written as

$$P(\lambda)=(-1)^n\left(\lambda^n-d_1\lambda^{n-1}+d_2\lambda^{n-2}-d_3\lambda^{n-3}\cdots+(-1)^n d_n\right) \tag{3.192}$$

Evstigneev and Yablonskii (1982) have proven that the coefficient of λ^k with k the exponential factor for the characteristic polynomial $P(\lambda)$ equals the sum of the weights of all the k spanning trees of the reaction graph if $k\neq0$ and equals zero if $k=0$. A k spanning tree for graph $G(x,y)$ is called an unconnected partial graph[1] containing all the nodes, that is, a set of rooted trees ("rooted forest"), all edges of which are directed toward the roots, that is, the given graph points x. A rooted tree can be degenerated, which means it consists of only one point. When speaking of trees, spanning trees, and graphs, we imply that they are oriented. The weight of the spanning tree is the product of weights of its edges. The weight of a degenerated component is assumed to be equal to unity.

Relationships for different coefficients are the following:

- $k = 1$. The coefficient at λ is equal to the sum of all the principal minors of order $(n-1)$. It was proven that in terms of graph theory every minor of this type equals the sum of the weights of spanning trees entering the point x.
- $1 < k < n - 1$. The coefficient at λ^k equals the sum of the principal minors of order $(n-k)$. Every term corresponds to an n-point graph without cycles and having k components.
- $k = n - 1$. The coefficient at λ^k equals the trace of the matrix $\mathbf{W}(\mathbf{c})$, that is, the sum of all elements on the main diagonal of the matrix, which are the reaction weights.
- $k = n$. The coefficient at λ^k equals the weight of the empty n-point graph having no points. By definition, its weight is unity.

As an example, for the catalytic isomerization reaction $A \rightleftarrows B$ with the mechanism given by Eq. (3.87), the reaction weights are given by $w_1^+ = k_1^+ c_A$; $w_1^- = k_1^-$; $w_2^+ = k_2^+$; $w_2^- = k_2^-$; $w_3^+ = k_3^+$; $w_3^- = k_3^- c_B$. The graph corresponding to this mechanism is shown in Fig. 3.4. The characteristic equation is

$$\lambda^2 + d_1\lambda + d_2 = 0 \tag{3.193}$$

The coefficient d_1 is the sum of the weights of all reaction weights:

$$d_1 = k_1^+ c_A + k_1^- + k_2^+ + k_2^- + k_3^+ + k_3^- c_B \tag{3.194}$$

The free term d_2 is given by

$$d_2 = k_1^+ c_A k_2^+ + k_1^+ c_A k_3^+ + k_2^+ k_3^+ + k_1^- k_2^- + k_2^- k_3^- c_B + k_3^- k_1^- c_B + k_1^+ k_2^- c_A + k_2^+ k_3^- c_B + k_3^+ k_1^- \tag{3.195}$$

If all steps are irreversible, these coefficients are reduced to

$$d_1 = k_1^+ c_A + k_2^+ + k_3^+ \tag{3.196}$$

$$d_2 = k_1^+ c_A \left(k_2^+ + k_3^+ \right) + k_2^+ k_3^+ \tag{3.197}$$

In accordance with Vieta's formulas, the roots of the characteristic polynomial can be determined from its coefficients. In this example,

$$\lambda_1 + \lambda_2 = -d_1 \tag{3.198}$$

$$\lambda_1 \lambda_2 = d_2 \tag{3.199}$$

---

[1] $H(y,v)$, with $y \leq x$ and $v \leq u$, is a partial graph of the graph $G(x,u)$.

Nomenclature

A	cross-sectional area (m^2)
$C_{c,si}$	cycle characteristic for cycle i containing step s
c_i	concentration of component i (mol m^{-3})
D_i	diffusion coefficient of component i (m^2 s^{-1})
d_i	coefficients in characteristic equation
E_a	activation energy (J mol^{-1})
F_i	molar flow rate of component i (mol s^{-1})
$\Delta_r H$	enthalpy change of reaction (J mol^{-1})
K_{eq}	equilibrium coefficient (–)
k	reaction rate coefficient ((mol m^{-3})$^{(1-n)}$s^{-1})
k_0	pre-exponential factor (same as k)
N	number (–)
n_{Ai}	number of moles of reactant i (mol)
n_{Bi}	number of moles of product i (mol)
$P(\lambda)$	characteristic polynomial
p	pressure (Pa)
q_V	volumetric flow rate (m^3 s^{-1})
R_g	universal gas constant (J mol^{-1} K^{-1})
R_i	net rate of production of component i (mol m^{-3} s^{-1})
r	reaction rate for a homogeneous reaction (mol m^{-3} s^{-1})
r_S	reaction rate per unit surface area of catalyst (mol m$_{cat}^{-2}$ s^{-1})
r_V	reaction rate per unit volume of catalyst (mol m$_{cat}^{-3}$ s^{-1})
r_W	reaction rate per unit mass of catalyst (mol kg$_{cat}^{-2}$ s^{-1})
S_{cat}	catalyst surface area (m$_{cat}^2$)
S_i	selectivity to component i (–)
T	temperature (K)
t	time (s)
u	superficial fluid velocity (m s^{-1})
V	reaction volume (m^3)
W	weight of spanning tree
W_{cat}	mass of catalyst in the reactor (kg$_{cat}$)
w	reaction weight (mol m$_{cat}^{-2}$ s^{-1})
x_i	conversion of component i (–)
z	axial reactor coordinate (m)

Greek symbols

α_i	absolute value of stoichiometric coefficient of reactant A_i (–)
α_j	absolute value of stoichiometric coefficient of intermediate X_j (–)
β_i	stoichiometric coefficient of product B_i (–)

β_j	stoichiometric coefficient of intermediate Y_j (–)
δ_{ik}	Kronecker delta (–)
ε	small parameter (–)
ε_b	void fraction of catalyst bed ($m_f^3 m^{-3}$)
κ_i	coupling parameter for cycle i containing step s
Γ_{X_j}	concentration of surface intermediate X_j ($mol\ m_{cat}^{-2}$)
Γ_t	total concentration of surface intermediates ($mol\ m_{cat}^{-2}$)
λ	characteristic root, eigenvalue
θ_{X_j}	normalized concentration of surface intermediate X_j (–)
ν_i	stoichiometric coefficient of component i (–)
ρ_{cat}	catalyst density ($kg_{cat}\ m_{cat}^{-3}$)
σ	Horiuti number (–)
τ	characteristic time (s)
τ	space time (s)

Subscripts

0	initial, inlet
cat	catalyst
eq	equilibrium
f	fluid
int	intermediate
s	step

Superscripts

+	of forward reaction
–	of reverse reaction
α_i	partial order of reaction in reactant A_i
α_j	partial order of reaction in intermediate X_j
β_i	partial order of reaction in product B_i
β_j	partial order of reaction in intermediate Y_j
n	(apparent) reaction order

References

Balandin, A.A., 1964. Catalysis and Chemical Kinetics. Academic Press, New York.
Clarke, B.L., 1980. Stability of complex reaction networks. In: Prigogine, I., Rice, S.A. (Eds.), Advances in Chemical Physics. Wiley, New York, pp. 1–215.
Ergun, S., 1952. Fluid flow through packed columns. CEP 48, 89–94.
Evstigneev, V.A., Yablonskii, G.S., 1982. Structured form of the characteristic equation of a complex chemical reaction (linear case). Theor. Exp. Chem. 18, 99–103.
Hudyaev, S.I., Vol'pert, A.I., 1985. Analysis in Classes of Discontinuous Functions and Equations of Mathematical Physics. Martinus Nijhoff Publishers, Dordrecht, The Netherlands.

Ivanova, A.N., 1979. Conditions of uniqueness of steady state related to the structure of the reaction mechanism. Kinet. Katal. 20, 1019–1023 (in Russian).

Ivanova, A.N., Tarnopolskii, B.L., 1979. On computer determining the critical regions in kinetic systems. Kinet. Katal. 20, 1541–1548 (in Russian).

King, E.L., Altman, C., 1956. A schematic method of deriving the rate laws for enzyme-catalyzed reactions. J. Phys. Chem. 60, 1375–1378.

Lazman, M.Z., Yablonsky, G.S., 2008. Overall reaction rate equation of single route catalytic reaction. In: Marin, G.B., West, D., Yablonsky, G.S. (Eds.), Advances in Chemical Engineering. Elsevier, Amsterdam, pp. 47–102.

Marin, G.B., Yablonsky, G.S., 2011. Kinetics of Chemical Reactions—Decoding Complexity. Wiley-VCH.

Mason, S.J., 1953. Feedback theory—some properties of signal flow graphs. Proc. Inst. Radio Eng. 41, 1144–1156.

Mason, S.J., 1956. Feedback theory—further properties of signal flow graphs. Proc. Inst. Radio Eng. 44, 920–926.

Schwab, G.M., 1982. Development of kinetic aspects in catalysis research. Crit. Rev. Solid State Mater. Sci. 10, 331–347.

Vol'pert, A.I., 1972. Differential equations on graphs. Math. USSR Sb. 17, 571–582.

Volkenstein, M.V., Goldstein, B.N., 1966a. A new method for solving the problems of the stationary kinetics of enzymological reactions. Biochim. Biophys. Acta 115, 471–477.

Volkenstein, M.V., Goldstein, B.N., 1966b. Method for derivation of enzyme kinetics equations. Biokhimiya 31, 541–547 (in Russian).

Yablonskii, G.S., Bykov, V.I., Gorban, A.N., 1983. Kinetic Models of Catalytic Reactions. Nauka, Novosibirsk.

Yablonskii, G.S., Bykov, V.I., Gorban, A.N., Elokhin, V.I., 1991. Kinetic Models of Catalytic Reactions. Comprehensive Chemical Kinetics, vol. 32. Elsevier, Amsterdam.

Yevstignejev, V.A., Yablonskii, G.S., 1979. On one fundamental correlation of steady-state catalytic kinetics. Kinet. Katal. 20, 1549–1555.

## CHAPTER 4

Physicochemical Principles of Simplification of Complex Models

4.1 Introduction

In chemical reactors an enormous variety of possible regimes, both steady state and non-steady state (transient) can be observed. Steady-state reaction rates can be characterized by maxima and hystereses. Non-steady-state kinetic dependences may exhibit many phenomena of complex behavior, such as "fast" and "slow" domains, ignition and extinction, oscillations and chaotic behavior. These phenomena can be even more complex when taking into account transport processes in three-dimensional media. In this case, waves and different spatial structures can be generated. For explaining these features and applying this knowledge to industrial or biochemical processes, models of complex chemical processes have to be simplified.

Simplification not only is a means for the easy and efficient analysis of complex chemical reactions and processes, but also is a necessary step in understanding their behavior. In many cases, "to understand" means "to simplify." Now the main question is: "Which reaction or set of reactions is responsible for the observed kinetic characteristics?" The answer to this question very much depends on the details of the reaction mechanism and on the temporal domain that we are interested in. Frequently, simplification is defined as a reduction of the "original" set of system factors (processes, variables, and parameters) to the "essential" set for revealing the behavior of the system, observed through real or virtual (computer) experiments. Every simplification has to be correct. As a basis of simplification, many physicochemical and mathematical principles/methods/approaches, or their efficient combination, are used, such as fundamental laws of mass conservation and energy conservation, the dissipation principle, and the principle of detailed equilibrium. Based on these concepts, many advanced methods of simplification of complex chemical models have been developed (Marin and Yablonsky, 2011; Yablonskii et al., 1991).

In the mathematical sense, simplification can be defined as "model reduction," that is, the rigorous or approximate representation of complex models by simpler ones. For example, in a certain domain of parameters or times, a model of partial differential equations ("diffusion-reaction" model) is approximated by a model of differential equations, or a model of differential equations is approximated by a model of algebraic equations, and so on.

83

See Gorban et al. (2010), Gorban and Karlin (2005), Gorban and Radulescu (2008), and Maas and Pope (1992).

In this chapter, both a set of physicochemical conceptual assumptions used in model simplification and a set of mathematical tools for this purpose are presented. First, we are going to explain physicochemical concepts of simplifications using models of chemical transformations. In this case, the transport processes are considered to be fast, and the models consist of ordinary differential equations (non-steady-state processes) or algebraic equations (steady-state processes).

4.2 Physicochemical Assumptions

Typically, assumptions are made on substances, on reactions and their parameters, on transport-reaction characteristics, and on experimental procedures.

4.2.1 Assumptions on Substances

(1) *Abundance of some substances* in comparison with others, so their amount/concentration can be assumed to be constant in the course of the process, either steady- or non-steady-state. For example, in aqueous-phase reactions, the water concentration is often taken as a parameter in kinetic reaction models.

(2) *Insignificant change of some substance amount/concentration* in comparison with its initial amount/concentration during a non-steady-state process. For example, in pulse-response experiments under vacuum conditions in a temporal analysis of products (TAP) reactor, the total number of active sites on the catalyst surface is much larger than the amount of gas molecules injected in one pulse. Therefore, the concentration of active catalyst sites may be assumed to remain equal during a pulse experiment.

(3) *Dramatic increase of the concentration/temperature* at the very beginning of a process in a batch reactor or at the inlet of a continuous-flow reactor, typically by a delta function or step function.

(4) *Complete conversion* of some substances in time during the process or at the very end (the final section) of the chemical reactor.

(5) *Gaussian distribution of the chemical composition* regarding some physicochemical properties, for example, the molecular weight of polymers.

(6) *Assumptions on intermediates of complex chemical reactions*:

    (a) Frequently, the concentrations of many intermediates are very small compared to the concentrations of others. At the limit, only one intermediate dominates. For heterogeneous catalysis, Boudart introduced the term "most abundant reaction intermediate" (*mari*), which is the only important surface intermediate on the catalyst surface under reaction conditions.

(b) Some intermediates can be considered to be in a quasi-steady state (QSS) or pseudo-steady state (PSS), see Sections 4.2.2 (assumption (5)) and 4.3.5 for a more detailed explanations. Here, a word on the prefixes "pseudo" and "quasi." In literature, they are both used interchangeably to mean "sort of," but in fact have somewhat different meanings. "Pseudo" is from Greek, meaning false or not real(ly), and is typically used for situations where deception is deliberate. "Quasi" is from Latin, meaning almost, as if, or as it were. It is often used to describe something that for the most part but not completely behaves like something else. Hence, we prefer to use quasi to describe this type of (non-)steady state.

4.2.2 Assumptions on Reactions and Their Parameters

(1) *Assumption of irreversibility of reaction steps.* Either all reaction steps are irreversible (strong irreversibility), or some reaction steps are irreversible (weak reversibility). Rigorously speaking, all reaction steps are reversible. If the rate of the forward reaction is much larger than that of the reverse reaction, we consider the reaction step to be irreversible. If in a sequence of steps, say in a heterogeneous catalytic cycle, at least one step is irreversible, the overall reaction can be considered to be irreversible.

(2) *Assumption of "rate-limiting or rate-determining step."* In a sequence of reaction steps, there usually are fast steps and slow steps. The kinetic parameters of the slow steps are much smaller than those of the fast steps, reversible or irreversible, and kinetic dependences are governed by these small parameters. If there is only a single slow step, this is called the rate-limiting or rate-determining step. However, this is not a rigorous definition of the rate-limiting-step concept, which remains a subject of permanent fierce discussions (Gorban et al., 2010; Gorban and Radulescu, 2008; Kozuch and Martin, 2011). In their paper, Kozuch and Martin express a provocative opinion on this subject.

(3) *Assumption of "quasiequilibrium" or "fast equilibrium."* If in a sequence of steps both the forward and reverse reactions of some reversible steps are much faster than other reaction steps, the assumption can be made that the forward and reverse reactions of such fast steps occur at approximately equal rates, that is, they are at equilibrium. Typically, this assumption is justified by the fact that the kinetic parameters of these fast steps are much larger than the kinetic parameters of the other, slow steps. For many chemical systems, the assumption of quasiequilibrium is complimentary to the assumption of a rate-limiting step; if one step is considered to be rate limiting, other, reversible steps can be assumed to be at equilibrium.

(4) *"Quasi-steady-state" assumption.* This assumption relates to reaction intermediates whose rate of change follows the time evolution of the concentrations of other species. According

to the QSS assumption, the rates of production and consumption of intermediates are approximately equal so their net rate of production is approximately equal to zero. The QSS assumption is typically used in the following situations:

(a)   Gas-phase chain reactions (eg, oxidation reactions) are propagated by free radicals, that is, species having an unpaired electron ($H\cdot$, $O\cdot$, $OH\cdot$, etc.). The kinetic parameters of reactions in which these short-lived, highly reactive free radicals participate are much larger than the kinetic parameters of reactions involving other species. Their concentration in the QSS is necessarily small.

(b)   Gas-solid catalytic reactions occur through catalytic surface intermediates. These are not necessarily short-lived but their concentrations are much smaller than the concentrations of reactants and products of the overall reaction. Therefore, the kinetic dependences of the surface intermediates are governed by the concentrations of the gaseous species. A similar reasoning holds for enzyme-catalyzed biochemical reactions, in which the number of active enzyme sites is small compared to the number of substrate and product molecules.

(5)   *Assumption of equality or similarity of chemical activity.* Based on a preliminary analysis, some groups of species with identical or similar chemical functions or activities can be distinguished, for example, a family of hydrocarbons of similar activity can be represented by just one hydrocarbon. This is the so-called lumping procedure (see also Section 4.3.8).

(6)   *Additional assumptions on parameters:*

(a)   *Assumption of equality of parameters* of some steps. For example, kinetic parameters of some adsorption steps or even coefficients of all irreversible reactions are equal.

(b)   *Assumption of "fast step."* The kinetic parameter of a certain step is assumed to be much larger than the kinetic parameters of other steps.

(c)   *Assumption regarding the hierarchy of kinetic parameters.* For example, in catalytic reactions adsorption coefficients are usually much larger than the kinetic parameters of reactions between different surface intermediates.

(7)   *Principle of critical simplification.* In accordance with this principle (Yablonsky et al., 2003), the behavior near critical points, for instance ignition or extinction points in catalytic combustion reactions, is governed by the kinetic parameters of only one reaction—adsorption for ignition and desorption for extinction—which is not necessarily the rate-limiting one.

4.2.3 Assumptions on Transport-Reaction Characteristics

(1)   *Assumption of continuity of flow.* When a fluid is in motion, it must move in such a way that mass is conserved.

(2) *Assumption of uniformity* of chemical composition and/or temperature and/or gas pressure in a chemical reactor.

(3) *Assumption of transport limitation*, that is, an assumption under which a model only comprising transport can be used (fast reaction and slow transport, in particular diffusion limitation).

(4) *Assumption of kinetic limitation,* that is, an assumption under which a model only including reaction can be used (fast advection or fast diffusion and slow reaction, kinetic limitation).

These assumptions are analyzed in detail in Chapter 5, which is devoted to transport-reaction problems. Special attention has to be paid to boundary and initial conditions in transport-reaction problems because these conditions reflect the physical realization of processes.

4.2.4 Assumptions on Experimental Procedures

(1) *Assumption of insignificant change* of the system characteristics during an experiment involving a small perturbation of the system: a *state-defining experiment*.

(2) *Assumption of controlled change* of the system characteristics during an experiment: a *state-altering experiment*.

(3) *Assumption of instantaneous change*: instantaneous injection of a reactant into a chemical reactor.

(4) *Assumption of linear change* of the controlled parameter: a linear temperature increase in the course of thermodesorption.

4.2.5 Combining Assumptions

It should be noted that some physicochemical assumptions are overlapping and some are complimentary. For example, assuming that some steps are fast automatically means assuming that other steps are slow. In the simplest case—the two-step mechanism—the assumption of a fast first step is identical to the assumption of a rate-limiting second step. Assumptions on the abundance of species and rate-limiting steps can be made both for reversible and irreversible reaction steps. In contrast, the quasiequilibrium assumption cannot be applied to a set of reaction steps that are all reversible. Sometimes many assumptions, not just one or two, are used for the development of a model. An example is the Michaelis-Menten model, which is well known in biocatalysis. In this model, for which the mechanism is shown in Fig. 4.1, the total amount of

Fig. 4.1
Mechanism of enzyme reaction.

active enzyme sites is much smaller than the amounts of liquid-phase substrate (S) and product (P). Because of that, QSS behavior of the enzyme species (free enzyme E and substrate-bound enzyme ES) is observed. It is assumed that

(1) the first step (E + S ⇌ ES) is reversible, while the second step (ES → P + E) is irreversible
(2) the kinetic parameters of the first step are much larger than those of the second step, that is, the first step is fast and the second step is slow.

Therefore, there are two simultaneous assumptions: the assumption of quasiequilibrium of the first step and the assumption that the second step is rate limiting.

4.3 Mathematical Concepts of Simplification in Chemical Kinetics

4.3.1 Introduction

In modeling, it is not enough to present assumptions or simplifications expressed in a verbal way. Physicochemical assumptions have to be translated into the language of mathematics. In 1963, Kruskal (1963) introduced a special term for this activity, "asymptotology." See Gorban et al. (2010) for a detailed analysis. Mathematical models have to be developed based on assumptions with clear physicochemical basis. Every physicochemical assumption has a domain of its correct application, and this domain has to be validated. Typically, this is done using the full model that includes the partial model the validity of which is tested. The partial model is generated asymptotically from the full model and the correctness of this asymptotic procedure has to be proven (Gorban and Radulescu, 2008). As stressed by Gorban et al. (2010) "often we do not know the rate constants for complex networks, and kinetics that is ruled by orderings rather than by exact values of rate constants may be very useful."

A mathematical analysis founded on the basic laws of physics, in particular laws of thermodynamics, may provide us with an understanding of "tricks" of which the physicochemical meaning was previously unclear or even with a formulation of new fundamental concepts. For example, the meaning of the QSS assumption introduced in chemistry at the very beginning of the 20th century was clarified only about 50 years after using the mathematical theory of singular perturbations and, even now, this knowledge is not sufficiently widespread. The lumping procedure, a commonly used approach to reduce the number of chemical species and reactions to be handled by grouping together species having similar chemical functions or activities into one pseudocomponent or lump, was theoretically grounded and realized by Wei and Prater (1962) and Wei and Kuo (1969). Complex chemical behavior that was discovered in chemical systems in the 1950–70s, such as bistability, oscillations, chaotic behavior, and so forth, has been understood only by transferring and adapting the concepts of the mathematical dynamic theory (stability, bifurcation, catastrophes, chaos, etc.). Maas and Pope (1992) efficiently used the mathematical technique of manifolds for understanding combustion processes. At the same time, many mathematical tools applied to chemical problems are still remaining "purely mathematical" not having a special chemical content, such as many methods of statistical analysis and sensitivity analysis. The process of conceptual interaction between chemical sciences and mathematics is still continuing and will never be finished.

We will now present an analysis of different types of simplifications.

4.3.2 Simplification Based on Abundance

This simplification (assumption (1) in Section 4.2.1) assumes that the concentration of at least one species is much larger than that of others. For example, in models of many reactions in aqueous solutions, the water concentration is considered to be constant. Similarly, in reactions in which precipitation occurs, the concentration of the solid phase is taken as the constant. These constant values are incorporated into the kinetic coefficients. In typical heterogeneous gas-solid reactions, the amount of reacting gas molecules is assumed to be much larger than the total amount of active catalyst sites. In this case, the concentration of the abundant gaseous species is included in the reaction rate coefficient as a constant (apparent kinetic coefficient). This simplifies the reaction model and often results in a linear model. Thus, for a surface catalytic process at steady state, the rate of adsorption for the reaction $A + Z \rightarrow AZ$ can be expressed as

$$r_{ads} = (k_{ads} c_A) \theta_Z = w_{ads} \theta_Z \tag{4.1}$$

Here, θ_Z is the normalized surface concentration of free active sites, which changes during the course of the surface reaction, and c_A is the constant concentration of the gaseous species A. Parameters k_{ads} and w_{ads} are the adsorption rate coefficient and the apparent adsorption rate coefficient or weight, respectively.

In contrast, in vacuum pulse-response experiments, also known as TAP experiments, the amount of gas molecules injected is much smaller than the total amount of active catalyst sites (Gleaves et al., 1988, 1997; Marin and Yablonsky, 2011) and the change in the normalized surface concentration of free active sites is insignificant. Hence, the reaction can be presented as follows:

$$r_{ads} = (k_{ads} \theta_Z) c_A = k'_{ads} c_A \tag{4.2}$$

and in this case the apparent adsorption rate coefficient, k'_{ads}, includes the concentration of free active sites as a parameter and not the concentration of the gaseous reactant.

Obviously, models (1) and (2) are linear, although for different reasons and in different ways: model (1) is linear regarding the concentration of free active sites and model (2) is linear regarding the concentration of the gaseous reactant.

4.3.3 Rate-Limiting Step Approximation

This approximation is based on assumption (2) in Section 4.2.2. The common opinion is that for not too complicated mechanisms peculiarities of transition regimes can be comprehended in detail. For example, for the sequence of two irreversible first-order reactions

$$A \xrightarrow{k_1} B \xrightarrow{k_2} C \qquad (4.3)$$

starting from pure A (no B or C present), the exact solutions for the concentrations of A, B, and C as a function of time in a batch reactor (or in a plug-flow reactor with t replaced by τ) are

$$c_A = c_{A0} e^{-k_1 t} \qquad (4.4)$$

$$c_B = c_{A0} \frac{k_1}{k_2 - k_1} \left(e^{-k_1 t} - e^{-k_2 t} \right) \qquad (4.5)$$

$$c_C = c_{A0} - c_A - c_B = c_{A0} \left(1 - \frac{k_2}{k_2 - k_1} e^{-k_1 t} - \frac{k_1}{k_1 - k_2} e^{-k_2 t} \right) \qquad (4.6)$$

A remarkable feature of reaction sequence Eq. (4.3) is a maximum in the concentration of B. Typically, there is a large difference between the magnitudes of the kinetic parameters k_1 and k_2 of the two steps, and two periods in temporal dependences are distinguished: (1) *a short initial period* governed by the kinetic parameter of the fast step and (2) *a long relaxation period*, which is determined by the kinetic parameter of the slow step.

If $k_2 \gg k_1$ (Fig. 4.2), the relaxation dependence during the short initial period is governed by the large kinetic parameter, k_2. The concentration maximum can be achieved very fast. During this initial period, $c_B \approx c_{A0} - c_A$. Then, during the second, long, period, the relaxation dependence and, especially, the tail dynamics are governed by the small kinetic parameter k_1 because the exponential $e^{-k_2 t}$ vanishes. During this period, $c_C \approx c_{A0} - c_A$ and $c_B \approx (k_1/k_2) c_A$. The ratio of the concentrations of B and A then equals $k_1/(k_2 - k_1)$, or just (k_1/k_2), since $k_2 \gg k_1$. In this case, the rate-limiting step is the first one. Since the expression $c_B = (k_1/k_2) c_A$ relates to the maximum concentration of B, this regime can be termed a "quasi-maximum" regime or even a "quasi-steady-state" regime, if the concentration of A changes very slowly. See subsequent text for a more detailed definition of the QSS regime.

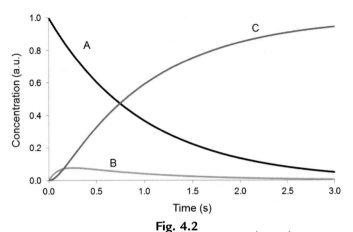

Fig. 4.2

Concentration dependences for the consecutive mechanism $A \xrightarrow{k_1} B \xrightarrow{k_2} C$; $k_1 = 1\,s^{-1}$; $k_2 = 10\,s^{-1}$.

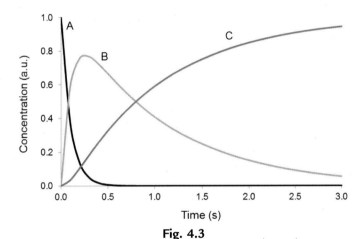

Fig. 4.3

Concentration dependences for the consecutive mechanism $A \xrightarrow{k_1} B \xrightarrow{k_2} C$; $k_1 = 10\,s^{-1}$; $k_2 = 1\,s^{-1}$.

If, on the other hand, $k_1 \gg k_2$ (Fig. 4.3) during the initial period the relaxation is governed by the large parameter, k_1. After that the exponential term $e^{-k_1 t}$ vanishes and the long relaxation dependence and the tail dynamics depend on the small kinetic parameter, k_2; the rate-limiting step is the second one. In contrast with the previous case, the ratio of the concentrations of B and A during the second period is not constant, but approaches infinity with $c_A = 0$ and $c_C = c_{A0} - c_B$.

Gorban and colleagues (Gorban et al., 2010; Gorban and Radulescu, 2008) developed a general theory of limitation in chemical reaction networks. The main new concept of this theory is an acyclic "dominant mechanism," which is distinguished within the complex mechanism. The opinion that for a not too complicated mechanism peculiarities of transition regimes can be comprehended in detail is true for the rate-limiting step approximation. Nevertheless, recently it was found that the mechanism of two consecutive irreversible reactions (4.3), one of the simplest in chemical kinetics, may exhibit an amazing variety of properties, in particular coincidences of different peculiarities of transition regimes such as double and triple intersections, extrema, and oscillation (Yablonsky et al., 2010). Mathematically, this can be explained by the interplay of the two exponential dependences that form part of the solution.

4.3.4 Quasi-equilibrium Approximation

The quasi-equilibrium approximation is based on assumption (3) in Section 4.2.2. One or more steps of the overall reaction are considered to be at equilibrium if their kinetic parameters, both forward and reverse, are much larger than the kinetic parameters of other steps. An example is the reaction sequence

$$A \underset{k_1^-}{\overset{k_1^+}{\rightleftharpoons}} B \overset{k_2}{\longrightarrow} C \tag{4.7}$$

with $k_1^+, k_1^- \gg k_2$ and starting from pure A. Fig. 4.4 shows the exact solutions for the concentrations of A, B, and C as a function of time. In this case, it is easy to show that two transient periods can be observed:

(1) *A short initial period* at the end of which an equilibrium (or rather a quasiequilibrium) of the first step is established, where the reaction rate of the forward reaction is approximately equal to the rate of the reverse reaction: $r_1^+ \approx r_1^-$ (During this initial period, hardly any C is formed and $c_B \approx c_{A0} - c_A$.).

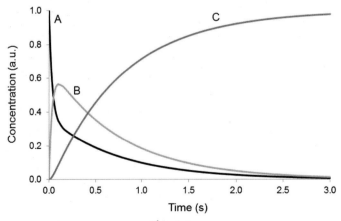

Fig. 4.4

Concentration dependences for the consecutive mechanism $A \underset{k_1^-}{\overset{k_1^+}{\rightleftharpoons}} B \overset{k_2}{\longrightarrow} C$; $k_1^+ = 20\,\text{s}^{-1}$; $k_1^- = 10\,\text{s}^{-1}$; $k_2 = 2\,\text{s}^{-1}$.

(2) *A long quasiequilibrium period* with initial concentrations that are the final concentrations of the initial period.

At the end of the initial period,

$$c_B \approx K_{eq,1} c_A \tag{4.8}$$

where $K_{eq,1} = k_1^+ / k_1^-$ is the equilibrium coefficient of the first step. Assuming that the concentration of C is negligible, $c_A + c_B = c_{A0}$.

Then, the final concentrations of A and B after the initial period can be approximated by

$$c_A = c_{A0} \left(\frac{1}{1 + K_{eq,1}} \right) \tag{4.9}$$

and

$$c_B = c_{A0} \left(\frac{K_{eq,1}}{1 + K_{eq,1}} \right) \tag{4.10}$$

These concentrations are taken as the initial concentrations for the second, quasi-equilibrium period, for which the kinetic dependence is quite simple:

$$\frac{dc_B}{dt} = -k_2 c_B \tag{4.11}$$

Integrating Eq. (4.11) with the initial condition Eq. (4.10), yields

$$c_B = c_{A0} \left(\frac{K_{eq,1}}{1 + K_{eq,1}} \right) e^{-k_2 t} \tag{4.12}$$

Then,

$$c_A = c_{A0} \left(\frac{1}{1 + K_{eq,1}} \right) e^{-k_2 t} \tag{4.13}$$

and

$$\frac{c_B}{c_A} = K_{eq,1} \tag{4.14}$$

Relationship Eq. (4.14) is approximately fulfilled within the quasiequilibrium domain.

The physicochemical meaning of this approximation is that in spite of the temporal change, the concentration ratio of some components in the mixture, in this example A and B, is governed by the equilibrium coefficient of the fast first step (see Eqs. 4.9, 4.10). However, the absolute concentration change in time is caused by the small kinetic parameter of the slow second step, see Eqs. (4.12), (4.13).

Constales et al. (2013) have described a wide variety of kinetic dependences that correspond to this two-step mechanism in which the first step is reversible.

4.3.5 Quasi-Steady-State Approximation

Every simplification has its own region of validity and this region has to be estimated at least qualitatively. The QSS approximation, which is based on assumption (5) in Section 4.2.2, is the simplification that is the most applied and the least understood. It can be called "the most complicated simplification." Despite the fact that this approximation has been in use since the 1910s, its physicochemical and mathematical meanings are still not well understood within the chemical community. Therefore, it makes sense to analyze a number of aspects of this approximation and illustrate the analysis with some examples.

First of all, the QSS is not a steady state; it is a special type of non-steady state. The popular version of the QSS approximation can be formulated as follows: During a chemical process, the concentrations of both species present in large amounts, usually the controllable and observed species, and species present in small amounts (intermediates such as radicals and surface intermediates), usually the uncontrollable and unobserved species, change in time. In the QSS approximation, the concentrations of the intermediates become functions of the concentrations of the observed abundant species; they "adapt" to the concentrations of the observed species as if they were steady-state concentrations.

The two-step consecutive reaction, $A \underset{k_1^-}{\overset{k_1^+}{\rightleftharpoons}} B \overset{k_2}{\longrightarrow} C$, can be used as the simplest example of QSS behavior. Let the kinetic parameters of the reactions in which B is consumed (k_1^- and k_2) be much larger than the kinetic parameter of the reaction in which B is formed (k_1^+). Fig. 4.5 shows an example of an exact solution for this situation. It can be proved that the QSS concentration of B can be approximated closely by the following asymptotic relationship:

$$c_B = \frac{k_1^+}{k_1^- + k_2} c_A \tag{4.15}$$

As the parametric ratio is small $\left(k_1^+ \ll k_1^-, k_2\right)$, B can be considered as a reactive molecule (eg, a free radical) with a very small concentration that follows the concentration of A.

Thus, in this simple example, three simplifications come together:

(1) The first step is rate limiting.
(2) In some temporal domain, an invariance is observed, where the ratio of the concentrations of B and A is approximately constant.
(3) Simultaneously this invariance is a fingerprint of the QSS.

Obviously, there is a similarity between the quasi-equilibrium relationship, Eq. (4.8), and the QSS relationship, Eq. (4.15). Both phenomena are caused by a difference in the magnitude of

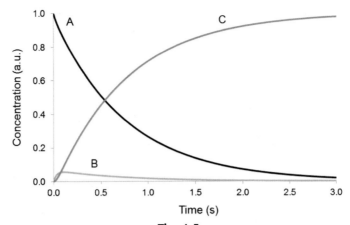

Fig. 4.5

Concentration dependences for the consecutive mechanism $A \underset{k_1^-}{\overset{k_1^+}{\rightleftharpoons}} B \xrightarrow{k_2} C$; $k_1^+ = 2\,s^{-1}$; $k_1^- = 10\,s^{-1}$; $k_2 = 20\,s^{-1}$.

the kinetic parameters. However, the physical meaning of these approximations is different. Relationship Eq. (4.8) reflects the fact that both forward and reverse reactions of the first step are fast, while Eq. (4.15) is a consequence of the fact that the rate of consumption of B is fast.

4.3.6 Mathematical Status of the Quasi-Steady-State Approximation

In previous sections, different simplifications and approximations have been introduced using first-order reactions involving linear models as examples. In many real situations, kinetic models corresponding to different physicochemical systems are nonlinear. As mentioned earlier, two typical scenarios where the QSS approximation can be used are gas- or liquid-phase reactions with free radicals as intermediates and catalytic or biocatalytic reactions involving catalytic surface intermediates or substrate-enzyme complexes. Within the traditional mathematical procedure for dealing with these intermediates, three steps can be distinguished:

(1) Write the non-steady-state model, that is, a set of ordinary differential equations for both the observed species and the unobserved intermediates.
(2) Replace the differential equations for the intermediates with the corresponding algebraic equations by setting their rates of production equal to their rates of consumption, so that the net rate of production is zero, which, in the case of catalytic surface intermediates, translates into putting:

$$\frac{d\theta_j}{dt} = 0 \qquad (4.16)$$

where θ_j is the normalized concentration of surface intermediate j, and then solving these equations, such that the concentrations of intermediates are expressed as a function of the concentrations of the observed species and the temperature. In fact, solving this set of equations is fairly easy for linear models corresponding to linear mechanisms, but for nonlinear models this may not be so simple.

(3)   Finally, construct expressions for the reaction rates of the observed species in terms of the reactant and product concentrations of the overall reaction only.

The idea to set the net rate of production of an intermediate equal to zero has its origins in the early 1900s and is almost always attributed to Bodenstein (1913). The QSS approximation is still the most popular approach for dealing with systems of complex chemical reactions involving radicals or (bio)catalysts. A complex reaction mechanism consisting of a combination of subsystems related to the observed variables x and unobserved variables y can be described by the following general model:

$$\begin{cases} \dfrac{d\mathbf{x}}{dt} = f(\mathbf{x}, \mathbf{y}) \\ \dfrac{d\mathbf{y}}{dt} = g(\mathbf{x}, \mathbf{y}) \end{cases} \qquad (4.17)$$

The subsystems in Eq. (4.17) are called subsystems of "slow" and "fast" motion, respectively.

The mathematical validity of the steady-state approximation can be illustrated by scaling the original set of equations and writing it in dimensionless form as

$$\begin{cases} \dfrac{d\bar{\mathbf{x}}}{d\tau} = f(\bar{\mathbf{x}}, \bar{\mathbf{y}}) \\ \varepsilon \dfrac{d\bar{\mathbf{y}}}{d\tau} = g(\bar{\mathbf{x}}, \bar{\mathbf{y}}) \end{cases} \qquad (4.18)$$

in which ε is the so-called small parameter ($\varepsilon \ll 1$). At the limit $\varepsilon \to 0$, Eq. (4.18) transforms into the so-called degenerated set of equations.

$$\begin{cases} \dfrac{d\bar{\mathbf{x}}}{d\tau} = f(\bar{\mathbf{x}}, \bar{\mathbf{y}}) \\ 0 \simeq g(\bar{\mathbf{x}}, \bar{\mathbf{y}}) \end{cases} \qquad (4.19)$$

The solution of the fast subsystem $g(\bar{\mathbf{x}}, \bar{\mathbf{y}}) = 0$ will yield a fast variable vector $\mathbf{y}$ as a function of the slow variable vector $\mathbf{x}$. Tikhonov's theorem provides conditions for which the solution of Eq. (4.18) approaches that of the degenerated set of equations, Eq. (4.19) (Vasil'eva and Butuzov, 1973).

Let $\mathbf{y} = \mathbf{y}_{ss}(\mathbf{x})$ be a continuous and continuously differentiable solution of $g(\bar{\mathbf{x}}, \bar{\mathbf{y}}) = 0$ in a certain domain $\mathbf{X}$ with $x \in \mathbf{X}$ and $\mathbf{y} = \mathbf{y}_{ss}(\mathbf{x})$ is an asymptotically stable global solution of the subsystem of fast motion, $\mathbf{y}(t) \to \mathbf{y}_{ss}(\mathbf{x})$ at $t \to \infty$. Then, if the solution $\mathbf{x} = \mathbf{x}(t)$ of the degenerated system Eq. (4.19) remains in the $\mathbf{X}$ domain at $0 \leq t \leq t_{\text{final}}$, for any $t_0 > 0$, the solution of the original set of equations [$\mathbf{x}(t)$, $\mathbf{y}_{ss}(\mathbf{x}(t))$] at $\varepsilon \to 0$ approaches that of the degenerated system uniformly on the segment [t_0, t_{final}]. The functions $\mathbf{x}(t)$ for the original and degenerated systems approach each other uniformly throughout the segment [0, t_{final}].

This statement can be presented qualitatively in a simpler way: The solution of the original system approaches the solution of the degenerated system if the subsystem of fast motion $g(\bar{\mathbf{x}}, \bar{\mathbf{y}}) = 0$ has a stable solution and the initial conditions are "attracted" by this solution.

Different actual systems generate the small parameter ε in different ways. For example, in homogeneous chain reactions, the small parameter is a ratio of rate coefficients. It arises because the reactions in which free radicals, which are unstable and thus are short-lived, participate are much faster than the other reactions.

In heterogeneous gas-solid catalytic systems, the small parameter is the ratio of the total amount of surface intermediates $n_{t,int}$ to the total amount of reacting gas molecules $n_{t,g}$ present in the reactor:

$$\varepsilon = \frac{n_{t,int}}{n_{t,g}} = \frac{\Gamma_t S_{cat}}{c_t V_g} \tag{4.20}$$

with Γ_t the total concentration of surface intermediates, S_{cat} the catalyst surface, c_t the total concentration of gas molecules, and V_g the gas volume.

In contrast to free radicals, surface intermediates may be relatively long-lived. Yablonskii et al. (1991) have indicated different scenarios for reaching QSS regimes in such systems.

If $\varepsilon \to 0$ and the system of fast motion has a unique and asymptotically stable global steady state at every fixed y_i, we can apply Tikhonov's theorem and, starting from a certain value of ε, use a QSS approximation.

Summing up the theoretical analysis of the QSS problem, we can distinguish two types of behavior:

- A QSS caused by a difference in kinetic parameters (*rate-parametric QSS*).
- A QSS caused by a difference in mass balances of species (*mass-balance QSS*), that is, a large difference in mass balances between gaseous or liquid reactants and products on the one side, and substrate-enzyme complexes or catalytic surface intermediates on the other.

An example of the rate-parametric QSS has been presented in Section 4.3.5 using the two-step consecutive reaction (4.7), $A \underset{k_1^-}{\overset{k_1^+}{\rightleftarrows}} B \overset{k_2}{\longrightarrow} C$. Many examples of the mass-balance QSS are presented by Yablonskii et al. (1991), in particular for a typical two-step catalytic mechanism: (1) $A + Z \rightleftarrows AZ$ and (2) $AZ \rightleftarrows B + Z$. See also the corresponding analysis in the excellent review on asymptotology by Gorban et al. (2010).

Two difficulties are linked with the use of the QSS approximation in heterogeneous catalytic reactions. The first is both pedagogical and theoretical: This approximation is an example of the right result obtained based on the wrong assumption, namely that $d\theta_j/dt = 0$. This is impossible for the fast surface intermediates. However, $\varepsilon d\theta_j/dt \approx 0$ is true and the rates of consumption of the surface intermediates are approximately equal to their rates of production. The second difficulty is that as far as we know, despite over 100 years of QSS approximation, its domain of validity for non-steady-state catalytic reactions has never been investigated systematically based on experimental data and comparing them with different models, both non-steady-state models and QSS models.

4.3.7 Simplifications and Experimental Observations

Summing up our analysis of all mentioned simplifications, the following problem can be posed: "How, based on simple experimental observations, can we distinguish which reaction is rate limiting and which is not, which reaction is irreversible and which is reversible, and, finally, which regime is established, a quasiequilibrium or QSS or just a regime with a rate-limiting step?" Answering these questions gives a primary diagnosis of the mechanism of a specific complex reaction. In fact, in the case of the relatively simple reaction,

$A \underset{k_1^-}{\overset{k_1^+}{\rightleftarrows}} B \overset{k_2}{\longrightarrow} C$, some answers to these questions are overlapping. Attention has to be paid

first to the data obtained in the course of the long (slow) period.

The following scenarios are possible:

- B is observed in negligible quantities. Thus, the approximate balance relation $c_A + c_C \approx c_{A0}$ is maintained. This means that step 1 is rate limiting and irreversible or virtually irreversible and step 2 is fast, so in effect the overall reaction $A \rightarrow C$ occurs (Fig. 4.2).
- A is observed in negligible quantities. Thus, the approximate balance relation is $c_B + c_C \approx c_{A0}$. This means that step 1 is fast and irreversible or virtually irreversible and step 2 is rate limiting (Fig. 4.3). In effect the overall reaction $B \rightarrow C$ occurs. Furthermore, the ratio of the concentrations of B and A, c_B/c_A, is not constant but approaches infinity after a short time. This is an additional argument in favor of the first step being irreversible and fast and the second step being slow.
- A and B are observed in relatively large quantities during the slow period. Their concentrations are changing, but the ratio of the concentrations of B and A is approximately constant (Fig. 4.4). Based on these observations, one can conclude that step 1 is fast and reversible and step 2 is rate limiting. The balance relation has the form: $c_A + c_B + c_C = c_{A0} \approx c_A (1 + K_{1,eq}) + c_C$.
- During the slow period the concentration of B is negligible. A and C are both observed, but the rates of their change are relatively small (Fig. 4.5). It can be concluded that step 1 is rate limiting and step 2 is fast. The balance relation has the following

approximate form: $c_A + c_C \approx c_{A0}$. In this case, the concentration of component B is characterized by QSS behavior regarding component A.

These features are reflected in Tables 4.1 and 4.2. The corresponding decision tree is presented in Fig. 4.6 as a sequence of Yes-No tests followed by the corresponding conclusions.

The following physicochemical considerations have to be taken into account in discussions on the similarities and differences of asymptotic regimes:

- All these regimes are provoked by large differences in certain properties of the reactions and processes involved, for example, differences in kinetic parameters or mass balances or both.
- In an asymptotic analysis, the time evolution of absolute values of concentrations is studied and linked to certain reaction features, such as a rate-limiting step, and species, for example highly reactive intermediates with kinetic parameters that are much larger than those of other species.
- At the same time, it is interesting to monitor the time evolution of concentration ratios, which for some asymptotic regimes can be invariances (unchanging ratios of kinetic parameters). If the parameters represent equilibrium coefficients, these in variances can be termed "thermodynamic invariances."

Table 4.1 Regimes for the reaction $A \xrightarrow{k_1} B \xrightarrow{k_2} C$

	$k_1 \ll k_2$	$k_1 \gg k_2$
First, fast period	k_2 dominates	k_1 dominates
	$c_B \approx c_{A0} - c_A$	$c_B \approx c_{A0} - c_A$
Second, slow period	k_1 dominates	k_2 dominates
	$\dfrac{c_B}{c_A} \approx \dfrac{k_1}{k_2}$	$\dfrac{c_B}{c_A} \to \infty$
	$c_C \approx c_{A0} - c_A$	$c_C \approx c_{A0} - c_A$
Regime characteristic	First step rate limiting during slow period; quasi-maximum or QSS regime	Second step rate limiting; limiting regime

Table 4.2 Regimes for the reaction $A \underset{k_1^-}{\overset{k_1^+}{\rightleftarrows}} B \xrightarrow{k_2} C$

	$k_2, k_1^- \gg k_1^+$	$k_1^+, k_1^- \gg k_2$
First, fast period	k_2, k_1^- dominate	k_1^+, k_1^- dominate
		$c_B \approx c_{A0} - c_A$
Second, slow period	k_1^+ dominates	k_2 dominates
	$\dfrac{c_B}{c_A} \approx \dfrac{k_1^+}{k_1^- + k_2}$	$\dfrac{c_B}{c_A} \approx K_{eq,1}$
Regime characteristic	First step rate limiting; QSS of B	Quasiequilibrium of the first step

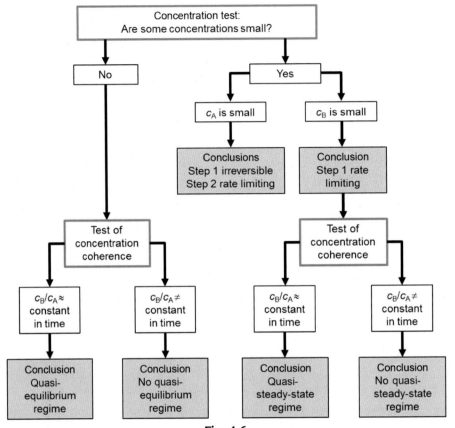

Fig. 4.6
Decision tree for reaction $A \rightleftarrows B \rightarrow C$.

Finally, it has to be noted that typically fast and slow regimes occur in kinetic behavior. In addition, the concept of the "turning point" (domain), which defines the transition between these two regimes, has to be introduced. Typically, the fast period is determined by the fast step(s), but strictly speaking the position of the turning point in time is determined by the fast parameters. The slow period is the obvious period of rate limitation. However, as a rule this limitation is accompanied by other phenomena such as quasi-equilibrium or QSS.

4.3.8 Lumping Analysis

In their classical paper, Wei and Prater (1962) introduced a so-called lumping procedure. They showed that for systems in which only monomolecular or pseudo-monomolecular reactions occur, a linear combination exists of concentrations that change in time independently. These linear combinations or "pseudocomponents" correspond to the left eigenvectors (row vectors) $\mathbf{l}$ of a kinetic matrix $\mathbf{K}$ as follows. If $\mathbf{lK} = \lambda \mathbf{l}$, then

$d(\mathbf{l}, \mathbf{c})/dt = \langle \mathbf{l}, \mathbf{c} \rangle \lambda$, where the standard inner product.[1] $\langle \mathbf{l}, \mathbf{c} \rangle$ is the concentration of a pseudocomponent. Wei and Prater also established a specially designed experiment for identifying these pseudocomponents.

The results obtained by Wei and Prater gave rise to a question: "How do we lump components into suitable pseudocomponents with proper accuracy?" Wei and Kuo determined conditions for exact (Wei and Kuo, 1969) and approximate (Kuo and Wei, 1969) lumping for monomolecular and pseudo-monomolecular reaction systems. They showed that under certain conditions, a large monomolecular system could be modeled well by a lower-order system.

Nomenclature

c_i	concentration of component i (mol m^{-3})
$\mathbf{K}$	kinetic matrix
K_{eq}	equilibrium coefficient (depends)
k	reaction rate coefficient (first-order reaction) (s^{-1})
k_{ads}	adsorption rate coefficient (s^{-1})
k'_{ads}	apparent adsorption rate coefficient (s^{-1})
$\mathbf{l}$	left eigenvector of kinetic matrix $\mathbf{K}$ (–)
$n_{t,g}$	total amount of reacting molecules in the gas phase (mol)
$n_{t,int}$	total amount of surface intermediates (mol)
r	reaction rate for a homogeneous reaction (mol m^{-3} s^{-1})
r_{ads}	rate of adsorption (mol m^{-3} s^{-1})
S_{cat}	catalyst surface area (m$^2_{cat}$)
t	time (s)
V_g	gas volume (m^3)
w_{ads}	apparent adsorption coefficient, weight (mol m^{-3} s^{-1})
x	observed variable
$\bar{x}$	dimensionless observed variable (–)
y	unobserved variable
$\bar{y}$	dimensionless unobserved variable (–)

Greek symbols

ε	small parameter (–)
Γ_t	total concentration of surface intermediates (mol m$^{-2}_{cat}$)
λ	eigenvalue (–)

---

[1] The inner product of two vectors having the same dimensionality is denoted here as $\mathbf{x}^T\mathbf{y}$ and yields a scalar:

$$\mathbf{x}^T\mathbf{y} = \sum_{i=1}^{n} x_i y_i = \mathbf{y}^T\mathbf{x}.$$ It is also referred to as the scalar product or dot product and denoted by $\mathbf{x} \cdot \mathbf{y}$.

θ_j normalized concentration of surface intermediate j (–)

τ dimensionless time (–)

Subscripts

0   initial, inlet

eq   equilibrium

g   gas

int   intermediate

ss   steady state

t   total

Superscripts

+   of forward reaction

–   of reverse reaction

References

Bodenstein, M., 1913. Eine Theorie der photochemischen Reaktionsgeschwindigkeiten. Z. Phys. Chem. (Leipzig) 85, 329–397.

Constales, D., Yablonsky, G.S., Marin, G.B., 2013. Intersections and coincidences in chemical kinetics: linear two-step reversible—irreversible reaction mechanism. Comput. Math. Appl. 65, 1614–1624.

Gleaves, J.T., Ebner, J.R., Kuechler, T.C., 1988. Temporal analysis of products (TAP)—a unique catalyst evaluation system with submillisecond time resolution. Catal. Rev. Sci. Eng. 30, 49–116.

Gleaves, J.T., Yablonskii, G.S., Phanawadee, P., Schuurman, Y., 1997. TAP-2: an interrogative kinetics approach. Appl. Catal. A 160, 55–88.

Gorban, A.N., Karlin, I.V., 2005. Invariant Manifolds for Physical and Chemical Kinetics. Springer, New York.

Gorban, A.N., Radulescu, O., 2008. Dynamic and static limitation in multiscale reaction networks, revisited. In: Marin, G.B., West, D.H., Yablonsky, G.S. (Eds.), Advances in Chemical Engineering—Mathematics in Chemical Engineering and Kinetics, vol. 34. Elsevier, London, pp. 103–173.

Gorban, A.N., Radulescu, O., Zinovyev, A.Y., 2010. Asymptotology of chemical reaction networks. Chem. Eng. Sci. 65, 2310–2324.

Kozuch, S., Martin, J.M.L., 2011. The rate-determining step is dead. Long live the rate-determining state! ChemPhysChem 12, 1413–1418.

Kruskal, M.D., 1963. Asymptotology. In: Dobrot, S. (Ed.), Mathematical Models in Physical Sciences. Prentice Hall, Englewood Cliffs, NJ, pp. 17–48.

Kuo, J.C.W., Wei, J., 1969. Lumping analysis in monomolecular reaction systems. Analysis of approximately lumpable system. Ind. Eng. Chem. Fundam. 8, 124–133.

Maas, U., Pope, S.B., 1992. Simplifying chemical kinetics: intrinsic low-dimensional manifolds in composition space. Combust. Flame 88, 239–264.

Marin, G.B., Yablonsky, G.S., 2011. Kinetics of Chemical Reactions—Decoding Complexity. Wiley-VCH, Weinheim, Germany.

Vasil'eva, A.B., Butuzov, V.F., 1973. Asymptotic Expansions of the Solutions of Singularly Perturbed Equations. Nauka, Moscow.

Wei, J., Kuo, J.C.W., 1969. A lumping analysis in monomolecular reaction systems. Analysis of exactly lumpable systems. Ind. Eng. Chem. Fundam. 8, 114–123.

Wei, J., Prater, C.D., 1962. The structure and analysis of complex reaction systems. In: Eley, D.D. (Ed.), Advances in Catalysis. Academic Press, New York, pp. 203–392.

Yablonskii, G.S., Bykov, V.I., Gorban, A.N., Elokhin, V.I., 1991. Kinetic models of catalytic reactions.
    In: Comprehensive Chemical Kinetics. vol. 32. Elsevier, Amsterdam.
Yablonsky, G.S., Constales, D., Marin, G.B., 2010. Coincidences in chemical kinetics: surprising news about simple
    reactions. Chem. Eng. Sci. 65, 2325–2332.
Yablonsky, G.S., Mareels, I.M.Y., Lazman, M., 2003. The principle of critical simplification in chemical kinetics.
    Chem. Eng. Sci. 58, 4833–4842.

## CHAPTER 5

Physicochemical Devices and Reactors

5.1 Introduction

Diffusion has a history in science of more than one and a half century (Philibert, 2005, 2006). Driven by the experimental observation of the German chemist Johann Döbereiner in 1823 that hydrogen gas diffuses faster than air, in the late 1820s and early 1830s the Scottish chemist Thomas Graham performed a systematic study on diffusion in gases and found that the rate of gas diffusion is inversely proportional to the square root of its molecular mass (Sherwood et al., 1971). The following milestones of diffusion theory and experiment can be distinguished:

- Fick's laws of diffusion (Fick, 1855a,b, 1995)
- Maxwell's theory of diffusion in gases (1860s)
- Einstein's theory of Brownian motion (Einstein, 1905)
- Knudsen diffusion in diluted gases discovered by Martin Knudsen (Knudsen, 1909, 1934; Feres and Yablonsky, 2004)
- Onsager's linear phenomenology with diffusion considered within the general context of linear nonequilibrium thermodynamics (Onsager, 1931a,b)
- Teorell's approach of the diffusion of ions through membranes (Teorell, 1935, 1937; Gorban et al., 2011)
- Prigogine's theory of "dissipative structures" generated by a combination of nonlinear chemical reactions and diffusion processes (Glansdorff and Prigogine, 1971; Nicolis and Prigogine, 1977)

In studies of heterogeneous systems, in particular those of gas-solid catalytic processes, attention was particularly paid to the interplay between diffusion and reaction in porous catalyst pellets. In 1939, Thiele introduced a dimensionless number, the Thiele modulus, to characterize this complex diffusion-reaction process in a porous medium (Thiele, 1939). The Thiele modulus quantifies the ratio of the reaction rate to the diffusion rate. Significant progress in the fundamental understanding of diffusion and adsorption in zeolites was achieved due to Kärger and Ruthven, whose book (Kärger and Ruthven, 1992) is considered by many to be the best contribution to the field of zeolite science and technology.

Keil (1999) authored a comprehensive review of theoretical and experimental studies in the field of heterogeneous catalysis. Excellent books and review articles devoted to mathematical diffusion and diffusion-reaction problems, in particular in porous media, have been published (see, for example Aris, 1975a,b; Crank, 1980; Vasquez, 2006; Gorban et al., 2011).

Summing up, diffusion-reaction problems have been popular topics in physical chemistry and in chemical and biochemical engineering for over one and a half century.

5.2 Basic Equations of Diffusion Systems

5.2.1 Fick's Laws

The basic equation of diffusion is Fick's first law. In accordance with this law, the diffusive flux (for ideal mixtures) is proportional to the negative concentration gradient, in multidimensional form:

$$\mathbf{J} = -D\,\mathrm{grad}\,c = -D\nabla c \tag{5.1}$$

where $\mathbf{J}$ is the diffusive flux vector, D is the diffusion coefficient, c is the concentration, and ∇ is the del or nabla operator, which in three dimensions (x, y, z) is expressed as $\nabla = (\partial/\partial x, \partial/\partial y, \partial/\partial z)$. According to Fick's second law, the time derivative of the concentration is the negative of the flux divergence:

$$\frac{\partial c}{\partial t} = -\mathrm{div}\,\mathbf{J} = D\Delta c \tag{5.2}$$

where the Laplace operator $\Delta = \nabla^2$

Interestingly, Fick introduced this law using an analogy with Fourier's law for heat transfer. So, Fick's law for mass transfer was coined *after* Fourier's law for heat transfer and not before.

5.2.2 Einstein's Mobility and Stokes Friction

Within Einstein's microscopic theory of motion of diluted particles in liquid systems, the diffusion coefficient is proportional to the absolute temperature:

$$D = mk_{\mathrm{B}}T \tag{5.3}$$

where $k_{\mathrm{B}} = R_{\mathrm{g}}/N_{\mathrm{A}}$ is Boltzmann's constant, with R_{g} the universal gas constant and N_{A} Avogadro's number, T is the absolute temperature and m is the so-called particle mobility, which equals the inverse of the Stokes friction coefficient f. The simplest case is a spherical particle with radius r_{p} in a Newtonian liquid with dynamic viscosity η:

$$m = \frac{1}{f} = \frac{1}{6\pi r_{\mathrm{p}}\eta} \tag{5.4}$$

Eqs. (5.3) and (5.4) combined represent the Stokes-Einstein diffusion coefficient:

$$D = \frac{k_B T}{6\pi r_p \eta} \tag{5.5}$$

Under a constant external force F_e, the velocity of the particle population equals mF_e.

5.2.3 Knudsen Diffusion and Diffusion in Nonporous Solids

For gas-solid systems with a low gas density in which gas molecules diffuse through long narrow pores, the mean free path of the molecules is much larger than the pore diameter. In this type of diffusion, so-called Knudsen diffusion, the transport properties are essentially determined by collisions of the gas molecules with the pore walls rather than by collisions with other gas molecules. Based on the kinetic theory of gases, in the Knudsen approach all rebounds are assumed to be governed by the cosine law of reflection. In a straight cylindrical pore of diameter d_{pore}, the Knudsen diffusion coefficient $D_{K,i}$ for component i with molecular mass M_i is given by

$$D_{K,i} = \frac{d_{pore}}{3}\sqrt{\frac{8R_g T}{\pi M_i}} \quad \text{for} \quad \frac{\lambda}{d_{pore}} > 5 \tag{5.6}$$

with λ the mean free path.

In solid bodies, the diffusion coefficient can be estimated using the Arrhenius dependence (Dushman and Langmuir, 1922)

$$D = D_0 \exp\left(-\frac{E_a}{R_g T}\right) \tag{5.7}$$

with E_a the activation energy.

5.2.4 Onsager's Approach

Within Onsager's theory (1931a,b), for multicomponent diffusion the flux of the ith component is given by the famous reciprocity relationship:

$$\mathbf{J}_i = \sum_j L_{ij}\mathbf{X}_j \tag{5.8}$$

where L_{ij} ($=L_{ji}$) are the so-called phenomenological coefficients that are part of the Onsager matrix $\mathbf{L}$ and $\mathbf{X}_j$ is the matrix of generalized thermodynamic forces (eg, a temperature, pressure, or chemical potential gradient). For pure diffusion, the jth thermodynamic force is the negative space gradient of the jth chemical potential μ_j, divided by the absolute temperature T. In ideal systems

$$\mu_j = R_g T \ln \frac{c_j}{c_{j,\text{eq}}} \qquad (5.9)$$

where $c_{j,\text{eq}}$ are equilibrium concentrations. Linearization near equilibrium yields

$$X_j = -\frac{1}{T} \operatorname{grad} \mu_j = -\frac{R_g}{c_{j,\text{eq}}} \operatorname{grad} c_j \qquad (5.10)$$

so

$$\mathbf{J}_i = -\sum_j L_{ij} \frac{R_g}{c_{j,\text{eq}}} \operatorname{grad} c_j = -\sum_j L_{ij} \frac{R_g}{c_{j,\text{eq}}} \nabla c_j \qquad (5.11)$$

and

$$\frac{\partial c_i}{\partial t} = -\operatorname{div} \mathbf{J}_i = R_g \sum_j L_{ij} \frac{\Delta c_j}{c_{j,\text{eq}}} \qquad (5.12)$$

Onsager's approach, by definition, is valid in the vicinity of equilibrium, and deviations of c_j from $c_{j,\text{eq}}$ are assumed to be small. The symmetry of the Onsager matrix, $L_{ij} = L_{ji}$, follows from the principle of microreversibility (see the classical monograph by de Groot and Mazur, 1962).

5.2.5 Teorell's Approach

The mobility-based concept used by Einstein was further applied by Teorell (Teorell, 1935, 1937), who constructed a theory on the diffusion of ions through membranes in aqueous solutions of univalent strong electrolytes (also see Gorban et al., 2011). Teorell assumed the solutions to be ideal and homogeneous on both sides of the membrane. The essence of his approach is captured in the following equation:

$$\text{Flux} = (\text{mobility}) \times (\text{concentration}) \times (\text{force per kilogram ion}) \qquad (5.13)$$

The force consists of two parts: (i) a diffusion force caused by a concentration gradient and (ii) an electrostatic force, caused by an electric potential gradient. With these two parts, Teorell's equation for the flux can be written as

$$J = mc\left(-R_g T \frac{1}{c}\frac{dc}{dx} + q\frac{d\varphi}{dx}\right) \qquad (5.14)$$

where q is the charge and φ is the electric potential.

It sometimes is convenient to represent the diffusion part of the force in a different way, with c_{eq} as reference equilibrium concentration (Gorban et al., 2011):

$$-R_g T \frac{1}{c}\frac{dc}{dx} = -R_g T \frac{d\ln\left(\dfrac{c}{c_{\text{eq}}}\right)}{dx} \qquad (5.15)$$

5.2.6 Nonlinear Diffusion Models

Gorban et al. (1980, 1986) proposed a model for surface diffusion of chemical species, which was later extended by Gorban and Sargsyan (1986). Different versions of this approach have been presented in the excellent paper by Gorban et al. (2011), in which an extended review is combined with many original results on these topics:

- mass-action law of diffusion
- mass action cell-jump formalism
- continuous diffusion equation
- generalized mass-action law for diffusion (thermodynamic approach)

Conceptually, the nonlinear mass-action law for surface diffusion can be introduced as follows. A system contains different surface species with concentrations $\Gamma_1, \Gamma_2, ..., \Gamma_{N_{int}}$. Empty sites with concentration Γ_0 are considered to be surface species as well. The sum of all surface concentrations, including those of the empty sites, is assumed to be constant, that is, the total amount of catalytic sites is conserved:

$$\sum_{j=0}^{N_{int}} \Gamma_j = \Gamma_t = \text{constant} \tag{5.16}$$

In a jump model of the surface diffusion process, this process is considered to proceed by jumplike substance exchange between different locations on the surface. Gorban et al. (1980) used this model for modeling of the oxidation of carbon monoxide over a platinum catalyst. One step of this reaction involves the reversible exchange of an adsorbed carbon monoxide molecule with the neighboring empty site:

$$PtCO + Pt \rightleftarrows Pt + PtCO \tag{5.17}$$

For the diffusive flux of species A_j ($j = 1, ..., N_{int}$), the jump model gives

$$J_j = -D_j(z\nabla\Gamma_j - \Gamma_j\nabla z) \tag{5.18}$$

Consequently, the corresponding diffusion equation is

$$\frac{\partial\Gamma_j}{\partial t} = -\text{div}J_j = D_j(z\nabla\Gamma_j - \Gamma_j\nabla z) \tag{5.19}$$

As a result of the conservation law for the active catalyst sites, the concentration of empty catalyst sites is equal to

$$\Gamma_0 = \Gamma_t - \sum_{j=1}^{N_{int}} \Gamma_j \tag{5.20}$$

and the system of N_{int} diffusion equations can be presented as

$$J_j = -D_j \left[\left(\Gamma_t - \sum_{j=1}^{N_{int}} \Gamma_j \right) \nabla \Gamma_j + \Gamma_j \nabla \left(\sum_{j=1}^{N_{int}} \Gamma_j \right) \right] \qquad (5.21)$$

$$\frac{\partial \Gamma_j}{\partial t} = -D_j \left[\left(\Gamma_t - \sum_{j=1}^{N_{int}} \Gamma_j \right) \Delta \Gamma_j + \Gamma_j \Delta \left(\sum_{j=1}^{N_{int}} \Gamma_j \right) \right] \qquad (5.22)$$

This system can be generalized under the assumption that all surface species can be exchanged with their closest neighbors by jump processes:

$$J_j = -\sum_k D_{jk} \left(\Gamma_k \nabla \Gamma_j - \Gamma_j \nabla \Gamma_k \right) \qquad (5.23)$$

$$\frac{\partial \Gamma_j}{\partial t} = \sum_k D_{jk} \left(\Gamma_k \nabla \Gamma_j - \Gamma_j \nabla \Gamma_k \right) \qquad (5.24)$$

5.3 Temporal Analysis of Products Reactor: A Basic Reactor-Diffusion System

5.3.1 Introduction

The temporal analysis of products (TAP) reactor is an experimental device for the systematic study of gas-solid diffusion-reaction systems. The TAP reactor can serve various purposes, such as

- precise characterization of diffusion properties and activity of solid materials.
- determination of the mechanism of gas-solid reactions over solid material, in particular the mechanism of heterogeneous catalytic reactions.
- formulation of recommendations on improved design of active materials and modification of operating regimes of industrial reactors.

The TAP reactor was created by John T. Gleaves in 1988 (Gleaves et al., 1988) and later modified (Gleaves et al., 1997). The TAP reactor has found a variety of applications in many areas of chemical kinetics and chemical engineering. The first 20 years of TAP application has been reflected by the special review of "Catalysis Today" (Kondratenko and Perez-Ramírez, 2007) and three other reviews (Berger et al., 2008; Gleaves et al., 2010; Yablonsky et al., 2005).

5.3.2 Typical TAP Reactor Configurations

Fig. 5.1 shows a typical TAP reactor system. It can be used for performing a battery of kinetic experiments, both transient and steady state, in a wide range of pressures (from 10^{-7} to 250 kPa) and temperatures (from 200 to 1200 K).

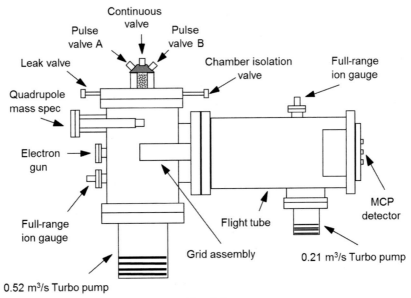

Fig. 5.1

Schematic of a TAP reactor coupled to a time-of-flight mass spectrometer. *Reprinted from Gleaves, J. T., Yablonsky, G., Zheng, X., Fushimi, R., Mills, P.L., 2010. Temporal analysis of products (TAP)—recent advances in technology for kinetic analysis of multi-component catalysts. J. Mol. Catal. A: Chem. 315, 108–134, Copyright (2010), with permission from Elsevier.*

However, the specific features of the TAP system are demonstrated in pulse-response experiments under vacuum conditions in the Knudsen-diffusion regime, in which the diffusion process is well defined. This means that—unlike the molecular diffusion coefficient—the Knudsen diffusion coefficient does not depend on the composition of the gas mixture. The diffusivity then is calculated as

$$D_{\text{eff},i} = \frac{\varepsilon_{\text{b}}}{\tau_{\text{b}}} D_{\text{K},i} = \frac{d_{\text{pore}}}{3} \sqrt{\frac{8R_{\text{g}}T}{\pi M_i}} \tag{5.25}$$

with

$$d_{\text{pore}} = \frac{2}{3} \frac{\varepsilon_{\text{b}}}{(1 - \varepsilon_{\text{b}})} d_{\text{p}} \tag{5.26}$$

where D_{eff} is the effective diffusion coefficient of the gas, ε_{b} is the bed voidage, τ_{b} is the tortuosity of the bed, and d_{p} is the diameter of the catalyst pellet.

Pulse intensities in vacuum experiments range from 10^{13} to 10^{17} molecules per pulse with a pulse width of 250 μs and a pulse frequency of between 0.1 and 50 pulses per second. Such a spectrum of time resolution is unique among kinetic methods. Possible experiments include high-speed pulsing, both single-pulse and multipulse response, steady-state isotopic transient kinetic analysis (SSITKA), temperature-programmed desorption (TPD), and temperature-programmed reaction (TPR).

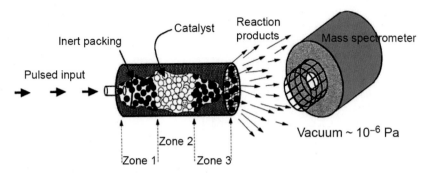

Fig. 5.2

Three-zone TAP reactor with quadrupole mass spectrometer; $L_{zone\ 1} = L_{zone\ 2} = L_{zone\ 3}$. *Reprinted from Shekhtman, S.O., Yablonsky, G.S., Chen, S., Gleaves, J.T., 1999. Thin-zone TAP-reactor—theory and application. Chem. Eng. Sci. 54, 4371–4378, Copyright (1999), with permission from Elsevier.*

5.3.2.1 Three-zone TAP reactor

The most common TAP microreactor is the three-zone TAP reactor (Fig. 5.2). In this reactor the active catalyst zone is sandwiched between two inert zones, which consist of packed beds of inert particles, typically quartz.

In the three-zone TAP reactor, it is difficult to maintain a uniform profile of the catalyst surface composition because of the gas concentration gradient, which is the driving force for the diffusion transport in the reactor. This nonuniformity of the catalyst surface composition becomes significant particularly in multipulse experiments.

5.3.2.2 Thin-zone TAP reactor

The thin-zone TAP reactor (TZTR) configuration (Fig. 5.3) was a proposal by Shekhtman et al. (1999). In a TZTR, the thickness of the catalyst zone is made very small compared to the total length of the reactor. Thus, the change of the gas concentration across the thin catalyst zone is

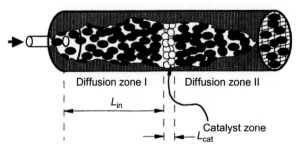

Fig. 5.3

Thin-zone TAP reactor. *Reprinted from Shekhtman, S.O., Yablonsky, G.S., Chen, S., Gleaves, J.T., 1999. Thin-zone TAP-reactor—theory and application. Chem. Eng. Sci. 54, 4371–4378, Copyright (1999), with permission from Elsevier.*

negligible and the catalyst composition can be considered to be uniform. In the thin-zone TAP approach, it is possible to separate chemical reaction from the well-defined Knudsen-diffusion process. Another advantage of the TZTR is that the catalyst zone can be more easily maintained at isothermal conditions than the catalyst zone in the conventional three-zone TAP reactor.

5.3.2.3 Single-particle TAP reactor

Zheng et al. (2008) reported results of atmospheric and vacuum pulse-response experiments on the catalytic oxidation of carbon monoxide. The catalyst sample used was a single polycrystalline platinum particle with a diameter of 400 μm, which was placed in a microreactor bed containing about 100,000 inert particles with diameters between 210 and 250 μm (Fig. 5.4).

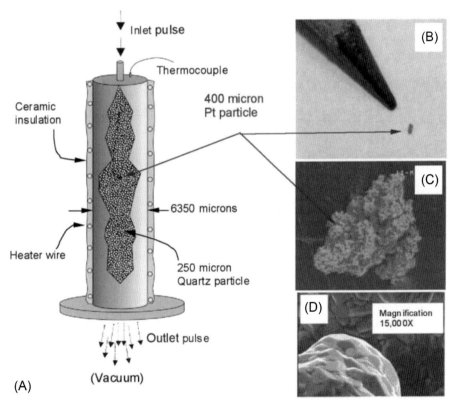

Fig. 5.4

(A) Schematic of TAP single-particle microreactor configuration. A Pt particle with a diameter of 400 μm is located in a sea of inert quartz particles with diameters between 210 and 250 μm; (B) image comparing a 400 μm Pt particle to a pencil point; (C) SEM image showing the complex surface structure of a polycrystalline Pt particle; (D) Larger magnification (15,000×) of the particle shown in (C), which demonstrates that the surface is nonporous. *Reprinted from Zheng, X., Gleaves, J.T., Yablonsky, G.S., Brownscombe, T., Gaffney, A., Clark, M., Han, S., 2008. Needle in a haystack catalysis. Appl. Catal. A 341, 86–92, Copyright (2008), with permission from Elsevier.*

This TAP single-particle microreactor is a further modification of the TZTR. The particle occupies less than 0.3% of the cross-sectional area of the TAP reactor, so the catalyst zone can be considered as a point source. Temperature and/or concentration gradients across the catalyst zone can be assumed to be negligible. Nevertheless, the conversion achieved in this single-particle microreactor is high because of the nature of the diffusion-reaction process.

5.3.3 Main Principles of TAP

The three main principles of TAP are

- well-defined Knudsen diffusion as a measuring stick for the characterization of chemical activity.
- insignificant change of the solid material during a single-pulse experiment and controlled change during a series of pulses.
- uniformity of the surface composition of the solid material across the active zone.

The first and second principles were formulated early on, in the pioneering paper by Gleaves et al. (1988) and further developed by Shekhtman et al. (1999). As mentioned, in TAP experiments, transport occurs in a well-defined diffusion regime, known as the Knudsen regime, the characteristics of which are independent of the gas composition. Such diffusion transport serves as a measuring stick for extracting kinetic information. During a pulse experiment, in the Knudsen regime a typical pulse contains 10^{13} molecules or approximately 10^{-10} mol. As an example, with a catalyst sample having a surface area of 10^4 m^2 kg^{-1}, a single pulse would address only $1/10^6$ part of the total surface of 10^{-4} kg sample. If the active surface area is a realistic fraction (say, more than 0.1%) of the total surface area, a single pulse will have a negligible effect on the active catalyst surface. Whenever the number of reactant molecules is significantly smaller than the number of active catalyst sites, the catalytic system is in nearly the same state before and after the experiment. This type of experiment performed in the TAP reactor is referred to as a *state-defining experiment* (SDE).

For distinguishing a SDE, the following experimental criteria can be used:

- insignificant change in the pulse response of a reactant or product during a series of pulses
- independence of the shape of a pulse-response curve on the pulse intensity

The first criterion implies that the reactivity of the catalyst surface does not change from pulse to pulse. This criterion can be applied to an irreversible process such as the reduction of a metal oxide catalyst by a hydrocarbon. The second criterion implies that the number of molecules in the input pulse is sufficiently small for a minor change of this number not to influence the shape of the exit flow response.

SDEs provide kinetic parameters corresponding to a given catalyst state. For example, in the case of the irreversible transformation of a hydrocarbon on a metal oxide catalyst, the corresponding kinetic coefficient depends on the state of the metal oxide surface and, in particular, on its oxidation state.

In a *state-altering experiment* (SAE), a multipulse procedure is performed. A large number of pulses are injected and the reactant and product responses are monitored. By performing TAP multipulse experiments, the interaction of a gas component with the surface at different catalyst compositions can be studied. State-defining and state-altering experiments can be performed in sequence. In this so-called *interrogative cycle*, a given catalyst state is probed using SDEs involving different gas molecules. Next, a SAE is performed, resulting in a new catalyst state. This state is probed again by different types of gas molecules in SDEs. The goal of this procedure, for which Gleaves et al. (1997) coined the term "interrogative kinetics," is to systematically probe a variety of different catalyst states and to understand how one state evolves into another.

The third basic principle of TAP, that of uniformity of the surface composition across the active zone, proposed by Shekhtman et al. (1999), has to be considered an absolute necessity in TAP studies because kinetic information corresponding to a certain catalyst composition cannot be obtained directly if nonuniform catalyst states are probed.

We consider the TAP reactor as a basic kinetic device for systematic studies of reaction-diffusion systems. In this chapter, we are going to (i) present and analyze models of different TAP configurations with a focus on their possibilities with respect to characterizing active materials and unraveling complex mechanisms, and (ii) demonstrate relationships between TAP models and other basic reactor models, that is, models for the ideal continuous stirred-tank reactor (CSTR), batch reactor (BR) and plug-flow reactor (PFR). In some situations, the TZTR can be considered a simple building block for constructing the various models.

5.4 TAP Modeling on the Laplace and Time Domain: Theory

5.4.1 The Laplace Transform

The first step in the successful analysis of TAP experiments is to abstract from the detailed physical model to obtain a practical equivalent mathematical formulation including only the *apparent* physical parameters.

For the one-dimensional model this amounts to

$$\varepsilon_b \frac{\partial c}{\partial t} = D_{eff} \frac{\partial^2 c}{\partial x^2} - k_a c \qquad (5.27)$$

Now we are facing the difficulty that this partial differential equation (PDE) is imposed on continuous domains in time and space, in fact on $t > 0$ (all positive values of time) and on $0 < x < L$ (all x coordinates inside a zone of constant apparent ε_b, D_{eff} and k_a in the reactor). Note that we exclude the boundary values $t=0$, $x=0$, and $x=L$ from these domains because the derivatives $\partial/\partial t$ and $\partial^2/\partial x^2$ are only directly defined for interior points.

Our task is now to reduce this differential equation to mathematical structures that can be expressed finitely. Two steps are required for this: (i) reducing the time dependence to the Laplace domain and (ii) integrating the relation expressed in x over the domain $0<x<L$.

The switch to the Laplace domain is realized as follows: when f is a function of at most exponential growth, so

$$|f(t)| \leq C_1 \exp(C_2 t) \quad \text{for } t > t_0 \tag{5.28}$$

with C_1 and C_2 real values and f a continuous function at least on separate intervals, the Laplace transform $\mathcal{L}\{f\}$ is defined through the integral equation

$$\mathcal{L}\{f\}(s) = \int_0^{+\infty} e^{-st} f(t) dt \tag{5.29}$$

where s denotes the Laplace variable, a value in the complex plane. From the assumed estimate, Eq. (5.28), of $|f(t)|$ in terms of an exponential, it can be shown that $\mathcal{L}\{f\}$ is an analytical function of s defined at least over the half-plane $\text{Re}(s) > C_2$.

The Laplace transform is an exceptionally useful computational technique because the equivalents of calculus operations in the Laplace domain are straightforward algebraic ones.

5.4.2 Key Results

The first key result in the present case is that the Laplace transform of a derivative with respect to t is readily expressed in terms of $\mathcal{L}\{f\}$, namely

$$\mathcal{L}\{f'\} = s\mathcal{L}\{f\} - f(0) \tag{5.30}$$

as follows directly from applying the definition, Eq. (5.28), and integrating by parts:

$$\int_0^{+\infty} e^{-st} f'(t) dt = e^{-st} f(t)\big|_0^{+\infty} - \int_0^{+\infty} \left(e^{-st}\right)' f(t) dt$$

$$= 0 - f(0) + s\int_0^{+\infty} e^{-st} f(t) dt = s\mathcal{L}\{f\} - f(0) \tag{5.31}$$

This allows us to get rid of the time derivative $\partial/\partial t$ on the left-hand side of Eq. (5.27). The term $\partial^2/\partial x^2$ is not affected by the Laplace transform in t, so that

$$\mathcal{L}\left\{\frac{\partial^2 f}{\partial x^2}\right\} = \frac{\partial^2}{\partial x^2} \mathcal{L}\{f\} \tag{5.32}$$

The second key result is that the Laplace transform uniquely simplifies accounting for different solicitations of a linear system over time. If the response of a system $\mathcal{L}\{f\} = g$ to $g(t)$ is $f(t)$

and the system is autonomous, it will respond to $g(t-\tau)$ by $f(t-\tau)$ when a time shift occurs.[1] Consequently, if we know the response $f_0(t)$ to a Dirac delta solicitation $\mathcal{L}\{f_0\} = \delta(t)$, and writing $g(t)$ as a superposition of Dirac delta expressions:

$$g(t) = \int_0^{+\infty} g(\tau)\delta(t-\tau)d\tau \tag{5.33}$$

the solution f to $\mathcal{L}\{f\} = g$ is given by

$$f(t) = \int_0^{+\infty} g(\tau)f_0(t-\tau)d\tau \tag{5.34}$$

as a result of linearity. This expression is so useful in practice that it is studied in its own right. It is called the *convolution product* of g and f_0, $g*f_0$. Note that since both factors in the convolution product are assumed to be causal, the integral can also be written as $\int_{-\infty}^{+\infty} \cdots$ or $\int_0^t \cdots$ without affecting its value. Calculating such convolution integrals might seem arduous but luckily the Laplace transform simplifies this considerably. If we write the definition of $\mathcal{L}\{f*g\}$ as

$$\mathcal{L}\{f*g\} = \int_0^{+\infty} e^{-st}(f*g)(t)dt = \int_{-\infty}^{+\infty} e^{-st}(f*g)(t)dt = \int_{-\infty}^{+\infty} e^{-st}\int_{-\infty}^{+\infty} f(\tau)g(t-\tau)d\tau dt \tag{5.35}$$

and exchange the order of integrals while reworking $e^{-st} = e^{-s\tau}e^{-s(t-\tau)}$, then Eq. (5.35) simplifies to

$$\mathcal{L}\{f*g\} = \int_{-\infty}^{+\infty} e^{-s\tau}f(\tau)d\tau \int_{-\infty}^{+\infty} e^{-su}g(u)du = (\mathcal{L}\{f\})(\mathcal{L}\{g\}) \tag{5.36}$$

that is, the Laplace transform of a (complicated) convolution is a (simple) product. This also means that systems of linear equations with convolution products, which would be difficult to analyze in the time domain, transform to systems of linear equations including ordinary products, for which the many resources of linear algebra are readily available.

We now apply these results to the problem at hand. After Laplace transformation with respect to t, the PDE becomes

$$\varepsilon_b(s\mathcal{L}\{c\}) - 0 = D_{\text{eff}}\frac{\partial^2 \mathcal{L}\{c\}}{\partial x^2} - k_a\mathcal{L}\{c\} \tag{5.37}$$

in which only derivatives with respect to the variable x remain, so that this is actually an ordinary differential equation (ODE). Furthermore, this ODE is linear with constant coefficients and homogeneous: from a basic analysis we know that its solutions can be written

---

[1] Note that all functions considered here are supposed to be causal, that is, their values vanish for all $t < 0$.

as linear combinations of exponential terms $\exp(\lambda x)$ where $\lambda \in \mathbb{C}$ must be chosen such that both sides of Eq. (5.36) are identical. Substitution leads to the characteristic equation

$$\varepsilon_b s = D_{\text{eff}} \lambda^2 - k_a \tag{5.38}$$

with roots

$$\lambda_\pm = \pm \sqrt{\frac{\varepsilon_b s + k_a}{D_{\text{eff}}}} \tag{5.39}$$

In general, s is allowed to be any complex value of sufficiently large real part for the defining integral of the Laplace transform to exist. In practice, however, due to the properties of analytic functions of complex variables, one can always assume that s is actually real, and large positive. All results obtained under these assumptions are guaranteed to be valid generally for all s in a right complex half-plane, but in cases such as these the positivity assumption brings the benefit that the square root in Eq. (5.39) need not be specified for nonreal or negative values of its argument. Similarly, although for $\varepsilon_b s = -k_a$ there is a complication due to the coincidence of the two characteristic roots (since then $\lambda_+ = \lambda_- = 0$), this case does not require further analysis.

Summarizing, the general solution of Eq. (5.37) can be expressed as

$$\mathcal{L}\{c\} = C_1 \exp\left(-x\sqrt{\frac{\varepsilon_b s + k_a}{D_{\text{eff}}}}\right) + C_2 \exp\left(x\sqrt{\frac{\varepsilon_b s + k_a}{D_{\text{eff}}}}\right) \tag{5.40}$$

where C_1 and C_2 are constants for x (but typically will depend on s and other parameters). This establishes the general framework within which the actual solution will necessarily fit.

To obtain results that are generally valid and can be used as computational building blocks for models with an arbitrary composition of zones, we now turn to the special boundary condition that expresses the physical contact between successive zones in the reactor (see Fig. 5.5). These contact or interface conditions are to be determined from a combination of physical insight and experimental results.

Z_1	Z_2	Z_3	Z_4	Z_5
L_1	L_2	L_3	L_4	L_5
$\varepsilon_{b,1}$	$\varepsilon_{b,2}$	$\varepsilon_{b,3}$	$\varepsilon_{b,4}$	$\varepsilon_{b,5}$
$k_{a,1}$	$k_{a,2}$	$k_{a,3}$	$k_{a,4}$	$k_{a,5}$
$D_{\text{eff},1}$	$D_{\text{eff},2}$	$D_{\text{eff},3}$	$D_{\text{eff},4}$	$D_{\text{eff},5}$

Fig. 5.5

Schematic of a five-zone TAP reactor. Each zone has its own length L_i and parameters $\varepsilon_{b,i}, k_{a,i}$ and $D_{\text{eff},i}$.

The first condition is that there is no jump in concentration, that is, $c(x-0)=c(x+0)$ everywhere, specifically at x values where two neighboring zones touch.

The second condition is similar, but applies to the diffusive flux, which is given by

$$J = -D_{\text{eff}} \frac{\partial c}{\partial x} \tag{5.41}$$

so $J(x-0)=J(x+0)$ everywhere, specifically at x values where two neighboring zones touch. At these points, however, D_{eff} need not be equal on both sides. For instance, in the schematic of Fig. 5.5 there is no reason why $D_{\text{eff},1}$ should equal $D_{\text{eff},2}$ or why $D_{\text{eff},2}$ should equal $D_{\text{eff},3}$, and so on; each zone may consist of a different material with unique properties. Consequently, the equality

$$D_{\text{eff},i} \left(\frac{\partial c}{\partial x} \right)_{\text{left}} = D_{\text{eff},i-1} \left(\frac{\partial c}{\partial x} \right)_{\text{right}} \tag{5.42}$$

entails that the derivative $\partial c/\partial x$ will show a jump at the interfaces between the zones.

Taking fully into account these jumps in material characteristics in a multizone model in the time domain is a pain; again the Laplace transform comes to the rescue and delivers computational elegance. In order to clarify this, we briefly return to our solution in the Laplace domain for a single zone, and express the concentration and flux values to the left of this zone in terms of the values to its right. The linearity of the equations guarantees that one can write

$$\begin{bmatrix} \mathcal{L}\{c\} \\ \mathcal{L}\{J\} \end{bmatrix}_{\text{left}} = \begin{bmatrix} M_{11} & M_{12} \\ M_{21} & M_{22} \end{bmatrix} \begin{bmatrix} \mathcal{L}\{c\} \\ \mathcal{L}\{J\} \end{bmatrix}_{\text{right}} \tag{5.43}$$

for certain matrix entries M_{ij}. In particular,

$$\begin{bmatrix} M_{11} \\ M_{21} \end{bmatrix} = \begin{bmatrix} \mathcal{L}\{c\} \\ \mathcal{L}\{J\} \end{bmatrix}_{\text{left}} \quad \text{when} \quad \begin{bmatrix} \mathcal{L}\{c\} \\ \mathcal{L}\{J\} \end{bmatrix}_{\text{right}} = \begin{bmatrix} 1 \\ 0 \end{bmatrix} \tag{5.44}$$

Expressing this in the general solution gives

$$\mathcal{L}\{c\} = \cosh \left(L \sqrt{\frac{\varepsilon_b s + k_a}{D_{\text{eff}}}} \right) \tag{5.45}$$

and

$$\mathcal{L}\{J\} = \sqrt{(\varepsilon_b s + k_a) D_{\text{eff}}} \ \sinh \left(L \sqrt{\frac{\varepsilon_b s + k_a}{D_{\text{eff}}}} \right) \tag{5.46}$$

Similarly,

$$\begin{bmatrix} M_{12} \\ M_{22} \end{bmatrix} = \begin{bmatrix} \mathcal{L}\{c\} \\ \mathcal{L}\{J\} \end{bmatrix}_{\text{left}} \quad \text{when} \quad \begin{bmatrix} \mathcal{L}\{c\} \\ \mathcal{L}\{J\} \end{bmatrix}_{\text{right}} = \begin{bmatrix} 0 \\ 1 \end{bmatrix} \tag{5.47}$$

which from the general solution leads to

$$\mathcal{L}\{c\} = \frac{\sinh\left(L\sqrt{\dfrac{\varepsilon_b s + k_a}{D_{eff}}}\right)}{\sqrt{(\varepsilon_b s + k_a)D_{eff}}} \tag{5.48}$$

and

$$\mathcal{L}\{J\} = \cosh\left(L\sqrt{\frac{\varepsilon_b s + k_a}{D_{eff}}}\right) \tag{5.49}$$

Written in more compact form, in the Laplace domain a zone corresponds to a matrix equation

$$\begin{bmatrix} \mathcal{L}\{c\} \\ \mathcal{L}\{J\} \end{bmatrix}_{left} = \mathbf{M}(s, \varepsilon_b, k_a, D_{eff}, L) \begin{bmatrix} \mathcal{L}\{c\} \\ \mathcal{L}\{J\} \end{bmatrix}_{right} \tag{5.50}$$

where the *zone transfer matrix* $\mathbf{M}$ is given by

$$\mathbf{M} = \begin{bmatrix} \cosh u & \dfrac{\sinh u}{w} \\ w \sinh u & \cosh u \end{bmatrix}; \quad u = L\sqrt{\frac{\varepsilon_b + k_a}{D_{eff}}}; \quad w = \sqrt{(\varepsilon_b + k_a)D_{eff}} \tag{5.51}$$

Notice that the determinant of $\mathbf{M}$, det $\mathbf{M} = 1$ as $\cosh^2 u - \sinh^2 u = 1$.

If, for instance, a TAP reactor consists of two adjacent zones, combining the zone transfer matrices leads to

$$\begin{bmatrix} \mathcal{L}\{c\} \\ \mathcal{L}\{J\} \end{bmatrix}_{inlet} = \underbrace{\mathbf{M}_1(s_1, \varepsilon_{b,1}, k_{a,1}, D_{eff,1}, L_1)\mathbf{M}_2(s_2, \varepsilon_{b,2}, k_{a,2}, D_{eff,2}, L_2)}_{\text{global transfer matrix}} \begin{bmatrix} \mathcal{L}\{c\} \\ \mathcal{L}\{J\} \end{bmatrix}_{outlet} \tag{5.52}$$

so that the global transfer matrix, linking the inlet and outlet concentration and flux values, is the product of these zone transfer matrices, ordered left to right, $\mathbf{M}_1\mathbf{M}_2$. If the reactor consists of three zones, the global transfer matrix is given by $\mathbf{M}_1\mathbf{M}_2\mathbf{M}_3$, etc. In general, we will express the global transfer matrix in terms of its elements as

$$\mathbf{M} = \begin{bmatrix} P_{11} & P_{12} \\ P_{21} & P_{22} \end{bmatrix} \tag{5.53}$$

Now this global transfer matrix can be used to express the boundary conditions at the inlet and outlet of the reactor. We assume that at $t=0$ at the inlet a sharp pulse of gas is injected into the system. Physically this means that $J_{inlet}(t) = \delta(t - 0^+)$. We also assume that the vacuum at the outlet is perfect, that is, $c_{outlet}(t) = 0$. Our interest lies with the outlet flux, J_{outlet}, which is observed experimentally. In the Laplace domain, keeping only the second equation in the matrix system yields

$$\mathcal{L}\{J_{\text{inlet}}\} = P_{21}\underbrace{\mathcal{L}\{c_{\text{outlet}}\}}_{=0} + P_{22}\mathcal{L}\{J_{\text{outlet}}\} \tag{5.54}$$

so that

$$\mathcal{L}\{J_{\text{outlet}}\} = \frac{1}{P_{22}} \tag{5.55}$$

which is the fundamental result of TAP reactor theory. It means that to completely analyze a multizone reactor setup in the Laplace domain, it suffices to form the zone transfer matrices, to multiply these matrices to obtain the global transfer matrix, and then take the value of the P_{22} component in the result.

5.5 TAP Modeling on the Laplace and Time Domain: Examples

5.5.1 Moment Definitions

Moments are experimental characteristics of the outlet flux that can readily be obtained in closed form from their Laplace transforms. Since $\mathcal{L}\{f\}(s) = \int_0^{+\infty} e^{-st} f(t) dt$ (Eq. 5.29), the so-called zeroth moment is

$$M_0 = \mathcal{L}\{f\}(0) = \int_0^{+\infty} f(t) dt \tag{5.56}$$

the first moment is

$$M_1 = -\frac{\partial \mathcal{L}}{\partial s}\{f\}(0) = \int_0^{+\infty} tf(t) dt \tag{5.57}$$

and so on. The nth moment is

$$M_n = (-1)^n \frac{\partial^n \mathcal{L}}{\partial s^n}\{f\}(0) = \int_0^{+\infty} t^n f(t) dt \tag{5.58}$$

5.5.2 One-Zone Reactor, Diffusion Only

For a one-zone TAP reactor with diffusion only, the global transfer matrix can be written as (with symbol * for complicated irrelevant entries)

$$\mathbf{M} = \begin{bmatrix} P_{11} & P_{12} \\ P_{21} & P_{22} \end{bmatrix} = \begin{bmatrix} * & * \\ * & \cosh\left(L\sqrt{\dfrac{\varepsilon_b s}{D_{\text{eff}}}}\right) \end{bmatrix} \tag{5.59}$$

so

$$\mathcal{L}\{J_{outlet}\} = \frac{1}{\cosh\left(L\sqrt{\dfrac{\varepsilon_b s}{D_{eff}}}\right)}$$ (5.60)

In this case, the zeroth moment is $M_0 = 1$ (all molecules pulsed in also leave the reactor since there is no reaction, $k_a = 0$) and $M_1 = \dfrac{L^2 \varepsilon_b}{2 D_{eff}}$. The average residence time, which is given by M_1/M_0, therefore equals $\dfrac{1}{2}\dfrac{\varepsilon_b L^2}{D_{eff}}$, a typical characteristic time for a diffusion process taking place with diffusivity D_{eff} over a length L. The Laplace transform $\mathcal{L}\{J_{outlet}\}$ therefore is a generating function for the moments of the outlet time distribution. It also coincides with the so-called characteristic function of the distribution if we set $s = -i\omega$, ω in $\mathbb{R}$. An alternative interpretation does not consider $\mathcal{L}\{J_{outlet}\}$ but its natural logarithm: in general we know from statistics that near $s = 0$,

$$\ln \mathcal{L}\{J_{outlet}\} = \ln M_0 - \frac{s}{1!}C_1 + \frac{s^2}{2!}C_2 - \frac{s^3}{3!}C_3 + \cdots$$ (5.61)

where the coefficients C_i are the so-called cumulants of the distribution: C_1 is the average, M_1/M_0, C_2 the variance, $M_2/M_0 - (M_1/M_0)^2$, C_3 a skewness, and so on. These cumulants help describe the outlet flux as a distribution of particle "escape times." Note that all moments and cumulants can be obtained exactly from the specifications of the reactor and the computation of the global transfer matrix described in Section 5.4. No discretization in time or space is involved; the results are mathematically exact.

The results in the Laplace domain can also be used to obtain results in the time domain. To this end, we first note that setting $s = i\omega$, with ω the Fourier frequency, in the Laplace transform definition yields

$$\mathcal{L}\{f\}(i\omega) = \int_0^{+\infty} e^{-i\omega t} f(t)dt = \int_{-\infty}^{+\infty} e^{-i\omega t} f(t)dt = \mathcal{F}\{f\}(\omega)$$ (5.62)

(the second step relying on the causal nature of f), so that the Fourier transform of f is readily obtained by this substitution. Now a very fast and robust numerical algorithm, aptly called the Fast Fourier Transform, is available for computing from a suitable sample of $\mathcal{F}\{f\}(\omega)$ values a corresponding set of $f(t_k)$ values in time.

A further method to return from the Laplace domain to the time domain is based on a special type of series expansion; in the present case expanding

$$\frac{1}{\cosh x} = \sum_{n=0}^{+\infty} \frac{(-1)^n (2n+1)\pi}{x^2 + \left(n + \frac{1}{2}\right)^2 \pi^2}$$ (5.63)

for $x = L\sqrt{\dfrac{\varepsilon_b s}{D_{eff}}}$ leads to

$$\mathcal{L}\{J_{\text{outlet}}\} = \sum_{n=0}^{+\infty} \frac{(-1)^n (2n+1)\pi}{\dfrac{L^2 \varepsilon_b}{D_{\text{eff}}} s + \left(n + \frac{1}{2}\right)^2 \pi^2} \tag{5.64}$$

Since

$$\mathcal{L}\left\{ \exp\left(-\frac{\left(n + \dfrac{1}{2}\right)^2 \pi^2 D_{\text{eff}}}{L^2 \varepsilon_b} t \right) \right\} = \frac{1}{s + \dfrac{\left(n + \frac{1}{2}\right)^2 \pi^2 D_{\text{eff}}}{L^2 \varepsilon_b}} \tag{5.65}$$

it follows from Eq. (5.64) that

$$J_{\text{outlet}} = \frac{\pi D_{\text{eff}}}{L^2 \varepsilon_b} \sum_{n=0}^{+\infty} (-1)^n (2n+1) \exp\left(-\frac{\left(n + \dfrac{1}{2}\right)^2 \pi^2 D_{\text{eff}}}{L^2 \varepsilon_b} t \right) \tag{5.66}$$

where the argument to the exponential function rapidly becomes very negative with increasing n. In this case, a different expansion yields

$$\frac{1}{\cosh x} = \frac{2}{e^x + e^{-x}} = 2\left(e^{-x} - e^{-3x} + e^{-5x} - e^{-7x} + \cdots \right) \tag{5.67}$$

so that

$$J_{\text{outlet}} = L \sqrt{\frac{\varepsilon_b}{\pi D_{\text{eff}} t^3}} \sum_{n=0}^{+\infty} (-1)^n (2n+1) \exp\left(-\frac{L^2 \varepsilon_b \left(n + \dfrac{1}{2}\right)^2}{D_{\text{eff}} t} \right) \tag{5.68}$$

5.5.3 One-Zone Reactor, Irreversible Adsorption

In the case of irreversible adsorption, the only difference from the diffusion-only case is that $\varepsilon_b s + k_a$ replaces $\varepsilon_b s$ in the expressions, that is, s must be replaced with $s + k_a/\varepsilon_b$. Consequently,

$$(J_{\text{outlet}})_{k_a \neq 0} = e^{-\frac{k_a}{\varepsilon_b} t} (J_{\text{outlet}})_{k_a = 0} \tag{5.69}$$

and the expressions in the time domain follow at once.

When the moments are concerned, M_0 no longer equals 1 but becomes

$$M_0 = \frac{1}{\cosh\left(L\sqrt{\dfrac{k_a}{D_{eff}}}\right)} < 1 \qquad (5.70)$$

meaning that only part of the injected gas leaves the reactor without being adsorbed. The conversion then equals $1 - M_0$ and increases with increasing k_a. Often a new dimensionless quantity $\varphi = L\sqrt{\dfrac{k_a}{D_{eff}}}$ is introduced to restrain the size of the expressions. The average residence time of gas molecules leaving the reactor is then given by

$$\frac{M_1}{M_0} = \underbrace{\frac{\tanh\varphi}{\varphi}}_{\text{reaction}} \cdot \underbrace{\frac{\varepsilon_b L^2}{2 D_{eff}}}_{\text{diffusion}} \qquad (5.71)$$

and is always smaller than the pure diffusion time: the molecules that escape are faster on average and thereby avoid adsorption.

5.5.4 One-Zone Reactor, Reversible Adsorption

The case of reversible adsorption has not yet been considered, but the flexibility of the analysis in the Laplace domain will allow us to reduce the problem to a known case as follows.

The original PDE (Eq. 5.27) now includes a desorption term:

$$\varepsilon_b \frac{\partial c}{\partial t} = D_{eff}\frac{\partial^2 c}{\partial x^2} - k_a c + k_d c_{cat}\theta \qquad (5.72)$$

where $\theta(x,t)$ is the so-called fractional surface coverage of adsorbed species. This coverage satisfies a differential equation of its own:

$$\frac{\partial\theta}{\partial t} = \frac{k_a}{c_{cat}}c - k_d\theta \qquad (5.73)$$

The resulting set of equations may be solved in the time domain by setting

$$c = \frac{c_{cat}}{k_a}\left(\frac{\partial\theta}{\partial t} + k_d\theta\right) \qquad (5.74)$$

and substitution in Eq. (5.72), but this would give rise to awkward terms of the form $\partial\theta^3/\partial x^2\partial t$ in the final equation. We observe, however, that in the Laplace domain the equation for θ is readily solved (assuming initial zero coverage):

$$s(\mathcal{L}\{\theta\} - 0) = \frac{k_a}{c_{cat}}\mathcal{L}\{c\} - k_d\mathcal{L}\{\theta\} \qquad (5.75)$$

so

$$\mathcal{L}\{\theta\} = \frac{k_a}{c_{cat}(s+k_d)}\mathcal{L}\{c\} \tag{5.76}$$

Substituting Eq. (5.76) in the Laplace-transformed PDE for c yields

$$\varepsilon_b(s\mathcal{L}\{c\} - 0) = D_{eff}\frac{\partial^2 \mathcal{L}\{c\}}{\partial x^2} - \left(k_a - \frac{k_a k_d}{s+k_d}\right)\mathcal{L}\{c\}$$
$$= D_{eff}\frac{\partial^2 \mathcal{L}\{c\}}{\partial x^2} - \frac{sk_a}{s+k_d}\mathcal{L}\{c\} \tag{5.77}$$

Thus, k_a merely has to be replaced with the expression $\dfrac{sk_a}{s+k_d}$ throughout, which for $k_d = 0$ clearly simplifies properly to k_a. The interface and boundary conditions only involve the diffusional parameter D_{eff} and are independent of k_a. Consequently, the zone transfer matrix is still of the form

$$\mathbf{M} = \begin{pmatrix} \cosh u & \dfrac{\sinh u}{w} \\ w\sinh u & \cosh u \end{pmatrix} \tag{5.78}$$

but now the more general expressions

$$u = L\sqrt{\frac{\varepsilon_b s + k_a s/(s+k_d)}{D_{eff}}} \tag{5.79}$$

and

$$w = \sqrt{(\varepsilon_b s + sk_a/(s+k_d))D_{eff}} \tag{5.80}$$

apply instead.

The essential value $\mathcal{L}\{J_{outlet}\}$ is still given by $1/P_{22}$, so that for a single-zone reactor with reversible adsorption

$$\mathcal{L}\{J_{outlet}\} = \frac{1}{\cosh\left(L\sqrt{\dfrac{\varepsilon_b s + sk_a/(s+k_d)}{D_{eff}}}\right)} \tag{5.81}$$

where $M_0 = 1$ (all gas molecules eventually escape from a reactor in which a reversible reaction takes place) and the average residence time is

$$\frac{M_1}{M_0} = \frac{1}{2}\frac{L^2}{D_{eff}}\left(\varepsilon_b + \frac{k_a}{k_d}\right) \tag{5.82}$$

The average residence time is the sum of the diffusional value and a reversible reaction time component of retardation. The results in the time domain are no longer obtainable from a simple shift in s; the roots of the denominator follow from a quadratic equation in s (see Gleaves et al., 1988 for details). The Fast Fourier technique remains though.

5.5.5 Three-Zone Reactor

Multiplying the three zone transfer matrices for P_{22} yields an expression that includes the sum of four products of hyperbolic functions, which is a little long but is easily obtained exactly using computer algebra. In a three-zone reactor, the middle zone typically is the only reactive zone. In the reversible case, $M_0 = 1$ as always and

$$\frac{M_1}{M_0} = \frac{1}{2}\frac{\varepsilon_{b,1}L_1^2}{D_{eff,1}}\left(1 + \overbrace{2\frac{L_2 D_{eff,1}}{L_1 D_{eff,2}} + 2\frac{L_3 D_{eff,1}}{L_1 D_{eff,3}}}^{\text{backmixing}}\right)$$

$$+ \frac{1}{2}\left(\varepsilon_{b,2} + \frac{k_a}{k_d}\right)\frac{L_2^2}{D_{eff,2}}\left(1 + \overbrace{2\frac{L_3 D_{eff,2}}{L_2 D_{eff,3}}}^{\text{backmixing}}\right) + \frac{1}{2}\frac{\varepsilon_{b,3}L_3^2}{D_{eff,3}}$$

(5.83)

in which the time contributions of the successive zones are split up, showing the presence of backmixing terms in the first and second zone components.

In the irreversible case ($k_d = 0$), the zeroth moment is given by

$$M_0 = \frac{1}{\cosh\varphi + \underbrace{\frac{L_3 D_{eff,2}}{L_2 D_{eff,3}}\varphi\sinh\varphi}_{\text{backmixing}}}$$

(5.84)

which is always less than the one-zone result due to backmixing from the third zone to the second.

5.5.6 Reactive Thin Zones

The previous results illustrate how irreversible adsorption leads to the survival of hyperbolic functions in the expressions obtained. This can be avoided by making the reactive zone very thin. If we let $L_2 \to 0$ in the three-zone reactor results, the contribution of the second zone will of course disappear entirely; but if we simultaneously increase k_a so that the product $k_a L_2$ remains fixed and finite, an interesting phenomenon arises: the zone transfer matrix tends to the nontrivial result

$$\mathbf{M} = \begin{bmatrix} 1 & 0 \\ k_a L_2 & 1 \end{bmatrix} \tag{5.85}$$

In the time domain this means that

$$c_{\text{left}} = c_{\text{right}} = c; \quad J_{\text{left}} = \underbrace{k_a L_2 c}_{\text{reaction rate}} + J_{\text{right}} \tag{5.86}$$

The zeroth moment no longer contains hyperbolic functions:

$$M_0 = \frac{1}{1 + \dfrac{(k_a L_2) L_3}{D_{\text{eff},3}}} \tag{5.87}$$

In the reversible case, the matrix tends to

$$\mathbf{M} = \begin{bmatrix} 1 & 0 \\ \dfrac{s(k_a L_2)}{s + k_d} & 1 \end{bmatrix} \tag{5.88}$$

since the reverse reaction is not affected by the limit in the zone length L_2.

5.5.7 Other Thin Zones

5.5.7.1 Capacity and resistor zones

The reactive thin zone is the one most frequently encountered in the literature, but several other ones are mathematically possible. The first is obtained by letting k_d increase in the thin reactive zone but in such a way that $k_a L_2/k_d$ remains finite. Then the matrix tends to

$$\mathbf{M} = \begin{bmatrix} 1 & 0 \\ s\left(\dfrac{k_a}{k_d} L_2\right) & 1 \end{bmatrix} \tag{5.89}$$

which, in analogy with electrical networks, we dub a *capacity zone*. In the time domain, this means that

$$c_{\text{left}}(t) = c_{\text{right}}(t) = \left(\frac{k_d}{k_a L_2}\right) \underbrace{\int_0^t \left(J_{\text{left}} - J_{\text{right}}\right) d\tau}_{\text{uptake}} \tag{5.90}$$

A different limit does not require any reaction ($k_a = k_d = 0$) but lets L_2 approach zero while L_2/D_{eff} remains fixed and finite. In this case the matrix tends to

$$\mathbf{M} = \begin{bmatrix} 1 & \dfrac{L_2}{D_{\text{eff}}} \\ 0 & 1 \end{bmatrix} \tag{5.91}$$

and in the time domain

$$J_{\text{left}} = J_{\text{right}}; \quad c_{\text{left}} = \frac{L_2}{D_{\text{eff}}} J_{\substack{\text{left} \\ \text{right}}} + c_{\text{right}} \tag{5.92}$$

which, in analogy with electrical networks, we dub a *thin resistor zone*. It is a "concentrated diffusive element."

5.5.7.2 Generalized boundary conditions

The standard boundary condition at the outlet is $c_{\text{outlet}} = 0$. In fact, this is an ideal limit of a prescribed velocity v at the outlet: $J_{\text{outlet}} = v c_{\text{outlet}}$; as $v \to +\infty$, this tends to $c_{\text{outlet}} = 0$, since J_{outlet} must remain finite. A remarkable fact is that this outlet condition can be reduced to the standard one by adding a thin resistor zone: if $c_{\text{extra}} = 0, J_{\text{outlet}} = J_{\text{extra}}$ and

$$c_{\text{outlet}} = \frac{1}{v} J_{\substack{\text{outlet} \\ \text{extra}}} + \overset{=0}{\overbrace{c_{\text{extra}}}} \text{ we indeed obtain } J_{\text{outlet}} = v c_{\text{outlet}}, \text{ the generalized outlet condition, by}$$

multiplying the global transfer matrix to the right by $\begin{bmatrix} 1 & \frac{1}{v} \\ 0 & 1 \end{bmatrix}$, so that its (2,2) element is

$\frac{1}{v} P_{21} + P_{22}$ and

$$\mathcal{L}\{J_{\text{outlet}}\} = \frac{1}{\frac{1}{v} P_{21} + P_{22}} \tag{5.93}$$

Similarly, different inlet conditions have been proposed, where instead of $J_{\text{inlet}} = \delta(t - 0^+)$ a premixing zone of length δ is considered, in which the concentration is determined by the injection flux, amounting to

$$\delta \frac{dc_{\text{pre-mix}}}{dt} = J_{\text{inlet}} - J_{\text{pre-mix}} \tag{5.94}$$

so that in the Laplace domain

$$s\delta \mathcal{L}\{c_{\text{pre-mix}}\} = J_{\text{inlet}} - \mathcal{L}\{J_{\text{pre-mix}}\} \tag{5.95}$$

which represents a thin capacity zone of matrix $\begin{bmatrix} 1 & 0 \\ s\delta & 1 \end{bmatrix}$. Accordingly, to take into account this model of a premixing zone, it suffices to multiply the global transfer matrix to the left with this matrix; the result's (2,2) element then reads $s\delta P_{12} + P_{22}$ and the outlet flux is corrected to

$$\mathcal{L}\{J_{\text{outlet}}\} = \frac{1}{s\delta P_{21} + P_{22}} \tag{5.96}$$

If both corrections are applied simultaneously, the outlet flux becomes

$$\mathcal{L}\{J_{\text{outlet}}\} = \frac{1}{\frac{s\delta}{v}P_{11} + s\delta P_{12} + \frac{1}{v}P_{21} + P_{22}} \tag{5.97}$$

5.6 Multiresponse TAP Theory

Up to now we have always considered a single gas of concentration c, possibly in combination with a single fractional coverage θ. The experimental capability of a TAP machine does, however, allow for the measurement of different gases with concentrations c_i, some of which might not even have been injected at all but are produced in the TAP reactor. The model's partial and ordinary differential equations must be generalized accordingly. To do this, we will switch to a vector $\mathbf{c}$ of gas concentrations and to a vector $\mathbf{\theta}$ of fractional coverages of surface species (not necessarily of the same dimension). The diffusion-reaction equation now becomes

$$\varepsilon_b \frac{\partial \mathbf{c}}{\partial t} = \mathbf{D}_{\text{eff}} \frac{\partial^2 \mathbf{c}}{\partial x^2} - \mathbf{K}_{\mathbf{GG}}\mathbf{c} + \mathbf{c}_{\text{cat}}\mathbf{K}_{\mathbf{GS}}\mathbf{\theta} \tag{5.98}$$

and

$$\frac{\partial \mathbf{\theta}}{\partial t} = \frac{\mathbf{K}_{\mathbf{SG}}}{\mathbf{c}_{\text{cat}}}\mathbf{c} - \mathbf{K}_{\mathbf{SS}}\mathbf{\theta} \tag{5.99}$$

where ε_b and $\mathbf{D}_{\text{eff}}$ are diagonal matrices of the zone characteristics, $\mathbf{K}_{\mathbf{GG}}$ is a matrix of gas reaction impact on gases (typically diagonal when only adsorption occurs), $\mathbf{K}_{\mathbf{SS}}$ is a matrix of the impact of surface species reactions on surface species, and $\mathbf{K}_{\mathbf{GS}}$ and $\mathbf{K}_{\mathbf{SG}}$ represent kinetic links between gases and surface species and vice versa. For the simple example of a reversible reaction in a one-gas context, $\mathbf{K}_{\mathbf{GG}} = [k_a]$, $\mathbf{K}_{\mathbf{GS}} = [k_d]$, $\mathbf{K}_{\mathbf{SG}} = [k_a]$ and $\mathbf{K}_{\mathbf{SS}} = [k_d]$. In a more complex example, the overall reaction $A \rightleftarrows B$, proceeding by elementary steps

$A + Z \underset{k_{d,1}}{\overset{k_{a,1}}{\rightleftarrows}} AZ \underset{k^-}{\overset{k^+}{\rightleftarrows}} BZ \underset{k_{a,2}}{\overset{k_{d,2}}{\rightleftarrows}} B + Z$, these matrices are given by

$$\mathbf{K}_{\mathbf{GG}} = \begin{bmatrix} k_{a,1} & 0 \\ 0 & k_{a,2} \end{bmatrix}, \quad \mathbf{K}_{\mathbf{GS}} = \begin{bmatrix} k_{d,1} & 0 \\ 0 & k_{d,2} \end{bmatrix}, \quad \mathbf{K}_{\mathbf{SG}} = \begin{bmatrix} k_{a,1} & 0 \\ 0 & k_{a,2} \end{bmatrix} \text{ and } \mathbf{K}_{\mathbf{SS}} = \begin{bmatrix} k_{d,1} & -k^- \\ -k^+ & k_{d,2} \end{bmatrix}.$$

Applying the Laplace transform first to the equations for the fractional surface coverage assuming zero initial coverage yields

$$s(\mathcal{L}\{\mathbf{\theta}\} - 0) = \frac{\mathbf{K}_{\mathbf{SG}}}{\mathbf{c}_{\text{cat}}}\mathcal{L}\{\mathbf{c}\} - \mathbf{K}_{\mathbf{SS}}\mathcal{L}\{\mathbf{\theta}\} \tag{5.100}$$

so that

$$(s + \mathbf{K}_{\mathbf{SS}})\mathcal{L}\{\mathbf{\theta}\} = \frac{\mathbf{K}_{\mathbf{SG}}}{\mathbf{c}_{\text{cat}}}\mathcal{L}\{\mathbf{c}\} \tag{5.101}$$

or

$$\mathcal{L}\{\mathbf{\theta}\} = (s + \mathbf{K}_{SS})^{-1} \frac{\mathbf{K}_{SG}}{\mathbf{c}_{cat}} \mathcal{L}\{\mathbf{c}\} \tag{5.102}$$

Substituting Eq. (5.102) into the PDE, Eq. (5.98) after Laplace transform leads to

$$\varepsilon_b s \mathcal{L}\{\mathbf{c}\} = \mathbf{D}_{eff} \frac{\partial^2 \mathcal{L}\{\mathbf{c}\}}{\partial x^2} - \mathbf{K}_{GG} \mathcal{L}\{\mathbf{c}\} + \mathbf{K}_{GS}(s + \mathbf{K}_{SS})^{-1} \mathbf{K}_{SG} \mathcal{L}\{\mathbf{c}\} \tag{5.103}$$

so that

$$\mathbf{D}_{eff} \frac{\partial^2 \mathcal{L}\{\mathbf{c}\}}{\partial x^2} = \left(\varepsilon_b s + \mathbf{K}_{GG} - \mathbf{K}_{GS}(s + \mathbf{K}_{SS})^{-1} \mathbf{K}_{SG} \right) \mathcal{L}\{\mathbf{c}\} = \mathbf{K}_S(s) \mathcal{L}\{\mathbf{c}\} \tag{5.104}$$

Now we can write this equation as a system in $\mathcal{L}\{\mathbf{c}\}$ and $\mathcal{L}\{\mathbf{J}\}$:

$$\frac{\partial}{\partial x} \begin{bmatrix} \mathcal{L}\{\mathbf{c}\} \\ \mathcal{L}\{\mathbf{J}\} \end{bmatrix} = \begin{bmatrix} 0 & -\mathbf{D}_{eff}^{-1} \\ -\mathbf{K}_S(s) & 0 \end{bmatrix} \begin{bmatrix} \mathcal{L}\{\mathbf{c}\} \\ \mathcal{L}\{\mathbf{J}\} \end{bmatrix} \tag{5.105}$$

so that, integrating in x from 0 to L,

$$\begin{bmatrix} \mathcal{L}\{\mathbf{c}\} \\ \mathcal{L}\{\mathbf{J}\} \end{bmatrix}_{left} = \exp \left(\begin{bmatrix} 0 & \mathbf{D}_{eff}^{-1} \\ \mathbf{K}_S(s) & 0 \end{bmatrix} L \right) \begin{bmatrix} \mathcal{L}\{\mathbf{c}\} \\ \mathcal{L}\{\mathbf{J}\} \end{bmatrix}_{right} \tag{5.106}$$

The calculation of the exponential will in practice require the determination of its eigenvalues. Since

$$\begin{bmatrix} 0 & \mathbf{D}_{eff}^{-1} L \\ \mathbf{K}_S(s) L & 0 \end{bmatrix}^2 = \begin{bmatrix} \mathbf{D}_{eff}^{-1} \mathbf{K}_S(s) L^2 & 0 \\ 0 & \mathbf{K}_S(s) \mathbf{D}_{eff}^{-1} L^2 \end{bmatrix} \tag{5.107}$$

and

$$\exp(x) = \frac{\sinh \sqrt{x^2}}{\sqrt{x^2}} x + \cosh \sqrt{x^2} \tag{5.108}$$

it follows that

$$\exp \begin{bmatrix} 0 & \mathbf{D}_{eff}^{-1} \\ \mathbf{K}_S(s) & 0 \end{bmatrix} = \begin{bmatrix} \cosh \sqrt{\mathbf{D}_{eff}^{-1} \mathbf{K}_S(s)} L & 0 \\ 0 & \cosh \sqrt{\mathbf{K}_S(s) \mathbf{D}_{eff}^{-1}} L \end{bmatrix}$$

$$+ \begin{bmatrix} 0 & \mathbf{D}_{eff}^{-1} \\ \mathbf{K}_S(s) & 0 \end{bmatrix} \begin{bmatrix} \dfrac{\sinh \sqrt{\mathbf{D}_{eff}^{-1} \mathbf{K}_S(s)} L}{\sqrt{\mathbf{D}_{eff}^{-1} \mathbf{K}_S(s)}} & 0 \\ 0 & \dfrac{\sinh \sqrt{\mathbf{K}_S(s) \mathbf{D}_{eff}^{-1}} L}{\sqrt{\mathbf{K}_S(s) \mathbf{D}_{eff}^{-1}}} \end{bmatrix} \tag{5.109}$$

where all results on $\mathbf{K_S}(s)\mathbf{D}_{\text{eff}}^{-1}$ can be readily obtained from the flux on $\mathbf{D}_{\text{eff}}^{-1}\mathbf{K_S}$ by the congruence relationship

$$f\left(\mathbf{K_S}(s)\mathbf{D}_{\text{eff}}^{-1}\right) = \mathbf{D}_{\text{eff}}f\left(\mathbf{D}_{\text{eff}}^{-1}\mathbf{K_S}(s)\right)\mathbf{D}_{\text{eff}}^{-1} \qquad (5.110)$$

This halves the computational work required.

Of course, if each gas lives separately in the reactor and all matrices, especially $\mathbf{K_S}(s)$, are diagonal, the comprehensive matrix with all gases will not offer new information. The nontrivial cases occur when adsorbed gases transform from one species to another, so that $\mathbf{K_{SS}}$ holds the kinetic information of the surface reaction system and $\mathbf{K_S}(s)$ is no longer diagonal.

For the inlet concentration, $\delta(t-0^+)$ is no longer sufficiently general because different gases might be injected in different quantities and even at different times. Therefore, we define a vector of inlet fluxes $\mathbf{J}(t)$ such that

$$\left(\mathbf{K_S}(s)\mathbf{D}_{\text{eff}}^{-1}\right) = -\mathbf{D}_{\text{eff}}\frac{\partial \mathbf{c}}{\partial x}\bigg|_{x=\text{inlet}} = \mathbf{J}(t) \qquad (5.111)$$

The zone transfer matrices will include all gases and their fluxes, so will be of dimension $(2N_g) \times (2N_g)$, and the global transfer matrix obtained as their product will have the same size, to be split up as $\begin{bmatrix} P_{11} & P_{12} \\ P_{21} & P_{22} \end{bmatrix}$ as before, with each P_{ij} an $N_g \times N_g$ matrix. The outlet condition can remain $\mathbf{c} = 0$.

From the lower block of equations, we then obtain

$$\mathcal{L}\{\mathbf{J}\} = P_{21} \cdot 0 + P_{22}\mathcal{L}\{\mathbf{J}_{\text{outlet}}\} \qquad (5.112)$$

and thus

$$\mathcal{L}\{\mathbf{J}_{\text{outlet}}\} = P_{22}^{-1}\mathcal{L}\{\mathbf{J}\} \qquad (5.113)$$

which is a clear generalization of the one-gas result.

Again, thin zones can be introduced. The resistor zones are of the form $\begin{bmatrix} 1 & \mathbf{v}^{-1} \\ 0 & 1 \end{bmatrix}$ where $\mathbf{v}$ is a diagonal matrix of formal velocities, so that

$$\mathbf{c}_{\text{left}} = \mathbf{c}_{\text{right}} + \mathbf{v}^{-1}\mathbf{J}_{\substack{\text{left} \\ \text{right}}} \qquad (5.114)$$

The active zones in their simplest form can be represented as $\begin{bmatrix} 1 & 0 \\ \mathbf{K_a}L & 1 \end{bmatrix}$ where $L \to 0$ but so that all elements of $\mathbf{K_a}L$ remain fixed and finite, and $\mathbf{K_a}$ is a diagonal matrix of reversible adsorption coefficients. However, the active zone can also exhibit much more complex surface reaction schemes. Generalizing the $\dfrac{sk_a}{s+k_d}$ result of the one-gas reversible adsorption to the case where

$\mathbf{K}(s)L$ remains finite as $L \to 0$, so that the zone transfer matrix tends to $\begin{bmatrix} 1 & 0 \\ \mathbf{K}(s)L & 1 \end{bmatrix}$.

As a special case, taking a further limit, $k_d \to +\infty$, leads to capacity zones of the form

$\begin{bmatrix} 1 & 0 \\ \left(\dfrac{k_aL}{k_d}\right)s & 1 \end{bmatrix}$. The treatment of alternative boundary conditions (inlet premixing zone,

prescribed outlet velocity) is then readily generalized.

As an example, for the reaction sequence $A+Z \underset{k_{d,1}}{\overset{k_{a,1}}{\rightleftharpoons}} AZ \underset{k^-}{\overset{k^+}{\rightleftharpoons}} BZ \underset{k_{a,2}}{\overset{k_{d,2}}{\rightleftharpoons}} B+Z$ the key result is

$$\frac{1}{\mathbf{M_{BA}}} = 1 + \frac{D_{\mathrm{eff,A3}}}{D_{\mathrm{eff,B3}}} \frac{k^- k_{d,1} k_{a,2}}{k^+ k_{a,1} k_{d,2}} + \frac{D_{\mathrm{eff,A3}}}{L_2 L_3} \frac{k^- k_{d,1} + k_{d,1} k_{d,2} + k_{d,2} k^+}{k_{d,2} k^+ k_{a,1}} \tag{5.115}$$

The result for $\dfrac{1}{\mathbf{M_{AB}}}$ is similar but with the roles of A and B switched. Furthermore,

$$\mathbf{M_{AA}} = 1 - \mathbf{M_{BA}}; \quad \mathbf{M_{BB}} = 1 - \mathbf{M_{BA}} \tag{5.116}$$

As always in such model problems, it is possible to interpret the separate contributions to these expressions in a more physical manner in terms of equilibrium coefficients and characteristic times of diffusion.

5.7 The Y Procedure: Inverse Problem Solving in TAP Reactors

The TAP theory developed up to now is linear and explains the many advantages of the Laplace-domain technique where obtaining exact and interpretable expressions is concerned. In the special case of TZTRs, it is possible to drop the requirement of linearity of the reaction and to reconstruct, purely from observed outlet fluxes, the temporal evolution of the concentrations and reaction rate in the thin reaction zones, without any a priori assumption of its linearity. The method used is called the Y procedure, in reference to the Cyrillic letter Я, an inverted R, for rate, which by a happy coincidence is also the first letter of the Roman surname of its inventor, G.S. Yablonsky. We consider the three-zone, TZTR (see Fig. 5.6), and write out the equations obtained from the zone transfer matrices as follows.

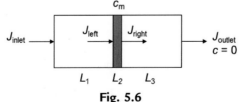

Fig. 5.6
Schematic of three-zone, thin-zone TAP reactor.

In the left zone, the inlet flux satisfies

$$\mathcal{L}\{J_{\text{inlet}}\} = v_1 s_1 \mathcal{L}\{c_m\} + c_1 \mathcal{L}\{J_{\text{left}}\} \tag{5.117}$$

so that

$$\mathcal{L}\{J_{\text{left}}\} = \frac{\mathcal{L}\{J_{\text{inlet}}\} - v_1 s_1 \mathcal{L}\{c_m\}}{c_1} \tag{5.118}$$

In the thin reaction zone, the rate of consumption R satisfies

$$L_2 \mathcal{L}\{R\} = \mathcal{L}\{J_{\text{left}}\} - \mathcal{L}\{J_{\text{right}}\} \tag{5.119}$$

by a balance argument.

In the right zone,

$$c_m = c_3 \cdot 0 + \frac{s_3}{v_3} \mathcal{L}\{J_{\text{outlet}}\} \tag{5.120}$$

and

$$J_{\text{right}} = v_3 s_3 \cdot 0 + c_3 \mathcal{L}\{J_{\text{outlet}}\} \tag{5.121}$$

Substituting Eqs. (5.118), (5.120), and (5.121) in Eq. (5.119) yields

$$
\begin{aligned}
c_1 L_2 \mathcal{L}\{R\} &= \mathcal{L}\{J_{\text{inlet}}\} - \left(c_1 c_3 + \frac{v_1}{v_3} s_1 s_3 \right) \mathcal{L}\{J_{\text{outlet}}\} \\
&= \left(c_1 c_3 + \frac{v_1}{v_3} s_1 s_3 \right) \left(\mathcal{L}\{J_{\text{outlet, diff}}\} - \mathcal{L}\{J_{\text{outlet}}\} \right)
\end{aligned}
\tag{5.122}
$$

where $J_{\text{outlet,diff}}$ denotes the outlet flux that would be produced in the absence of reaction, by a diffusion system only. This equation shows how, in principle and for exactly known data, R can be recovered. Note that on the left-hand side of the equation, R is multiplied with L_2, so that the effects will diminish as L_2 becomes too small, unless a similar limiting sequence is used as in the modeling of the linear thin zone, where although $L_2 \rightarrow 0$, $\mathbf{K_a}L$ remains fixed and finite. Furthermore, $c_1 L_2 R$ is obtained from a difference between two fluxes, where the relative error can be amplified due to near cancellation of terms if the effect of the thin zone is not sufficiently marked. Finally, the factors that occur have a net amplifying influence on higher frequencies, as can be seen when analyzing the case $s = i\omega$ for the Fourier transform analogue of the Laplace-domain equation. In practice, a damping factor must be inserted to prevent the experimental magnification (asymptotically of type $\exp(\lambda\sqrt{\omega})$) of high-frequency error components. Nevertheless, state-of-the-art TAP machines can perform the Y procedure reliably, with the most limiting difficulty being the interference of the electrical mains with the measurements, at their characteristic frequency of 50 or 60 Hz.

5.8 Two- and Three-Dimensional Modeling

Up to now we have considered the diffusion-reaction equation only in a one-dimensional spatial setting. If a correct formulation is to be given in space, the second derivative term $\partial^2 c/\partial x^2$ must be replaced by the Laplacian $\partial^2 c/\partial x^2 + \partial^2 c/\partial y^2 + \partial^2 c/\partial z^2 = \Delta c$, taking into account the diffusion along all three coordinate axes, and Eq. (5.27) generalizes to

$$\varepsilon_b \frac{\partial c}{\partial t} = D_{\text{eff}} \Delta c - k_a c \tag{5.123}$$

Clearly, this equation needs to be completed with boundary conditions corresponding to the cylindrical shape of the TAP reactor. This requires the use of cylindrical coordinates (r, θ, z) instead of the Cartesian coordinates (x, y, z). Note that the cylindrical z coordinate is longitudinal and corresponds to the Cartesian x coordinate used so far. Substituting the coordinate transformation equations and applying the chain rule we obtain the following expression for the Laplacian in cylindrical coordinates:

$$\Delta c = \frac{1}{r} \frac{\partial}{\partial r}\left(r \frac{\partial c}{\partial r} \right) + \frac{1}{r^2} \frac{\partial^2 c}{\partial \theta^2} + \frac{\partial^2 c}{\partial z^2} \tag{5.124}$$

In what follows, it is assumed that the TAP reactor cylinder has a radius $R > 0$. The boundary condition at $r = R$ (ie, the cylinder's mantle) is that no flux can escape it; since the diffusive flow is given by the gradient of c this boundary condition can be written as

$$\left. \frac{\partial c}{\partial r} \right|_{r=R} = 0 \tag{5.125}$$

The other boundary conditions (at the inlet, at the outlet, and at the interfaces between successive zones) remain the same as for the one-dimensional model. Solving Eq. (5.123) may seem exhausting, but the application of a standard mathematical technique, namely *separation of variables*, is very helpful. To understand its working, first consider that a function of several variables $f(x, y, z)$ can always be expanded in a local Taylor series, for example, near $(0, 0, 0)$ as

$$\begin{aligned} f(x, y, z) = a_{000} &+ a_{100}x + a_{010}y + a_{001}z + a_{200}x + a_{110}xy \\ &+ a_{020}y^2 + a_{011}yz + a_{101}xz + a_{002}z^2 + \cdots \end{aligned} \tag{5.126}$$

under suitable smoothness assumptions. This is a kind of "infinite" degree polynomial with building blocks that are expressions of the form $x^m y^n z^p$ with $m, n, p \in \mathbb{N}$. Now we could try to substitute an expression (5.126) for the unknown function c in Eq. (5.123) and group like terms together, but in practice this would lead to an infinitely intricate set of equations in the unknown coefficients $a_{mnp}(t)$ and their derivatives. Since we are aiming at true, mathematical solutions (as opposed to more numerical approximations), this is not acceptable. The road to success for the present problem lies in a suitable generalization of Eq. (5.126), while retaining one crucial property: we want to find special solutions of

Eq. (5.123) for which the spatial dependency splits into three independent factors, each handling one of the three spatial coordinates r, θ, z. Hence, we wish to seek solutions of Eq. (5.123) that can be represented as (typically infinite) sums of products $f_1(z)f_2(r)f_3(\theta)$ multiplied by a coefficient that may depend on time. We will choose $f_1, f_2,$ and f_3 in such a way that the representation makes geometric sense and that the PDE becomes almost trivial to solve.

5.8.1 The Two-Dimensional Case

For clarity of exposition, we will first assume that c is not dependent on the angle θ, but only on the longitudinal coordinate z and the transversal distance to the cylinder's axis, r. The PDE, Eq. (5.123), then reads

$$\varepsilon_b \frac{\partial c}{\partial t} = D_{eff}\left(\frac{1}{r}\frac{\partial}{\partial r}\left(r\frac{\partial c}{\partial r}\right) + \frac{\partial^2 c}{\partial z^2}\right) - k_a c \tag{5.127}$$

and we wish to find solutions of the form $c = a(t)f_1(z)f_2(r)$. Only the term $\dfrac{1}{r}\dfrac{\partial}{\partial r}\left(r\dfrac{\partial c}{\partial r}\right) + \dfrac{\partial^2 c}{\partial z^2}$ shows a dependence on r, so if we can coax c so that it is merely a multiple of c itself, the dependence on r will be removed entirely. This requires that

$$\frac{1}{r}\frac{\partial}{\partial r}\left(r\frac{\partial c}{\partial r}\right) = \mu c \tag{5.128}$$

for some constant μ; substituting $a(t)f_1(z)f_2(r)$ in this equation and simplifying yields

$$\frac{1}{r}\frac{d}{dr}(rf_2(r)) = \mu f_2(r) \tag{5.129}$$

since $a(t)$ and $f_1(z)$, which can be assumed not to vanish in an identical manner, cancel out. The resulting ODE for $f_2(r)$ has the general solution

$$f_2(r) = C_1 J_0\left(\sqrt{-\mu}r\right) + C_2 Y_0\left(\sqrt{-\mu}r\right) \tag{5.130}$$

where J_0 and Y_0 are so-called Bessel functions, special functions from mathematical analysis that cannot be expressed finitely in terms of functions such as logarithms or trigonometric functions. Inspecting the graphs of these functions (they are readily available in numerical and computer algebra packages), it can be seen that Y_0 is singular (tends to infinity) as $r \to 0$, whereas J_0 remains finite. Since our candidate functions serve to represent physical concentrations, it is clear that we will have to set $c_2 = 0$ in this and work only with J_0. It is also convenient here to write λ for $\sqrt{-\mu}$ so that $f_2(r) = J_0(\lambda r)$ will lead to

$$\varepsilon_b \frac{\partial c}{\partial t} = D_{eff}\frac{\partial^2 c}{\partial z^2} - \left(k_a + D_{eff}\lambda^2\right)c \tag{5.131}$$

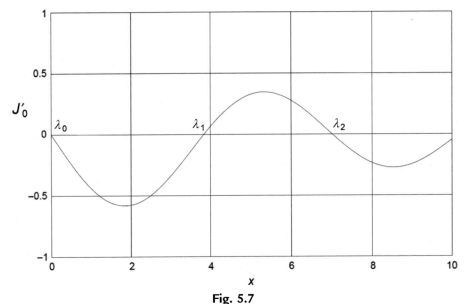

Fig. 5.7

Graph of Bessel function J_0' versus x with roots λ_k.

where $c = a(t)f_1(z)f_2(r)$ as before. To determine λ, the new boundary condition, Eq. (5.125), comes in most usefully:

$$\frac{\partial}{\partial r}(a(t)f_1(z)f_2(r)) = a(t)f_1(z)f_2'(r) \tag{5.132}$$

hence the requirement is that $J_0'(\lambda R) = 0$. Fig. 5.7 shows a plot of $J_0'(x)$ as a function of x. Roots occur at values $\lambda_0 = 0$, $\lambda_1 = 3.831705970$, $\lambda_2 = 7.015586670$, and so on for an infinity of possible values. The corresponding values of λ are then $\dfrac{\lambda_0}{R}, \dfrac{\lambda_1}{R}, \dfrac{\lambda_2}{R}, \cdots$ and the PDE corresponding to λ_k ($k = 0, 1, 2, \ldots$) is

$$\varepsilon_b \frac{\partial c}{\partial t} = D_{\text{eff}}\frac{\partial^2 c}{\partial z^2} - \left(k_a + \frac{D_{\text{eff}}}{R^2}\lambda_k^2\right)c \tag{5.133}$$

The remarkable fact about this equation is that it is identical to the one-dimensional PDE, Eq. (5.27), except that k_a, the adsorption rate coefficient, is increased with a term equal to $\dfrac{D_{\text{eff}}\lambda_k^2}{R^2}$.

The interpretation of this fact is as follows: when considering the 2D-solution to Eq. (5.123) for which

$$f_2(r) = J_0\left(\frac{\lambda_k r}{R}\right) \tag{5.134}$$

the rate coefficient of the forward reaction increases by a "pseudoreaction" term $\frac{\lambda_\mu^2}{\tau_r}$, where

$\tau_r = \frac{R^2}{D_{eff}}$ is a transversal diffusion characteristic term, and λ_k is the corresponding root of $J_0'(x)$, the kth one when counting 0 as λ_0.

Physically, the origin of this effect is that the diffusion reduces the solution with $k = 1, 2, \ldots$ (the ones that are not identically constant) ever more rapidly as k increases. This is no coincidence as our simplifying requirement that $\frac{1}{r}\frac{\partial}{\partial r}\left(r\frac{\partial c}{\partial r}\right) = \mu c$ already expresses that the left-hand side, representing transversal diffusion (up to a constant factor D_{eff}), reduces to the right-hand side, a linear term typical of a first-order reaction. For each k separately, the PDE (Eq. 5.133) can be treated using zone transfer matrices just as before, except that the pseudoreaction term must be added to the adsorption rate coefficient k_a. When inert zones are to be modeled, the same coefficient must be added to their value of $k_a = 0$, so that in the model an inert zone formally becomes a zone of irreversible adsorption in the $k = 1, 2, \ldots$ components of the two-dimensional model. The results of the analysis in the Laplace domain then carry through, including the closed-form expression in the time domain that can be found in some cases.

5.8.2 The Three-Dimensional Case

Reuse of the previous results is a recurring theme in mathematical modeling, and one that will again apply to the three-dimensional versus the two-dimensional case much in the way that the two-dimensional case has been reduced to the one-dimensional case in the previous subsection.

In the three-dimensional case, we must allow for a dependence factor $f_3(\theta)$ in the product $c = a(t)f_1(z)f_2(r)f_3(\theta)$, where θ is the transversal polar coordinate. Since θ is an angle, there are no "boundary" conditions as such, but we need periodicity, viz. $f_3(\theta + 2\pi) = f_3(\theta)$ (in radians). The diffusional term $\frac{1}{r^2}\frac{\partial^2 c}{\partial \theta^2}$ now has to be reduced as much as possible: we cannot make r disappear from this term, but we want to reduce the second derivate $\frac{\partial^2 c}{\partial \theta^2}$ to a multiple of c, that is, we wish that $f_3''(\theta) = \nu f_3(\theta)$ for some constant ν. The general solution of this ODE is

$$f_3(\theta) = C_1 \cos\left(\sqrt{-\nu}\theta\right) + C_2 \sin\left(\sqrt{-\nu}\theta\right) \tag{5.135}$$

and it is clear that periodicity can only be obtained if $\sqrt{-\nu}$ is some integer n, so $\nu = -n^2$ $(n = 0, 1, 2, \ldots)$. The case $n = 0$ is the one where there is no dependence on θ, which is the one that has been treated in the previous subsection. We now assume that $n = 1, 2, \ldots$ to include a dependence on θ. If $c = a(t)f_1(z)f_2(r)f_3(\theta)$ as before, with such an f_3, after expansion of $\partial/\partial r$ Eq. (5.123) reads

$$\varepsilon_b \frac{\partial c}{\partial t} = D_{eff} \underbrace{\left(\frac{\partial^2 c}{\partial r^2} + \frac{1}{r}\frac{\partial c}{\partial r} - \frac{n^2}{r^2}c}_{r \text{ part}} + \underbrace{\frac{\partial^2 c}{\partial z^2}}_{z \text{ part}} \right) - k_a c \tag{5.136}$$

so that our f_2 will have to be picked correctly to simplify the whole of the r part, that is,

$$f_2'' = \frac{1}{r}f_2' - \frac{n^2}{r^2}f_2 = \mu f_2 \tag{5.137}$$

where compared to the two-dimensional case, the new term is $-\frac{n^2}{r^2}f_2$. This ODE has the general solution given by

$$f_2(r) = C_1 J_n\left(\sqrt{-\mu}r\right) + C_2 Y_n\left(\sqrt{-\mu}r\right) \tag{5.138}$$

where J_n and Y_n are Bessel functions of order n ($n = 1, 2, \ldots$). Again, the mathematical definitions are such that Y_n diverges as its argument tends to 0, but that J_n remains finite. For each n the analysis can be carried out as in the two-dimensional case, with J_n replacing J_0 and the roots $\lambda_{n,k}$ of $J_n'(x) = 0$ replacing the λ_k.

The contribution of pseudoreactions then leads to

$$\varepsilon_b \frac{\partial c}{\partial t} = D_{eff} \frac{\partial^2 c}{\partial z^2} - \left(k_a + \frac{D_{eff}}{R^2}\lambda_{n,k}^2 \right)c \tag{5.139}$$

which again reduces to the one-dimensional PDE.

5.8.3 Summary

The general three-dimensional solution of the TAP PDE, Eq. (5.123), is the superposition of the following components:

- the one-dimensional solution $c_I(z,t)$ of the one-dimensional equation:

$$\varepsilon_b \frac{\partial c_I}{\partial t} = D_{eff} \frac{\partial^2 c_I}{\partial z^2} - k_a c_I \tag{5.140}$$

- the two-dimensional solution $c_{II}(z,t)J_0\left(\lambda_k \frac{r}{R}\right)$ with λ_k the kth root of J_0' ($k = 1, 2, \ldots$) and c_{II} satisfying

$$\varepsilon_b \frac{\partial c_{II}}{\partial t} = D_{eff} \frac{\partial^2 c_{II}}{\partial z^2} - \left(k_a + \frac{D_{eff}}{R^2}\lambda_k^2 \right)c_{II} \tag{5.141}$$

- the three-dimensional solution $c_{III}(z,t)J_n\left(\lambda_{n,k}\frac{r}{R}\right)$ with $\lambda_{n,k}$ the kth root of $J_n'(x)$ ($n = 1, 2, \ldots$, $k = 1, 2, \ldots$) and c_{III} satisfying

$$\varepsilon_b \frac{\partial c_{\mathrm{III}}}{\partial t} = D_{\mathrm{eff}} \frac{\partial^2 c_{\mathrm{III}}}{\partial z^2} - \left(k_a + \frac{D_{\mathrm{eff}}}{R^2} \lambda_{n,k}^2 \right) c_{\mathrm{III}} \tag{5.142}$$

All of the PDEs in z and t obtained this way can be treated separately using the zone transfer matrix symbolism.

In Constales et al. (2001b) these results have been applied to typical temporal experimental data from a TAP procedure to show that for a TAP reactor with an aspect ratio L/R larger than ten, all two- and three-dimensional effects can be neglected, as was to be expected considering the size of the pseudoreaction terms $\frac{D_{\mathrm{eff}}}{R^2} \lambda_{n,k}^2$. This may seem like a negative result, but it is a very important one to fully justify the TAP reactor design, which aims at the closest possible approximation of the one-dimensional linear case; the possibility of using exact results for two- and three-dimensional models is an extra bonus.

More generally, the methodology that was used here in a practical way—separation of variables to reduce transversal diffusion to formal reaction, reduction of two and three dimensions to one dimension—can be generalized to other reactor configurations, such as nonideal PFRs and cascades of CSTRs and also to more complex arrangements.

5.8.4 Radial Diffusion in Two and Three Dimensions

The zone transfer matrix calculated in the Laplace domain for a single zone in the one-dimensional model exactly expresses its behavior as a component of any system built from such a zone, not only as part of a TAP reactor. We now study what happens if we were to consider concentric situations in two and three dimensions instead. Let d denote the dimension ($d = 1$, 2, or 3). Then the spatial PDE, Eq. (5.123), takes the following form for radial situations

$$\varepsilon_b \frac{\partial c}{\partial t} = D_{\mathrm{eff}} \left(\frac{\partial^2 c}{\partial r^2} + \frac{d-1}{r} \frac{\partial c}{\partial r} \right) - k_a c \tag{5.143}$$

This corresponds to the one-dimensional case for $d = 1$. In its zone transfer matrix, the cosh and sinh functions occurred naturally considering the formal boundary conditions $\begin{bmatrix} \mathcal{L}c \\ \mathcal{L}J \end{bmatrix} = \begin{bmatrix} 1 \\ 0 \end{bmatrix}$ and $\begin{bmatrix} 0 \\ 1 \end{bmatrix}$, respectively. The new term $\frac{d-1}{r} \frac{\partial c}{\partial r}$ in Eq. (5.143) affects its solution in the Laplace domain: instead of the cosh and sinh functions, we find

$$C_1 r^{1-\frac{d}{2}} I_{\frac{d}{2}-1} \left(\sqrt{\frac{\varepsilon_b s + k_a}{D_{\mathrm{eff}}}} r \right) + C_2 r^{1-\frac{d}{2}} K_{\frac{d}{2}-1} \left(\sqrt{\frac{\varepsilon_b s + k_a}{D_{\mathrm{eff}}}} r \right) \tag{5.144}$$

where I and K are the so-called modified Bessel functions of order $d/2 - 1$. The case $d = 1$ can be recovered from this using special properties of $I_{-1/2}$ and $K_{-1/2}$, but for other values of d the

Bessel symbols are unavoidable. To make the first formal boundary condition, we need $\mathcal{L}\{c\}$ to be proportional to

$$g = \frac{r^{1-\frac{d}{2}}I_{\frac{d}{2}-1}\left(\sqrt{\frac{\varepsilon_b s + k_a}{D_{\text{eff}}}}r\right)}{\dfrac{d}{dr}\left(r^{1-\frac{d}{2}}I_{\frac{d}{2}-1}\left(\sqrt{\frac{\varepsilon_b s + k_a}{D_{\text{eff}}}}r\right)\right)\bigg|_{r=r_2}} - \frac{r^{1-\frac{d}{2}}K_{\frac{d}{2}-1}\left(\sqrt{\frac{\varepsilon_b s + k_a}{D_{\text{eff}}}}r\right)}{\dfrac{d}{dr}\left(r^{1-\frac{d}{2}}K_{\frac{d}{2}-1}\left(\sqrt{\frac{\varepsilon_b s + k_a}{D_{\text{eff}}}}r\right)\right)\bigg|_{r=r_2}} \tag{5.145}$$

so that $\partial g/\partial r$ at r_2 simplifies to $1-1=0$.

Then $\dfrac{g|_{r=r_1}}{g|_{r=r_2}}$ satisfies the formal boundary conditions and is the (1,1) entry of the zone transfer matrix. Subsequently, the (2,1) entry is found by deriving this with respect to r_1 and multiplying by $-D_{\text{eff}}$ to calculate the diffusive flux.

Similarly,

$$h = \frac{r^{1-\frac{d}{2}}I_{\frac{d}{2}-1}\left(\sqrt{\frac{\varepsilon_b s + k_a}{D_{\text{eff}}}}r\right)}{\left(r^{1-\frac{d}{2}}I_{\frac{d}{2}-1}\left(\sqrt{\frac{\varepsilon_b s + k_a}{D_{\text{eff}}}}r\right)\right)\bigg|_{r=r_2}} - \frac{r^{1-\frac{d}{2}}K_{\frac{d}{2}-1}\left(\sqrt{\frac{\varepsilon_b s + k_a}{D_{\text{eff}}}}r\right)}{\left(r^{1-\frac{d}{2}}K_{\frac{d}{2}-1}\left(\sqrt{\frac{\varepsilon_b s + k_a}{D_{\text{eff}}}}r\right)\right)\bigg|_{r=r_2}} \tag{5.146}$$

will yield $h=0$ at $r=r_2$ so that $\dfrac{h|_{r=r_1}}{-D_{\text{eff}}h'|_{r=r_2}}$ will be the (1,2) entry of the matrix and $\dfrac{h'|_{r=r_1}}{h'|_{r=r_2}}$ its (2,2) entry.

A more compact way of determining the transfer function is based on the general solution, Eq. (5.144), in matrix terms

$$\begin{bmatrix} f \\ -D_{\text{eff}}\dfrac{\partial f}{\partial r} \end{bmatrix} = r^{1-\frac{d}{2}}\mathbf{M}(r)\begin{bmatrix} C_1 \\ C_2 \end{bmatrix} \tag{5.147}$$

with

$$\mathbf{M}(r) = \begin{bmatrix} I_{\frac{d}{2}-1}\left(\sqrt{\frac{\varepsilon_b s + k_a}{D_{\text{eff}}}}r\right) & K_{\frac{d}{2}-1}\left(\sqrt{\frac{\varepsilon_b s + k_a}{D_{\text{eff}}}}r\right) \\ -\sqrt{D_{\text{eff}}(\varepsilon_b s + k_a)}I_{\frac{d}{2}-1}\left(\sqrt{\frac{\varepsilon_b s + k_a}{D_{\text{eff}}}}r\right) & \sqrt{D_{\text{eff}}(\varepsilon_b s + k_a)}K_{\frac{d}{2}-1}\left(\sqrt{\frac{\varepsilon_b s + k_a}{D_{\text{eff}}}}r\right) \end{bmatrix} \tag{5.148}$$

Then,

$$\begin{bmatrix} f \\ -D_{\text{eff}}\dfrac{\partial f}{\partial r} \end{bmatrix}_{r=r_1} = \underbrace{\mathbf{M}(r_1)\mathbf{M}^{-1}(r_2)}_{\text{zone transfer matrix}}\begin{bmatrix} f \\ -D_{\text{eff}}\dfrac{\partial f}{\partial r} \end{bmatrix}_{r=r_2} \tag{5.149}$$

As an example, for $d=3$, working out the (2,2) element of the zone transfer matrix of Eq. (5.149) leads to

$$P_{22} = \frac{r_2}{r_1}\left(\cosh\left(\sqrt{\frac{\varepsilon_b S}{D_{\text{eff}}}}(r_2-r_1)\right)\right) + \frac{\sinh\left(\sqrt{\frac{\varepsilon_b S}{D_{\text{eff}}}}(r_2-r_1)\right)}{r_1\sqrt{\frac{\varepsilon_b S}{D_{\text{eff}}}}} \tag{5.150}$$

so that the distribution of diffusing particles injected at $r=r_1$ that reach $r=r_2$ has this element $1/P_{22}$ as Laplace transform, up to the factor r_2^2/r_1^2 to take into account the relative surface of the spheres at $r=r_1$ and $r=r_2$:

$$\frac{1}{P_{22}}\frac{r_2^2}{r_1^2} = 1 - \underbrace{\frac{1}{6}\frac{\varepsilon_b(r_2-r_1)^2}{D_{\text{eff}}}(2r_1+r_2)\,s}_{\text{residence time}} + \cdots \tag{5.151}$$

In the special case where $r_1 \to 0$, we find

$$\frac{1}{P_{22}}\frac{r_2^2}{r_1^2} \to \frac{\sqrt{\frac{\varepsilon_b S}{D_{\text{eff}}}}\,r_2}{\sinh\left(\sqrt{\frac{\varepsilon_b S}{D_{\text{eff}}}}\,r_2\right)} \tag{5.152}$$

If, for instance, one is simulating a diffusing particle inside an intricate domain and the ball with radius r_2 centered at the particle's position is completely contained in this domain, a random jump to the ball's surface can be taken after a time generated by the random variable whose Laplace transform is given by Eq. (5.152). The cumulative distribution of Eq. (5.152) in the case $\left(\sqrt{\frac{\varepsilon_b}{D_{\text{eff}}}}\,r_2\right) = 1$ can be obtained from the inverse Laplace technique illustrated in the TAP case and equals

$$\frac{2}{\sqrt{\pi t}}\sum_{k=0}^{+\infty}\exp\left(-\frac{\left(k+\frac{1}{2}\right)^2}{t}\right) \tag{5.153}$$

5.9 TAP Variations

5.9.1 Infinite Exit Zone

Now, we further illustrate the flexibility of the transfer matrix method to calculate the effect of replacing the vacuum exit by an infinite zone of fixed characteristics $\varepsilon_b, D_{\text{eff}}$ and k_a. We consider J_{outlet} to be the flux at the exit of the reactor, which coincides with the entry of the infinite zone (see Fig. 5.8).

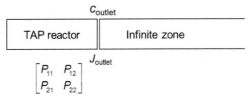

Fig. 5.8
Schematic of TAP reactor with infinite exit zone.

The general solution of the diffusion equation in the Laplace domain,

$$\varepsilon_b(s\mathcal{L}\{c\}) = D_{eff}\frac{\partial^2\mathcal{L}\{c\}}{\partial x^2} - k_a\mathcal{L}\{c\} \tag{5.154}$$

is given by

$$\mathcal{L}\{c\} = C_1\exp\left(-x\sqrt{\frac{\varepsilon_b s + k_a}{D_{eff}}}\right) + C_2\exp\left(x\sqrt{\frac{\varepsilon_b s + k_a}{D_{eff}}}\right) \tag{5.155}$$

When facing an infinite zone that extends into $x \to +\infty$, only the first term in this expansion remains finite. We must therefore force the solution to be expressible in the first term only. For this, we merely have to calculate its diffusive flux,

$$-D_{eff}\frac{\partial}{\partial x}\exp\left(-x\sqrt{\frac{\varepsilon_b s + k_a}{D_{eff}}}\right) = \sqrt{(\varepsilon_b s + k_a)D_{eff}}\left(-x\sqrt{\frac{\varepsilon_b s + k_a}{D_{eff}}}\right) \tag{5.156}$$

so that the requirement simplifies to

$$J_{outlet} = \sqrt{(\varepsilon_b s + k_a)D_{eff}}\,c_{outlet} \tag{5.157}$$

This is similar to the outlet condition $J_{outlet} = vc_{outlet}$ for a prescribed outlet velocity v, except that its role is played by the expression $\sqrt{(\varepsilon_b s + k_a)D_{eff}}$, which is a function of the Laplace variable s. Proceeding as usual to extract J_{outlet} from the global transfer matrix,

$$\begin{bmatrix} * \\ J_{inlet} \end{bmatrix} = \begin{bmatrix} P_{11} & P_{12} \\ P_{21} & P_{22} \end{bmatrix}\begin{bmatrix} \dfrac{J_{outlet}}{\sqrt{(\varepsilon_b s + k_a)D_{eff}}} \\ J_{outlet} \end{bmatrix} \tag{5.158}$$

leads to the closed-form expression

$$J_{outlet} = \frac{1}{P_{22} + \dfrac{P_{21}}{\sqrt{(\varepsilon_b s + k_a)D_{eff}}}} \tag{5.159}$$

where the denominator includes, along with the usual P_{22} term, a term $\dfrac{P_{21}}{\sqrt{(\varepsilon_b s + k_a)D_{eff}}}$, which represents backmixing from the infinite zone into the TAP reactor. As a check, consider the case of a single-zone reactor with the same characteristics as the infinite zone. Then

$$J_{outlet} = \cfrac{1}{\cosh\left(L\sqrt{\cfrac{\varepsilon_b s + k_a}{D_{eff}}}\right) + \cfrac{\sqrt{(\varepsilon_b s + k_a)D_{eff}}\sinh\left(L\sqrt{\cfrac{\varepsilon_b s + k_a}{D_{eff}}}\right)}{\sqrt{(\varepsilon_b s + k_a)D_{eff}}}}$$

$$= \exp\left(-L\sqrt{\frac{\varepsilon_b s + k_a}{D_{eff}}}\right)$$

(5.160)

as could have been expected from the uniformity of the reactor. Now

$$\ln J_{outlet} = -\sqrt{\frac{k_a L^2}{D_{eff}}} - \frac{1}{2}\frac{\varepsilon_b L}{\sqrt{k_a D_{eff}}}s + \cdots = -\sqrt{Da_{II}} - \underbrace{\frac{1}{2}\left(\frac{\varepsilon_b L^2}{D_{eff}}\right)\frac{1}{\sqrt{Da_{II}}}s}_{\text{residence time}} + \cdots$$

(5.161)

where $Da_{II} = \dfrac{k_a L^2}{D_{eff}}$ is the second Damköhler number.

5.9.2 Internal Injection

If a TAP reactor is run with the injection point not at the "left" end, but internal to the cylinder, the formulation of the transfer matrix leads to further elegant equations. We write P_{ij} for the global transfer matrix elements of the reactor to the left of the injection point and Q_{ij} for the elements to the right of this point (see Fig. 5.9).

Expressing all relationships as usual in the Laplace domain gives

$$\begin{bmatrix} \mathcal{L}\{c_1\} \\ 0 \end{bmatrix} = \begin{bmatrix} P_{11} & P_{12} \\ P_{21} & P_{22} \end{bmatrix} \begin{bmatrix} \mathcal{L}\{c_{inj}\} \\ \mathcal{L}\{J_{left}\} \end{bmatrix}$$

(5.162)

$$\begin{bmatrix} \mathcal{L}\{c_{inj}\} \\ \mathcal{L}\{J_{right}\} \end{bmatrix} = \begin{bmatrix} Q_{11} & Q_{12} \\ Q_{21} & Q_{22} \end{bmatrix} \begin{bmatrix} 0 \\ \mathcal{L}\{J_{outlet}\} \end{bmatrix}$$

(5.163)

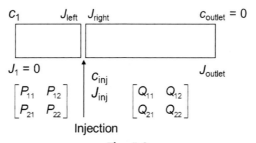

Fig. 5.9
Schematic of TAP reactor with internal injection.

and

$$\mathcal{L}\{J_{\mathrm{inj}}\} = \mathcal{L}\{J_{\mathrm{right}}\} = \mathcal{L}\{J_{\mathrm{left}}\} \tag{5.164}$$

Solving this set of five equalities for the unknowns: $c_1, c_{\mathrm{inj}}, J_{\mathrm{left}}, J_{\mathrm{right}}$, and J_{outlet}, for the outlet flux yields

$$\mathcal{L}\{J_{\mathrm{outlet}}\} = \frac{\mathcal{L}\{J_{\mathrm{inj}}\}}{Q_{22} + \dfrac{P_{21}Q_{12}}{P_{22}}} \tag{5.165}$$

where the denominator again includes a backmixing term.

A variation of this occurs when the "left" side of the reactor is not closed, but open to the vacuum exterior; instead of J_1 being equal to zero and c_1 to be determined, then c_1 equals zero and J_1 is to be determined from

$$\begin{bmatrix} 0 \\ \mathcal{L}\{J_1\} \end{bmatrix} = \begin{bmatrix} P_{11} & P_{12} \\ P_{21} & P_{22} \end{bmatrix} \begin{bmatrix} \mathcal{L}\{c_{\mathrm{inj}}\} \\ \mathcal{L}\{J_{\mathrm{left}}\} \end{bmatrix}, \tag{5.166}$$

The expression for the outlet flux then is given by

$$\mathcal{L}\{J_{\mathrm{outlet}}\} = \frac{\mathcal{L}\{J_{\mathrm{inj}}\}}{Q_{22} + \dfrac{P_{11}Q_{12}}{P_{12}}} \tag{5.167}$$

To get a feel for the different implications of these two results, let us assume that both reactor parts are inert with such characteristics, s, that

$$\begin{bmatrix} P_{11} & P_{12} \\ P_{21} & P_{22} \end{bmatrix} = \begin{bmatrix} Q_{11} & Q_{12} \\ Q_{21} & Q_{22} \end{bmatrix} = \begin{bmatrix} \cosh\sqrt{s} & \dfrac{\sinh\sqrt{s}}{\sqrt{s}} \\ \sqrt{s}\sinh\sqrt{s} & \cosh\sqrt{s} \end{bmatrix} \tag{5.168}$$

Then with the assumption that $J_{\mathrm{inj}} = 1$ the "left-closed" equation, Eq. (5.165), yields

$$\mathcal{L}\{J_{\mathrm{outlet}}\} = \frac{1}{\cosh\sqrt{s} + \dfrac{\sinh^2\sqrt{s}}{\cosh\sqrt{s}}} = \frac{\cosh\sqrt{s}}{\cosh(2\sqrt{s})} = 1 - \frac{3}{2}s + \cdots \tag{5.169}$$

while the "left-open" equation, Eq. (5.167), results in

$$\mathcal{L}\{J_{\mathrm{outlet}}\} = \frac{1}{\cosh\sqrt{s} + \dfrac{\cosh\sqrt{s}\dfrac{\sinh\sqrt{s}}{\sqrt{s}}}{\dfrac{\sinh\sqrt{s}}{\sqrt{s}}}} = \frac{1}{2\cosh(\sqrt{s})} = \frac{1}{2}\left(1 - \frac{1}{4}s + \cdots\right) \tag{5.170}$$

So, in the first case, all molecules exit to the right with an average residence time of 3/2, while in the second case, only half of the molecules exit to the right with an average residence time of 1/4.

5.9.3 Nonideal Plug-Flow Reactor via Transfer Matrix

In a nonideal PFR, besides transport by convection some diffusion occurs in the axial direction. For such a reactor, a model equation of the form

$$\varepsilon_b \frac{\partial c}{\partial t} = D_{eff} \frac{\partial^2 c}{\partial x^2} - v \frac{\partial c}{\partial x} - k_a c \tag{5.171}$$

is obtained. Its treatment mimics the TAP case, but presents some additional computational complexity. A Laplace transform of Eq. (5.171) yields

$$\varepsilon_b s \mathcal{L}\{c\} = D_{eff} \frac{\partial^2 \mathcal{L}\{c\}}{\partial x^2} - v \frac{\partial \mathcal{L}\{c\}}{\partial x} - k_a \mathcal{L}\{c\} \tag{5.172}$$

Now we need the solutions to the characteristic equation of this ODE in space:

$$D_{eff} \lambda^2 + v\lambda - (k_a + \varepsilon s) = 0 \tag{5.173}$$

with roots

$$\lambda_\pm = \frac{1}{2D_{eff}} \left(-v \pm \sqrt{v^2 - 4D_{eff}(\varepsilon_b s + k_a)} \right) \tag{5.174}$$

The transfer matrix must now take into account the presence of the convective term in the flux expression

$$J = vc - D_{eff} \frac{\partial c}{\partial x} \tag{5.175}$$

The first column of the transfer matrix for a reactor zone of length L with characteristics ε_b, D_{eff}, and v is calculated from a linear superposition of $\exp(\lambda_\pm (L-x))$ that has zero flux and unit concentration at $x=L$. This is readily seen to be

$$\frac{\dfrac{\exp(\lambda_+(L-x))}{v+D_{eff}\lambda_+} - \dfrac{\exp(\lambda_-(L-x))}{v+D_{eff}\lambda_-}}{\dfrac{1}{v+D_{eff}\lambda_+} - \dfrac{1}{v+D_{eff}\lambda_-}} \tag{5.176}$$

Similarly, the second column is obtained from a linear combination that has a concentration of zero and a flux of one at $x=L$:

$$\frac{\exp(\lambda_+(L-x)) - \exp(\lambda_-(L-x))}{(v+D_{eff}\lambda_+) - (v+D_{eff}\lambda_-)} \tag{5.177}$$

After working out the details, the zone transfer matrix is given by

$$
\begin{bmatrix}
\dfrac{\dfrac{e^{\lambda_+ L}}{v+D_{\text{eff}}\lambda_+}-\dfrac{e^{\lambda_- L}}{v+D_{\text{eff}}\lambda_-}}{\dfrac{1}{v+D_{\text{eff}}\lambda_+}-\dfrac{1}{v+D_{\text{eff}}\lambda_-}} & \dfrac{e^{\lambda_+ L}-e^{\lambda_- L}}{D_{\text{eff}}(\lambda_+ -\lambda_-)} \\[4mm]
\dfrac{e^{\lambda_+ L}-e^{\lambda_- L}}{\dfrac{1}{v+D_{\text{eff}}\lambda_+}-\dfrac{1}{v+D_{\text{eff}}\lambda_-}} & \dfrac{(v+D_{\text{eff}}\lambda_+)e^{\lambda_+ L}-(v+D_{\text{eff}}\lambda_-)e^{\lambda_- L}}{D_{\text{eff}}(\lambda_+ -\lambda_-)}
\end{bmatrix}
\tag{5.178}
$$

The technique of building the global transfer matrix is the same as for TAP reactors, but the boundary conditions are typically generalized to $J = vc$, that is, no diffusive flux at the outlet. This is the condition studied before in a generalized outlet condition for TAP reactors and leads to an outlet flux of the form

$$
J_{\text{outlet}} = \frac{1}{P_{22}+\dfrac{P_{21}}{v}}
\tag{5.179}
$$

5.10 Piecewise Linear Characteristics

5.10.1 Introduction

In our treatment of TAP reactors using the transfer function, so far we have always assumed that the characteristics ε_b, D_{eff}, and k_a are constant in each separate zone. In practice, if these were to exhibit a continuous dependence on the longitudinal coordinate x, this could be approximated by slicing the domain up and selecting average values of the characteristics in each resulting slice. The availability of computer algebra software has brought a wealth of special functions from advanced mathematics within reach of less specialized scientists, and in this respect it is useful to note that closed-form solutions exist even for piecewise linear characteristics (linear in each zone). For instance, the linear characteristic PDE

$$
(\varepsilon_b +\tilde{\varepsilon}_b x)s\mathcal{L}\{c\} = \frac{\partial}{\partial x}\left(\left(D_{\text{eff}}+\tilde{D}_{\text{eff}}x\right)\frac{\partial \mathcal{L}\{c\}}{\partial x}\right)-\left(k_a +\tilde{k}_a x\right)
\tag{5.180}
$$

has a general closed-form solution that can be expressed in terms of the functions

$$
\exp\left(-\sqrt{\frac{\tilde{\varepsilon}_b s+\tilde{k}_a}{\tilde{D}_{\text{eff}}}}x\right)\begin{Bmatrix} M \\ U \end{Bmatrix}
$$

$$
\left(\frac{1}{2}\left(1-\frac{D_{\text{eff}}\left(\tilde{\varepsilon}_b s+\tilde{k}_a\right)-\tilde{D}_{\text{eff}}(\varepsilon_b s+k_a)}{\tilde{D}_{\text{eff}}^{3/2}\sqrt{\tilde{\varepsilon}_b s+\tilde{k}_a}}\right),1,2\frac{\sqrt{\tilde{\varepsilon}_b s+\tilde{k}_a}}{\tilde{D}_{\text{eff}}^{3/2}}\left(D_{\text{eff}}+\tilde{D}_{\text{eff}}x\right)\right)
\tag{5.181}
$$

where m and u denote Kummer's function $M(\mu,\nu,z)$ and Tricomi's function $U(\mu,\nu,z)$, solutions of Kummer's ordinary differential equation

$$z\frac{d^2y}{dz^2} + (\nu - z)\frac{dy}{dz} - \mu y = 0 \tag{5.182}$$

These results are presumably not directly useful for practical applications, except maybe to check the accuracy of software developed for modeling variable-characteristic reactors mathematically by providing an exact theoretical value independently.

5.10.2 Conical Reactors

For reactors that are not cylindrical but conical, equations can still be established for the case with piecewise-constant characteristics. The diffusion equation is a mass balance that must now take into account the shape and the fact that the surface area of a transversal slice is proportional to x^2 for a conical reactor. Between x and $x+dx$, the gas content is $\varepsilon_b x^2 dx\, c(x)$, while the diffusive flux at x is given by $-x^2 D_{eff}\dfrac{\partial c(x)}{\partial x}$ and the reaction loss in the volume considered is $x^2 dx - k_a c(x)$. Hence,

$$\varepsilon_b x^2 \frac{\partial c}{\partial t} = D_{eff}\frac{\partial}{\partial x}\left(x^2\frac{\partial c}{\partial x}\right) - k_a x^2 c \tag{5.183}$$

which after rearrangement becomes

$$\varepsilon_b \frac{\partial c}{\partial t} = D_{eff}\frac{1}{x^2}\frac{\partial}{\partial x}\left(x^2\frac{\partial c}{\partial x}\right) - k_a c \tag{5.184}$$

This is the three-dimensional radial equation presented before, but reduced to the cone in its domain. The general solution is constructed from $\dfrac{\exp\left(\pm\sqrt{\dfrac{\varepsilon_b s + k_a}{D_{eff}}}x\right)}{x}$ and the transfer matrix can be calculated accordingly.

5.10.3 Surface Diffusion

The treatment of surface diffusion can be reduced formally to the multiresponse case as follows. Assume that the vector with fractional surface coverages $\boldsymbol{\theta}$ satisfies the equation

$$\frac{\partial \boldsymbol{\theta}}{\partial t} = D_\theta \frac{\partial^2 \boldsymbol{\theta}}{\partial x^2} + \text{reaction terms} \tag{5.185}$$

Then for the zone under consideration, this is a gas-type reaction-diffusion equation and $\boldsymbol{\theta}$ can be incorporated formally into the vector $\mathbf{c}$ of gas concentrations and its diffusive flux,

$\mathbf{J_\theta} = -D_\theta \dfrac{\partial \mathbf{\theta}}{\partial t}$, can be incorporated into the vector of gas fluxes. Nevertheless, then the problem arises that the coverages $\mathbf{\theta}$ are restricted to the zone considered and cannot diffuse into neighboring zones. If we establish the transfer matrix with the fractional surface coverages and diffusive fluxes of the surface intermediates incorporated in the vectors of the gas concentrations and gas fluxes, after rearrangement of the concentrations and fluxes from the order $\mathbf{c}, \mathbf{\theta}, \mathbf{J_c}, \mathbf{J_\theta}$ to the order $\mathbf{c}, \mathbf{J_c}, \mathbf{\theta}, \mathbf{J_\theta}$, a matrix relationship of the form

$$\begin{bmatrix} \mathbf{c} \\ \mathbf{J_c} \\ \mathbf{\theta} \\ \mathbf{J_\theta} \end{bmatrix}_{\text{left}} = \begin{bmatrix} \alpha & \beta & \gamma \\ \delta & \varepsilon & \kappa \\ \lambda & \mu & \nu \end{bmatrix} \begin{bmatrix} \mathbf{c} \\ \mathbf{J_c} \\ \mathbf{\theta} \\ \mathbf{J_\theta} \end{bmatrix}_{\text{right}} \tag{5.186}$$

is obtained. Our task is then to set $\mathbf{J_\theta} = 0$ both left and right and to eliminate $\mathbf{\theta}$ on both sides so as to obtain a transfer function in $\mathbf{c}$ and $\mathbf{J_c}$ only, which can then be plugged into the product of transfer matrices constituting the global transfer matrix. From the last row, it follows that

$$0 = \lambda \begin{bmatrix} \mathbf{c} \\ \mathbf{J_c} \end{bmatrix}_{\text{right}} + \mathbf{\mu}\mathbf{\theta}_{\text{right}} + \mathbf{\nu} \cdot 0 \tag{5.187}$$

so that

$$\mathbf{\theta}_{\text{right}} = -\mathbf{\mu}^{-1}\lambda \begin{bmatrix} \mathbf{c} \\ \mathbf{J_c} \end{bmatrix}_{\text{right}} \tag{5.188}$$

From the two top rows, we find

$$\begin{bmatrix} \mathbf{c} \\ \mathbf{J_c} \end{bmatrix}_{\text{left}} = \alpha \begin{bmatrix} \mathbf{c} \\ \mathbf{J_c} \end{bmatrix}_{\text{right}} + \beta\mathbf{\theta}_{\text{right}} = \underbrace{(\alpha - \beta\mathbf{\mu}^{-1}\lambda)}_{\text{transfer matrix}} \begin{bmatrix} \mathbf{c} \\ \mathbf{J_c} \end{bmatrix}_{\text{right}} \tag{5.189}$$

5.10.4 Independence of the Final Profile of Catalyst Surface Coverage

The general theory of TAP experiments is linear because the device itself and its use of pulse experiments causing an insignificant change to the solid material's surface were designed with a simple linear model in mind. Under some circumstances, however, exact mathematical results can also be obtained for nonlinear model equations, albeit only at $t = +\infty$. The plainest example of this is the irreversible adsorption reaction $A + Z \rightarrow AZ$, where Z denotes an active catalyst site. The rate of this elementary reaction is given by

$$r = k_a c_A N_Z \tag{5.190}$$

which in the linear model is represented as

$$r = (k_a N_Z) c_A = k_{a,\text{app}} c_A \tag{5.191}$$

because it is assumed that the total number of molecules in each pulse is negligible compared to the number of active catalyst sites, N_Z involved. If this assumption is not made, and if the law of mass action is applied, the "time" rate expression must be proportional to $\theta_Z c_A$ or $(1-\theta_{AZ})c_A$, with $\theta_Z (= N_Z/(N_Z+N_{AZ})$ the fractional surface coverage with free active catalyst sites, $\theta_{AZ} (= N_{AZ}/(N_Z+N_{AZ})$ the fractional surface coverage with adsorbed A, and $\theta_Z+\theta_{AZ}=1$. The model set of PDEs then reads

$$
\begin{cases}
\varepsilon_b \dfrac{\partial c_A}{\partial t} = D_{\text{eff}}\dfrac{\partial^2 c_A}{\partial x^2} - k_a(1-\theta_{AZ})c_A \\[2mm]
c_{\text{cat}}\dfrac{\partial \theta_{AZ}}{\partial t} = k_a(1-\theta_{AZ})c_A
\end{cases}
\tag{5.192}
$$

This is of second order and cannot be solved in terms of Laplace transforms because the product $\theta_{AZ}c_A$ cannot be expressed conveniently in terms of θ_{AZ} and c_A when transformed to the Laplace domain. Nevertheless, the second equation can be rearranged,

$$
\frac{1}{(1-\theta_{AZ})}\frac{\partial \theta_{AZ}}{\partial t} = \frac{k_a}{c_{\text{cat}}}c_A
\tag{5.193}
$$

and then integrated (assuming $\theta_{AZ}(0)=0$):

$$
- \ln(1-\theta_{AZ}) = \frac{k_a}{c_{\text{cat}}}\int_0^t c_A dt
\tag{5.194}
$$

or, solving for θ_{AZ} as a function of t:

$$
\theta_{AZ} = 1 - \exp\left(-\frac{k_a}{c_{\text{cat}}}\int_0^t c_A dt\right)
\tag{5.195}
$$

The integral $\int_0^t c_A dt$ is the area beneath the curve of c_A from 0 to t. This does not help much for finite t, but for $t = +\infty$ we can use it for substitution in the time integral of both sides of the equations for $\partial c_A/\partial t$:

$$
\varepsilon_b \underbrace{\int_0^{+\infty}\frac{\partial c_A}{\partial t}dt}_{\substack{=c_A(+\infty)-c_A(0)\\=0-0=0}} = D_{\text{eff}}\frac{\partial^2}{\partial x^2}\int_0^{+\infty}c_A dt - c_{\text{cat}}\int_0^{+\infty}\underbrace{\frac{k_a}{c_{\text{cat}}}(1-\theta_{AZ})c_A\,dt}_{\dfrac{\partial \theta_{AZ}}{\partial t}}
\tag{5.196}
$$

Writing $M(x)$ for the dimensionless expression $\dfrac{k_a}{c_{\text{cat}}}\displaystyle\int_0^t c_A dt$ and then using Eq. (5.195), we obtain

$$
0 = \frac{D_{\text{eff}}}{k_a}\frac{d^2M}{dx^2} - \theta_{AZ}(+\infty) = \frac{D_{\text{eff}}}{k_a}\frac{d^2M}{dx^2} - (1-\exp(-M))
\tag{5.197}
$$

This ODE for $M(x)$ can be integrated using the standard method of introducing an auxiliary expression $V(x)$ for $\dfrac{dM}{dx}$ and rewriting $\dfrac{d^2M}{dx^2} = \dfrac{dV}{dx}$ as $\left(\dfrac{dV}{dM}\right)\cdot\left(\dfrac{dM}{dx}\right) = V\dfrac{dV}{dM}$.

Then we integrate the equation obtained by this substitution in Eq. (5.197),

$$\frac{D_{\text{eff}}}{k_a} V \frac{dV}{dM} = 1 - \exp(-M) \tag{5.198}$$

to obtain

$$\frac{1}{2}\frac{D_{\text{eff}}}{k_a}(V^2 - V_1^2) = M + \exp(-M) - (M_1 + \exp(-M_1)) \tag{5.199}$$

where M_1 and V_1 denote the values of $M(x)$ and $V(x)$ at the left side of the zone. In the case of a single-zone reactor, the diffusive flux $J_{A,\text{inlet}} = -D_{\text{eff}}\frac{\partial c_A}{\partial x}\Big|_{\text{inlet}}$ is prescribed, and integrating from 0 to $+\infty$ yields

$$\frac{k_a}{c_{\text{cat}}}\int_0^{+\infty} J_{\text{inlet}}dt = -D_{\text{eff}}V_1 \tag{5.200}$$

from which the value of V_1 can be obtained. Solving for $V = \frac{dM}{dx}$ gives

$$\frac{dM}{dx} = -\sqrt{V_1^2 + \frac{2k_a}{D_{\text{eff}}}(M + \exp(-M) - (M_1 + \exp(-M_1)))} \tag{5.201}$$

so that

$$\int_{M(x)}^{M_1} \frac{dM}{\sqrt{V_1^2 + \frac{2k_a}{D_{\text{eff}}}(M + \exp(-M) - (M_1 + \exp(-M_1)))}} = x \tag{5.202}$$

This equation clearly holds for $x=0$ where $M(x)=M_1$, and considering the outlet of the reactor at $x=L$ where $c_A=0$ and thus $M=0$,

$$\int_0^{M_1} \frac{dM}{\sqrt{V_1^2 + \frac{2k_a}{D_{\text{eff}}}(M + \exp(-M) - (M_1 + \exp(-M_1)))}} = L \tag{5.203}$$

is the equation that allows us to determine M_1. This explicit solution is somewhat intricate, but the essential property is that when the integral from $t=0$ to $+\infty$ is considered, an ODE is obtained that determines $M(x)$ and hence $\theta(x, +\infty)$ *uniquely* with boundary conditions that depend only on the *total amount* $\int_0^{+\infty} J_{\text{inlet}}dt$ pulsed in. It is exceptional for such properties to hold; if we were to model, for instance, two competing adsorption reactions, $A+Z_1 \to AZ_1$ and $A+Z_2 \to AZ_2$, or if further reactions occurred after adsorption, for example, $A+Z \to AZ \to BZ$, the final profile of the catalyst coverage would be dependent not only on the total amount injected but also on the evolution of $J_{\text{inlet}}(t)$ in time. The key properties that lead to a dependence on the total amount injected are only that:

- at any position, the final catalytic composition is defined uniquely by the integral amount of gas concentration at that position, that is, $\theta_j(x)$ is a function of $M(x)$;
- the integral amount of gas concentration is defined uniquely by the total amount of molecules admitted, that is, that $M(x)$ satisfies the ODE where J_{inlet} only enters through the value $\int_0^{+\infty} J_{\text{inlet}} dt$.

For more detailed numerical and exact results of this type, refer to Constales et al. (2015) and Phanawadee et al. (2013).

5.11 First- and Second-Order Nonideality Corrections in the Modeling of Thin-Zone TAP Reactors

5.11.1 Thin-Zone TAP Reactor

Consider a three-zone TAP reactor in which an irreversible (adsorption) reaction occurs, and for which the middle zone is intended to be thin (the TZTR). The transfer matrix $\mathbf{M}_1$ for the first zone is given by

$$\mathbf{M}_1 = \begin{bmatrix} \cosh\left(L_1\sqrt{\dfrac{\varepsilon_{b,1}s}{D_{\text{eff},1}}}\right) & \dfrac{\sinh\left(L_1\sqrt{\dfrac{\varepsilon_{b,1}s}{D_{\text{eff},1}}}\right)}{\sqrt{\varepsilon_{b,1}D_{\text{eff},1}s}} \\[4ex] \sqrt{\varepsilon_{b,1}D_{\text{eff},1}s}\,\sinh\left(L_1\sqrt{\dfrac{\varepsilon_{b,1}s}{D_{\text{eff},1}}}\right) & \cosh\left(L_1\sqrt{\dfrac{\varepsilon_{b,1}s}{D_{\text{eff},1}}}\right) \end{bmatrix} \tag{5.204}$$

The transfer matrix for the third zone, $\mathbf{M}_3$, is similar to this matrix but with subscripts "3" instead of "1." The second zone's transfer matrix, $\mathbf{M}_2$, takes into account the adsorption rate coefficient and is given by

$$\mathbf{M}_2 = \begin{bmatrix} \cosh\left(L_2\sqrt{\dfrac{\varepsilon_{b,2}s+\left(\dfrac{\kappa}{L_2}\right)}{D_{\text{eff},2}}}\right) & \dfrac{\sinh\left(L_2\sqrt{\dfrac{\varepsilon_{b,2}s+\left(\dfrac{\kappa}{L_2}\right)}{D_{\text{eff},2}}}\right)}{\sqrt{\varepsilon_{b,2}s+\left(\dfrac{\kappa}{L_2}\right)D_{\text{eff},2}}} \\[6ex] \sqrt{\varepsilon_{b,2}s+\left(\dfrac{\kappa}{L_2}\right)D_{\text{eff},2}}\,\sinh\left(L_2\sqrt{\dfrac{\varepsilon_{b,2}s+\left(\dfrac{\kappa}{L_2}\right)}{D_{\text{eff},2}}}\right) & \cosh\left(L_2\sqrt{\dfrac{\varepsilon_{b,2}s+\left(\dfrac{\kappa}{L_2}\right)}{D_{\text{eff},2}}}\right) \end{bmatrix}$$

$$\tag{5.205}$$

where κ (in m s^{-1}) represents a reactivity parameter of the ideal thin zone that is being approximated (see Constales et al., 2001a for details). We are interested in the limit $L_2 \to 0$, which is the TZTR case. In this limit

$$\lim_{L_2 \to 0} \mathbf{M}_2 = \begin{bmatrix} 1 & 0 \\ \kappa & 1 \end{bmatrix} \tag{5.206}$$

In the Laplace domain, the observed outlet flux can be calculated from

$$J_3 = \frac{1}{(\mathbf{M}_1 \mathbf{M}_2 \mathbf{M}_3)_{22}} \tag{5.207}$$

and in the TZTR limit we have

$$\begin{aligned} \frac{1}{J_{\text{TZTR}}} &= \cosh\left(L_1 \sqrt{\frac{\varepsilon_{b,1} s}{D_{\text{eff},1}}}\right) \cosh\left(L_3 \sqrt{\frac{\varepsilon_{b,3} s}{D_{\text{eff},3}}}\right) \\ &+ \sqrt{\frac{\varepsilon_{b,1} D_{\text{eff},1}}{\varepsilon_{b,3} D_{\text{eff},3}}} \sinh\left(L_1 \sqrt{\frac{\varepsilon_{b,1} s}{D_{\text{eff},1}}}\right) \sinh\left(L_3 \sqrt{\frac{\varepsilon_{b,3} s}{D_{\text{eff},3}}}\right) \\ &+ \frac{\kappa}{\sqrt{\varepsilon_{b,3} D_{\text{eff},3} s}} \cosh\left(L_1 \sqrt{\frac{\varepsilon_{b,1} s}{D_{\text{eff},1}}}\right) \sinh\left(L_3 \sqrt{\frac{\varepsilon_{b,3} s}{D_{\text{eff},3}}}\right) \end{aligned} \tag{5.208}$$

As $L_2 \to 0, J_3 \to J_{\text{TZTR}}$ with $J_{\text{TZTR}} - J_3 = O(L_2)$, but unfortunately, due to the absence of a term proportional to

$$\sqrt{s} \sinh\left(L_1 \sqrt{\frac{\varepsilon_{b,1} s}{D_{\text{eff},1}}}\right) \cosh\left(L_3 \sqrt{\frac{\varepsilon_{b,3} s}{D_{\text{eff},3}}}\right) \tag{5.209}$$

the parameters of the limit TZTR (L_1, etc.) proportionally to L_2 cannot be corrected so as to ensure a faster $O(L_2^2)$ approach of the limit.

In practice, this relates to the question of what to do with the nonzero thickness of the middle zone: should it just be ignored or should it be split up and one half added to L_1 and the other to L_3? From our asymptotic viewpoint where $L_2 \to 0$, neither of these choices will improve the convergence of the approximation qualitatively, simply because a TZTR is too coarse an approximation of a real three-zone reactor with a relatively thin active zone.

This has motivated the introduction of the thin-sandwiched TAP reactor (TSTR), an ideal reactor offering a far superior approximation from the asymptotic viewpoint, while retaining the property of concentration uniformity in the reactive zone that is essential for application of the Y procedure (see Yablonsky et al., 2002).

5.11.2 Thin-Sandwiched TAP Reactor Matrix Modeling

Consider an inert zone whose thickness L_i will conceptually be decreasing to zero, while keeping D_{eff} proportional to L_i with proportionality factor $1/\rho$ (where ρ is a "diffusional resistance"). The corresponding transfer matrix will then tend to the limit

$$\begin{bmatrix} 1 & \rho \\ 0 & 1 \end{bmatrix} \tag{5.210}$$

as $L_i \to 0$. This is called a "thin resistor zone" (see also Section 5.5.7).

A TSTR consists of inert first and last zones, just like a TZTR, but the middle zone is no longer merely a thin reactive zone, but a "sandwich" consisting of a thin reactive zone between two thin resistor zones (in this case, of equal diffusional resistance values ρ, because we wish to retain the symmetry of the middle zone). The transfer matrix of this TSTR is given by

$$\begin{bmatrix} 1 & \rho \\ 0 & 1 \end{bmatrix} \begin{bmatrix} 1 & \rho \\ \kappa & 1 \end{bmatrix} \begin{bmatrix} 1 & \rho \\ 0 & 1 \end{bmatrix} \tag{5.211}$$

and its corresponding outlet flux is represented by

$$\begin{aligned}
\frac{1}{J_{\text{TSTR}}} =\; & (1+\kappa\rho)\cosh\left(L_1\sqrt{\frac{\varepsilon_{b,1}S}{D_{\text{eff},1}}}\right)\cosh\left(L_3\sqrt{\frac{\varepsilon_{b,3}S}{D_{\text{eff},3}}}\right) \\
& + (1+\kappa\rho)\sqrt{\frac{\varepsilon_{b,1}D_{\text{eff},1}}{\varepsilon_{b,3}D_{\text{eff},3}}}\sinh\left(L_1\sqrt{\frac{\varepsilon_{b,1}S}{D_{\text{eff},1}}}\right)\sinh\left(L_3\sqrt{\frac{\varepsilon_{b,3}S}{D_{\text{eff},3}}}\right) \\
& + \frac{\kappa}{\sqrt{\varepsilon_{b,3}D_{\text{eff},3}S}}\cosh\left(L_1\sqrt{\frac{\varepsilon_{b,1}S}{D_{\text{eff},1}}}\right)\sinh\left(L_3\sqrt{\frac{\varepsilon_{b,3}S}{D_{\text{eff},3}}}\right) \\
& + \rho(2+\kappa\rho)\sqrt{\varepsilon_{b,1}D_{\text{eff},1}}\sinh\left(L_1\sqrt{\frac{\varepsilon_{b,1}S}{D_{\text{eff},1}}}\right)\cosh\left(L_3\sqrt{\frac{\varepsilon_{b,3}S}{D_{\text{eff},3}}}\right).
\end{aligned} \tag{5.212}$$

The fourth term in this expression will enable us to approximate the TSTR limit with the discrepancy reduced to $O(L_2^2)$.

In a *first-order approximation*, the values corrected with $O(L_2)$ are:

$$\hat{L}_1 = L_1 + \frac{\varepsilon_{b,2}}{\varepsilon_{b,1}}\frac{\varepsilon_{b,1}D_{\text{eff},1}}{\varepsilon_{b,1}D_{\text{eff},1}+\varepsilon_{b,3}D_{\text{eff},3}}L_2 \tag{5.213}$$

$$\hat{L}_3 = L_3 + \frac{\varepsilon_{b,2}}{\varepsilon_{b,3}}\frac{\varepsilon_{b,3}D_{\text{eff},3}}{\varepsilon_{b,1}D_{\text{eff},1}+\varepsilon_{b,3}D_{\text{eff},3}}L_2 \tag{5.214}$$

$$\hat{\kappa} = \kappa + \frac{\kappa^2}{6D_{\text{eff},2}}L_2 \tag{5.215}$$

$$\hat{\rho} = \left(1 - \frac{2\varepsilon_{b,2}D_{\text{eff},2}}{\varepsilon_{b,1}D_{\text{eff},1}+\varepsilon_{b,3}D_{\text{eff},3}}\right)\frac{L_2}{2D_{\text{eff},2}} \tag{5.216}$$

These values will ensure an $O(L_2^2)$ discrepancy between the TZTR with parameters L_1, etc., and the TSTR with parameters L_1, etc.

There is considerable freedom in choosing which parameters to correct, but this choice appears the most justified because the value κ here only affects $\hat{\kappa}$ and not ρ or the other TSTR parameters. Note also that the equations for $\hat{L}_1$ and $\hat{L}_3$ amount to a weighted doling out of the "void length" $\varepsilon_{b,2}L_2$ over the void lengths of the inert zones, $\varepsilon_{b,1}L_1$ and $\varepsilon_{b,3}L_3$, with weights $\varepsilon_{b,1}D_{\text{eff},1}$ and $\varepsilon_{b,3}D_{\text{eff},3}$. It is possible for $\hat{\rho}$ to be negative, which would not be physically meaningful if a single zone were considered, but is no problem here for the modeling of a complete TSTR.

In general, no *second-order correction* can ensure $O(L_2^3)$ convergence, but in the special case where $\varepsilon_{b,1}D_{\text{eff},1} = \varepsilon_{b,3}D_{\text{eff},3}$ (most notably, whenever the first and last zones have the same physical characteristics, even if their lengths differ), some terms in the general solution's second derivative with respect to L_2 at 0 vanish, and then the corrected values (after elimination of $D_{\text{eff},3}$) will ensure an $O(L_2^3)$ discrepancy between the three-zone TAP reactor and the TSTR reactor. Note that all parameters are now influenced by κ:

$$\hat{L}_1 = L_1 + \frac{\varepsilon_{b,2}}{\varepsilon_{b,1}}L_2 + \frac{1}{\varepsilon_{b,1}}\left(3\frac{\varepsilon_{b,2}D_{\text{eff},2}}{\varepsilon_{b,1}D_{\text{eff},1}} - 2\right)\frac{\varepsilon_{b,2}\kappa L_2^2}{24D_{\text{eff},2}} \tag{5.217}$$

$$\hat{L}_3 = L_1 + \frac{\varepsilon_{b,2}}{\varepsilon_{b,3}}L_2 + \frac{1}{\varepsilon_{b,3}}\left(3\frac{\varepsilon_{b,2}D_{\text{eff},2}}{\varepsilon_{b,1}D_{\text{eff},1}} - 2\right)\frac{\varepsilon_{b,2}\kappa L_2^2}{24D_{\text{eff},2}} \tag{5.218}$$

$$\hat{\kappa} = \kappa + \frac{\kappa^2}{6D_{\text{eff},2}}L_2 + \frac{\kappa^3}{120D_{\text{eff},2}^2}L_2^2 \tag{5.219}$$

$$\hat{\rho} = \left(1 - \frac{\varepsilon_{b,2}D_{\text{eff},2}}{\varepsilon_{b,1}D_{\text{eff},1}}\right)\frac{L_2}{2D_{\text{eff},2}} - \left(1 - 2\frac{\varepsilon_{b,2}D_{\text{eff},2}}{\varepsilon_{b,1}D_{\text{eff},1}} + 3\left(\frac{\varepsilon_{b,2}D_{\text{eff},2}}{\varepsilon_{b,1}D_{\text{eff},1}}\right)^2\right)\frac{\kappa L_2^2}{24D_{\text{eff},2}^2} \tag{5.220}$$

Nomenclature

c_{cat}	catalyst concentration (mol m^{-3})
c_i	concentration of component i (mol m^{-3})
D	diffusion coefficient (m^2 s^{-1})
D_0	preexponential factor (m^2 s^{-1})
D_{eff}	effective diffusion coefficient (m^2 s^{-1})
D_{K}	Knudsen diffusion coefficient (m^2 s^{-1})
d_{p}	diameter of catalyst pellet (m)
d_{pore}	pore diameter (m)
E_{a}	activation energy (J mol^{-1})
F_{e}	external force (kg m s^{-2})
f	Stokes's friction coefficient (kg s^{-1})
I, K	modified Bessel functions

J	diffusive flux vector (mol m^{-2} s^{-1})
J_0	Bessel function
k_a	adsorption rate coefficient (s^{-1})
k_B	Boltzmann's constant (J K^{-1})
k_d	desorption rate coefficient (s^{-1})
L	reactor length (m)
L_i	length of zone i (m)
M_i	molecular mass of component i (kg mol^{-1})
M_n	nth moment
m	particle mobility (s k g^{-1})
N_A	Avogadro's number (mol)
N_{int}	total number of intermediates (–)
N_Z	number of active catalyst sites (–)
q	charge (C)
R	cylinder radius (m)
R	net rate of production (mol m^{-3} s^{-1})
R_g	universal gas constant (J mol^{-1} K^{-1})
r	transversal distance to cylinder's axis (m)
r	rate of reaction (mol m^{-3} s^{-1})
r_p	particle radius (m)
s	Laplace variable (s^{-1})
T	temperature (K)
t	time (s)
v	velocity (m s^{-1})
X_j	jth thermodynamic force (depends)
x	distance from reactor inlet (m)
x, y, z	Cartesian coordinates (m)
Y_0	Bessel function
z	longitudinal coordinate (m)

Greek symbols

Γ_j	concentration of surface intermediate j (mol m$_{cat}^{-2}$)
δ	length of premixing zone (m)
ε_b	bed voidage (–)
θ	angle in cylindrical coordinate system
θ_j	fractional surface coverage with component j (–)
η	dynamic viscosity of liquid (kg m^{-1} s^{-1})
κ	reactivity parameter (m s^{-1})
λ	mean free path (m)
μ	constant in Eq. (5.128) (–)

μ_j chemical potential of component j (J mol^{-1})

τ time shift (s)

τ_b tortuosity of bed (–)

φ electric potential (J C^{-1})

Subscripts

cat   catalyst

eq   equilibrium

g   gas

t   total

References

Aris, R., 1975a. The Mathematical Theory of Diffusion and Reaction in Permeable Catalysts. Vol. 1: The Theory of the Steady State. Clarendon Press, Oxford.

Aris, R., 1975b. The Mathematical Theory of Diffusion and Reaction in Permeable Catalysts. Vol. 2: Questions of Uniqueness, Stability and Transient Behaviour. Clarendon Press, Oxford.

Berger, R.J., Kapteijn, F., Moulijn, J.A., et al., 2008. Dynamic methods for catalytic kinetics. Appl. Catal. A 342, 3–28.

Constales, D., Yablonsky, G.S., Marin, G.B., Gleaves, J.T., 2001a. Multi-zone TAP-reactors theory and application: I. The global transfer matrix equation. Chem. Eng. Sci. 56, 133–149.

Constales, D., Yablonsky, G.S., Marin, G.B., Gleaves, J.T., 2001b. Multi-zone TAP-reactors theory and application: II. The three-dimensional theory. Chem. Eng. Sci. 56, 1913–1923.

Constales, D., Yablonsky, G.S., Phanawadee, P., Pongboutr, N., Limtrakul, J., Marin, G.B., 2015. When the final catalyst activity profile depends only on the total amount of admitted substance: theoretical proof. AIChE J. 61, 31–34.

Crank, J., 1980. The Mathematics of Diffusion, second ed. Oxford University Press, London.

de Groot, S.R., Mazur, P., 1962. Nonequilibrium Thermodynamics. NorthHolland, Amsterdam.

Dushman, S., Langmuir, L., 1922. The diffusion coefficient in solids and its temperature coefficient. Phys. Rev. 20, 113.

Einstein, A., 1905. Über die von der molekularkinetischen Theorie der Wärme geforderte Bewegung von in ruhenden Flüssigkeiten suspendierten Teilchen. Ann. Phys. 322, 549–560.

Feres, R., Yablonsky, G., 2004. Knudsen's cosine law and random billiards. Chem. Eng. Sci. 59, 1541–1556.

Fick, A., 1855a. Über diffusion. Ann. Phys. 170, 59–86.

Fick, A., 1855b. V. On liquid diffusion. Philos. Mag. Ser. 4 10, 30–39.

Fick, A., 1995. On liquid diffusion. J. Membr. Sci. 100, 33–38. Reprinted from Fick, A. (1855a).

Glansdorff, P., Prigogine, I., 1971. Thermodynamic Theory of Structure, Stability, and Fluctuations. Wiley, New York, NY.

Gleaves, J.T., Ebner, J.R., Kuechler, T.C., 1988. Temporal analysis of products (TAP)—a unique catalyst evaluation system with submillisecond time resolution. Catal. Rev. Sci. Eng. 30, 49–116.

Gleaves, J.T., Yablonskii, G.S., Phanawadee, P., Schuurman, Y., 1997. TAP-2: an interrogative kinetics approach. Appl. Catal. A 160, 55–88.

Gleaves, J.T., Yablonsky, G., Zheng, X., Fushimi, R., Mills, P.L., 2010. Temporal analysis of products (TAP)— recent advances in technology for kinetic analysis of multi-component catalysts. J. Mol. Catal. A Chem. 315, 108–134.

Gorban, A.N., Bykov, V.I., Yablonskii, G.S., 1980. Macroscopic clusters induced by diffusion in catalytic oxidation reactions. Chem. Eng. Sci. 35, 2351–2352.

Gorban, A.N., Bykov, V.I., Yablonskii, G.S., 1986. Essays on Chemical Relaxation. Nauka, Novosibirsk.

Gorban, A.N., Sargsyan, H.P., 1986. Mass action law for nonlinear multicomponent diffusion and relations between its coefficients. Kinet. Catal. 27, 527.

Gorban, A.N., Sargsyan, H.P., Wahab, H.A., 2011. Quasichemical models of multicomponent nonlinear diffusion. Math. Model. Nat. Phenom. 6, 184–262.

Kärger, J., Ruthven, D.M., 1992. Diffusion in Zeolites and Other Microporous Solids. Wiley, New York, NY.

Keil, F., 1999. Diffusion und Chemische Reaktionen in der Gas/Feststoff-Katalyse. Springer, Berlin.

Knudsen, M., 1909. Die Gesetze der Molekularströmung und der inneren Reibungsströmung der Gase durch Röhren. Ann. Phys. 333, 75–130.

Knudsen, M., 1934. The Kinetic Theory of Gases: Some Modern Aspects. Methuen, London.

Kondratenko, E.V., Perez-Ramírez, J., 2007. The TAP Reactor in Catalysis: Recent Advances in Theory and Practice. Catal. Today 121, 159–281.

Marin, G.B., Yablonsky, G.S., 2011. Kinetics of Chemical Reactions—Decoding Complexity. Wiley-VCH, Weinheim.

Nicolis, G., Prigogine, I., 1977. Self-Organization in Non-Equilibrium Systems: From Dissipative Structures to Order Through Fluctuations. Wiley, New York, NY.

Onsager, L., 1931a. Reciprocal relations in irreversible processes. I. Phys. Rev. 37, 405–426.

Onsager, L., 1931b. Reciprocal relations in irreversible processes. II. Phys. Rev. 38, 2265–2279.

Phanawadee, P., Pongboutr, N., Yablonsky, G.S., Constales, D., Jarungmanorom, C., Soikham, W., Limtrakul, J., 2013. Independence of active substance profiles from the pulse response experimental procedure. AIChE J. 59, 3574–3577.

Philibert, J., 2005. One and a half century of diffusion: Fick, Einstein, before and beyond. Diffus. Fundam. 2, 1.1–1.10.

Philibert, J., 2006. One and a half century of diffusion: Fick, Einstein, before and beyond. Diffus. Fundam. 4, 6.1–6.19.

Shekhtman, S.O., Yablonsky, G.S., Chen, S., Gleaves, J.T., 1999. Thin-zone TAP-reactor—theory and application. Chem. Eng. Sci. 54, 4371–4378.

Sherwood, J.N., Chadwick, A.V., Muir, W.M., et al., 1971. Diffusion Processes: Proceedings of the Thomas Graham Memorial Symposium, University of Strathclyde. Gordon and Breach, London.

Teorell, T., 1935. Studies on the "diffusion effect" upon ionic distribution. I. Some theoretical considerations. Proc. Natl. Acad. Sci. U. S. A. 21, 152–161.

Teorell, T., 1937. Studies on the "diffusion effect" upon ionic distribution. II. Experiments on ionic accumulation. J. Gen. Physiol. 21, 107–122.

Thiele, E.W., 1939. Relation between catalytic activity and size of particle. Ind. Eng. Chem. 31, 916–920.

Vasquez, J.L., 2006. The Porous Medium Equations: Mathematical Theory. Oxford University Press, Oxford.

Yablonsky, G.S., Constales, D., Gleaves, J.T., 2002. Multi-scale problems in the quantitative characterization of complex catalytic materials. Syst. Anal. Model. Simul. 42, 1143–1166.

Yablonsky, G.S., Olea, M., Marin, G.B., 2005. Temporal analysis of products: basic principles, applications, and theory. J. Catal. 216, 120–134.

Zheng, X., Gleaves, J.T., Yablonsky, G.S., Brownscombe, T., Gaffney, A., Clark, M., Han, S., 2008. Needle in a haystack catalysis. Appl. Catal. A 341, 86–92.

## CHAPTER 6

Thermodynamics

6.1 Introduction

6.1.1 Thermodynamic Systems

Thermodynamics is the science concerned with the study of the relations between temperature, heat, work, and the energy content of a system at equilibrium. A system is something that can be distinguished from the surrounding medium. In thermodynamics, the main types of systems defined are isolated, closed, and open systems. An *isolated* system is a system in which there is no exchange of heat and matter with the surroundings. In a *closed* system, there is no exchange of matter with the surroundings, but heat exchange generally does occur. In an *open* system, exchange of matter with the surroundings takes place and heat exchange may occur or not. Certainly, these definitions are idealizations of real situations.

Expensive chemicals are often produced in small amounts using closed systems. For example, reactors in the pharmaceutical industry (eg, for the production of drugs) are typically closed systems (batch reactors). On the other hand, relatively cheap bulk chemicals, such as ammonia and methanol, are usually produced in open systems (continuous-flow reactors) in the presence of a solid catalyst bed (fixed-bed reactors). Rigorously, these gas-solid catalytic reactors can be considered to be *semiopen* systems because there is exchange between the gas-phase reactants and products with the surroundings, but the catalyst stays in the reactor. However, in some reactor types, the catalyst moves through the reactor along with the reactants, for example, in the methanol-to-propylene process using a so-called moving-bed reactor.

Obviously, a closed system can be considered as a particular case of an open system, that is, an open system without exchange of matter between the system and the surroundings.

For closed systems, idealized interactions with the surroundings can be of different types. The most important ideal thermodynamic processes are

- isothermal processes: occur at constant temperature
- isobaric processes: occur at constant pressure
- isochoric processes: occur at constant volume

Advanced Data Analysis and Modelling in Chemical Engineering. http://dx.doi.org/10.1016/B978-0-444-59485-3.00006-0

Thermodynamic systems may be ideal or nonideal. An *ideal* thermodynamic system can be defined as a system in which molecule sizes can be considered to be negligible and in which interactions between molecules—repulsion or attraction—can be neglected. In a *nonideal* thermodynamic system such interactions have to be taken into account.

6.1.2 Thermodynamic State Functions

Thermodynamic state functions are characteristics of a system that only depend on its current state. They do not depend on the history of the system, that is, on the way in which the system achieved that particular state. Typical state functions that are used in chemistry and chemical engineering are

- mass
- chemical composition
- pressure
- volume
- temperature
- energy
  - o enthalpy
  - o internal energy
  - o Gibbs free energy
- entropy

Changes in state functions depend only on their initial and final values and are independent of the path between them. In contrast, the final values of heat and work depend on the specific path between the two states. These characteristics are process quantities and not state functions. Excellent textbooks with detailed information on thermodynamic functions are those by Smith et al. (2005) and Atkins (2010).

Relationships between state functions are called equations of state. An example is the relationship between the temperature, pressure, and volume at a given amount of substance:

$$f(p, V, T) = 0 \tag{6.1}$$

where p is the absolute pressure, V is the volume, and T is the absolute temperature. One of the simplest equations of state is the ideal gas law, which is valid under the assumption that there is no interaction (attraction or repulsion) between molecules,

$$pV = nR_gT \quad \text{or} \quad f(p, V, T) = pV - nR_gT = 0 \tag{6.2}$$

where $n = \sum_{i=1}^{N_c} n_i$ is the total number of moles in the system with n_i the number of moles of component i and R_g is the universal gas constant.

In nonideal systems, interactions between molecules cannot be neglected. Furthermore, the sizes of the molecules must be taken into account. Two of the most popular equations of state for nonideal systems are the Van der Waals equation and the Redlich-Kwong equation.

The key thermodynamic characteristic of a component i in a mixture is its chemical potential, defined as

$$\mu_i = \left(\frac{\partial G}{\partial n_i}\right)_{T,p,n_{j\neq i}} \tag{6.3}$$

where G is the total Gibbs free energy of the mixture. In an ideal gas mixture, the chemical potential is expressed as

$$\mu_i = \mu_i^o + R_g T \ln \frac{n_i}{n} = \mu_i^o + R_g T \ln \frac{p_i}{p^o} \tag{6.4}$$

where μ_i^o is the chemical potential of component i at standard conditions ($p^o = 1$ bar), which is independent of the composition, and p_i is the partial pressure of component i. The first equality in Eq. (6.4) is true for any ideal thermodynamic system, moreover it may be used as the definition of an ideal system (Gorban et al., 2006). Examples of other ideal systems are an ideal solution, an ideal adsorbed layer (substances adsorbed on the surface of a solid phase), and an ideal plasma.

The chemical potential must satisfy the following conditions (Yablonskii et al., 1991):

- the condition of symmetry:

$$\frac{\partial \mu_i}{\partial n_j} = \frac{\partial \mu_j}{\partial n_i} \tag{6.5}$$

- the condition of positive values: $\mu_i > 0$
- the quadratic form,

$$\sum_{i,j=1}^{N_c} \sum x_i \frac{\partial \mu_i}{\partial n_j} x_j \tag{6.6}$$

in which x_i and x_j are arbitrary variables related to components i and j, is nonnegatively determined in the real coordinate space $\mathbb{R}^n$
- the quadratic form, Eq. (6.6), is positively determined in any hyperplane with a positive normal $\mathbf{v}$ (all $\mathbf{v} > 0$)

The first and second conditions ensure the existence of the thermodynamic Gibbs free energy function or, using the mathematical term, the convex Lyapunov function for kinetic equations. The Lyapunov function is a strictly positive function with a nonpositive derivative. The one exception to this definition is that at the equilibrium point, the Lyapunov function equals zero. In physicochemical sciences, the Gibbs free energy is an extremely important Lyapunov function for understanding the stability of equilibria.

An alternative representation of the chemical potential for a component of an ideal gas mixture is

$$\mu_i(\mathbf{c}, T) = R_g T \ln \frac{c_i}{c^o} + u_i - C_{V,i} T \ln T + R_g T \delta_i \tag{6.7}$$

where $\mathbf{c}$ is a vector of concentrations, c^o is the standard concentration (1 kmol m^{-3}), u_i is the energy of the normal state of component i, $C_{V,i}$ is the molar heat capacity of component i at constant volume, and δ_i is an arbitrary constant. Eq. (6.7) makes it possible to obtain an explicit form for the Gibbs free energy. For example, for a thermally isolated system at constant volume and constant internal energy U

$$G = \sum_{i=1}^{N_c} \mu_i n_i = \sum_{i=1}^{N_c} n_i R_g T \left(\ln \frac{n_i}{V} + \delta_i - 1 \right) - C_V T (\ln T + 1) \tag{6.8}$$

where the total heat capacity of the mixture is given by

$$C_V = \sum_{i=1}^{N_c} C_{V,i} n_i \tag{6.9}$$

and

$$C_V T = U - \sum_{i=1}^{N_c} u_i n_i \tag{6.10}$$

The nonideal chemical potential can be expressed as

$$\mu_i = \mu_i^o + R_g T \ln a_i \tag{6.11}$$

where a_i is the activity of component i, which is the product of the concentration and an activity coefficient:

$$a_i = \gamma_i \frac{c_i}{c^o} \tag{6.12}$$

where the division by the standard concentration is necessary to ensure that both the activity and the activity coefficient are dimensionless. Generally, activity coefficients γ_i depend on temperature, pressure, and composition.

In terms of the Marcelin-de Donder kinetics, the reaction rate is expressed as a function of chemical potentials (Feinberg, 1972):

$$r(\mathbf{c}, T) = r^{\circ}(\mathbf{c}, T) \left[\exp \left(\sum_i \frac{\alpha_i \mu_i}{R_g T} \right) - \exp \left(\sum_i \frac{\beta_i \mu_i}{R_g T} \right) \right] \tag{6.13}$$

where $r^{\circ}(\mathbf{c}, T)$ is a positive function.

In the case of an ideal gas, Eq. (6.4), the Marcelin-de Donder equations represent traditional mass-action law kinetics with the accompanying power-law terms (polynomials). In the case of a nonideal gas, Eq. (6.11), Marcelin-de Donder kinetics generates more complex models, in which every term can be mixed, including power-law factors and exponential factors.

6.1.3 Steady State and Equilibrium

Many physical and chemical processes occurring in chemical reactors determine the temporal evolution of the reaction mixture. In an open chemical system, the *steady state* is a state in which the composition and other characteristics (temperature, pressure, volume, etc.) are considered to be constant in time. In contrast, in a *nonsteady-state* regime, the system characteristics change during the time of observation.

Besides these two regimes, another regime, with a temporally periodic change of the chemical composition (chemical oscillation or self-oscillation), may also be observed. A famous example of this phenomenon is the Belousov-Zhabotinsky reaction. Another example of complex kinetic behavior in open chemical systems is the occurrence of *multiple steady states* due to the fact that for some components of the reaction mixture the rate of consumption and rate of production can be balanced at more than one point. This type of behavior has become the subject of detailed theoretical and computational analyses (Marin and Yablonsky, 2011; Yablonskii et al., 1991). Despite the fact that there are many experimental data concerning such complex behavior, the steady-state regime with characteristics that are constant in time still is the most observed phenomenon.

Chemical equilibrium is a particular case of the steady state, that is, the steady state for a closed system. For a reversible reaction consisting of a single elementary step, chemical equilibrium means that the rate of the forward reaction is the same as the rate of the reverse reaction. Defining the equilibrium of a complex chemical system is a little more complicated. In that case, many different elementary steps occur in the chemical reactor, with every step consisting of a forward and a reverse reaction. For the whole system to be at equilibrium, every mechanistic step must be at equilibrium.

6.1.4 Reversibility and Irreversibility

Reversibility and irreversibility are central concepts of classical thermodynamics. Rigorously speaking, in physics and chemistry all processes are reversible. However, for many chemical processes the probability of the reverse process is negligible. For instance, processes such as the combustion of coal or hydrocarbons and enzyme reactions are in fact irreversible. On the other hand, even if the overall chemical reaction is irreversible, it always comprises at least one reversible step. For example

- the irreversible oxidation of carbon monoxide on platinum-containing catalysts includes the reversible adsorption of carbon monoxide.
- in accordance with the Michaelis-Menten mechanism, every enzyme reaction includes the reversible adsorption of the enzyme, $E + S \rightleftarrows ES$.
- every gas-phase chain reaction contains a reversible step in which radicals are generated, for example, in the oxidation of hydrogen: $H_2 \rightleftarrows 2H\cdot$.

Yablonsky et al. (2011b) discuss different meanings of the term "reversible." First of all, all processes are governed by the second law of thermodynamics in accordance with which the entropy of an isolated system may increase or stay the same, but never decreases. From this point of view, all processes involving entropy growth are not reversible. Second, processes should satisfy the principle of detailed balance and the Onsager relations, based on the requirement of microreversibility (see Section 6.1.5). From this point of view, all processes are reversible, or better, "time-reversible." In the last sense, one very much related to the common sense of chemists and chemical engineers, reversibility is the existence of inverse processes. If the transition $A \rightarrow B$ exists, then the transition $B \rightarrow A$ also exists. This condition is significantly weaker than that of microreversibility.

Time-reversibility of irreversible processes sounds paradoxical and requires some explanation (Yablonsky et al., 2011b). The most direct interpretation of time-reversing is to go back in time. This means taking a solution of the dynamic equations $x(t)$ and checking whether $x(-t)$ is also a solution. For microscopic dynamics (the Newton or Schrödinger equations), we expect this to be the case. Nonequilibrium statistical physics combines this idea with the description of macroscopic or mesoscopic kinetics by an ensemble of elementary processes (reactions). The microscopic reversing of time at this level turns into "reversing of arrows": the reaction $\sum \alpha_i A_i \rightarrow \sum \beta_i B_i$ transforms into $\sum_i \beta_i B_i \rightarrow \sum_i \alpha_i A_i$ and vice versa. The equilibrium ensemble should be invariant to this transformation. This immediately leads to the concept of *detailed balance*: each process is balanced by its reverse process (see Section 6.1.5). A "time-reversible kinetic process" is an irreversible process with underlying time-reversible microdynamics.

6.1.5 Equilibrium and Principle of Detailed Balance

The principle of detailed balance can be formulated in a simple way as follows: at equilibrium conditions the rate of every forward reaction should be balanced by the rate of the corresponding reverse reaction. It has to be stressed that this is not about the total reaction rate of every chemical component in the system. The principle of detailed balance is stronger than the statement, "for every component under equilibrium conditions the rate of its consumption must be balanced by the rate of its production." This principle was introduced in 1931 by Onsager (1931a,b), but in fact was known long before Onsager's results, as it was introduced for the Boltzmann equation in 1872 (Boltzmann, 1964). In 1901, Wegscheider (1901) was the first to touch upon the problem of detailed balance when analyzing a particular cyclic mechanism ($A_1 \rightleftarrows A_2$; $A_3 \rightleftarrows A_4$; $A_2 + A_3 \rightleftarrows A_1 + A_4$). Wegscheider also showed that irreversible cycles of the type $A_1 \rightarrow A_2 \rightarrow \cdots \rightarrow A_n \rightarrow A_1$ are impossible. He found explicit relations between the kinetic coefficients that are generated by the principle of detailed balance. For example, for the cyclic mechanism $A_1 \underset{k_1^-}{\overset{k_1^+}{\rightleftarrows}} A_2 \underset{k_2^-}{\overset{k_2^+}{\rightleftarrows}} A_3 \underset{k_3^-}{\overset{k_3^+}{\rightleftarrows}} A_1$ the following equations are valid in accordance with this principle:

$$k_1^+ c_{A_1} = k_1^- c_{A_2}; \; k_2^+ c_{A_2} = k_2^- c_{A_3}; \; k_3^+ c_{A_2} = k_3^- c_{A_1} \tag{6.14}$$

Multiplying these equations, one obtains

$$k_1^+ k_2^+ k_3^+ = k_1^- k_2^- k_3^- \quad \text{or} \quad K_{eq,1} K_{eq,2} K_{eq,3} = 1 \tag{6.15}$$

Einstein used this principle for the linear kinetics of emission and absorption of radiation (Einstein, 1916).

Onsager (1931a,b) formulated his famous reciprocal relations, for which he received the Nobel Prize in chemistry in 1968, connecting them to the principle of detailed balance. He made an explicit reference to Wegscheider's analysis: "chemists apply a very interesting approach." The connection between the principle of detailed balance and Onsager's reciprocal relations has been clarified by van Kampen (1973). These relations were also extended to various types of coordinate transformations, possibly including time derivatives and integration in time (Stöckel, 1983). Recently, Astumian (2008) derived reciprocal relations for nonlinear coupled transport processes between pairs of reservoirs that are energetically coupled at mesoscopic contact points. Now, an elegant geometric framework has been elaborated for Onsager's relations and their generalizations (Grmela, 2002).

Onsager's relations are widely used for extracting kinetic information about reciprocal processes from experiments and for the validation of this information, see, for example, Ozer and Provaznik (2005); one can measure how process A affects process B and from this extract the reciprocal information of how process B affects process A. These relations have been tested experimentally for many physical systems, but not for chemical ones. A remarkable review by Miller (1960a,b) on the experimental verification of Onsager's relations is still often cited. An analysis of many phenomena (thermoelectricity, isothermal diffusion, etc.) showed the validity of the corresponding reciprocal relations, but with respect to chemical reactions, Miller's point was, "The experimental studies of this phenomenon ... have been inconclusive and the question is still open from an experimental point."

Currently, the principle of detailed balance is explained as a macroscopic form of the principle of microscopic reversibility. As mentioned in Section 6.1.4, the microscopic reversing of time at the kinetic level turns into the reversing of arrows in the chemical reaction equation, that is, elementary chemical processes transform into their reverse processes.

Tolman (1938) and Boyd (1974) have described a derivation of the principle of detailed balance from the principle of microscopic reversibility. In the presence of an external magnetic field, it is possible that the equilibrium is not a detailed balance. De Groot and Mazur (1962) have formulated a modification of the principle of detailed balance for this case.

6.2 Chemical Equilibrium and Optimum

6.2.1 Introduction

The understanding of chemical equilibrium and optimum problems is extremely important from an academic and practical point of view, particularly in reactor design and control. However, the relationship between these two problems is not well understood. Historically, equilibrium-optimum considerations have been proclaimed in the famous Le Chatelier's principle. In chemistry, this principle is used to influence reversible chemical reactions. For example, the equilibrium conversion of an exothermic reaction, that is, a reaction liberating heat, is more favorable at lower temperature, so cooling of the reaction mixture shifts the equilibrium to the product side. Le Chatelier's principle is part of the curriculum of university students in chemistry and chemical engineering. Unfortunately, the relation between this principle and the analysis of equilibria and optima often is not presented clearly. In particular, there is no explicit explanation of how to apply Le Chatelier's principle, which has been formulated for closed systems at equilibrium (so at zero value of the net overall reaction rate), to continuous-flow reactors, in which the reaction rate certainly is not zero. This section is based on an article by Yablonsky and Ray (2008), which aims at bridging this gap between the concepts of equilibrium and optimum.

6.2.2 Le Chatelier's Principle

A closed system is at chemical equilibrium when the rates of the forward and reverse reactions are equal, so that the concentrations of reactants and products do not change over time. The system will remain in this equilibrium state, unless something is done to disturb it. Changes that may disturb the system are called stresses. Le Chatelier's principle states that when a system at equilibrium is subjected to a stress, then the equilibrium will shift in such a way as to relieve that stress.

Using Le Chatelier's principle, we are able to predict the result of applying a stress to a closed system at equilibrium and thus to manipulate a reversible reaction. The four most common ways to apply a stress are

- adding a reactant or subtracting a product.
- changing the volume of the system.
- changing the pressure of the system.
- changing the temperature of the system.

The first three stresses do not affect the value of the equilibrium coefficient, but changing the temperature does. The equilibrium coefficient, K_{eq}, is constant at a given temperature, independent of the original concentrations, reactor volume, or pressure. A very small K_{eq} means that a reaction is not likely to happen, while a very large K_{eq} means that the reaction can occur easily.

Responding to any of these four changes, the equilibrium of the system will shift. For example, if for any reaction $A + B \rightleftarrows C + D$ an additional amount of reactant A or B is added, the rate of the forward reaction increases and the equilibrium shifts to the product side, C and D. Then, as the added reactant is being consumed, the rate of the forward reaction slows. When the rates of the forward and reverse reactions are equal again, the system has returned to a different, equilibrium state with different concentrations of reactants and products. This is why for some reactions, an excess of one reactant, usually the cheapest one, is used to drive the reaction to the product side.

Changing the temperature of a system at equilibrium causes the system to attempt to relieve this stress by shifting to the side of the reactants or products. The direction of this shift depends on whether the forward reaction is exothermic or endothermic. Although heat is not actually a reactant or product of the chemical reaction, it is convenient to treat it as such in the representation of a reaction. An endothermic reaction is a reaction that consumes heat and can be represented as

$$A + B + heat \rightleftarrows C + D \tag{6.16}$$

The equilibrium coefficient for this reaction is defined as

$$K_{eq} = \frac{c_C c_D}{c_A c_B} = \exp\left(-\frac{\Delta_r G^o}{R_g T}\right) \tag{6.17}$$

where $\Delta_r G^o$ is the standard Gibbs free energy change for this reaction.

Once heat is placed into the reaction equation, the direction of the equilibrium shift upon a temperature change can be predicted. If the temperature of the system increases, the equilibrium coefficient also increases. As a result, the system shifts to the product side, while dissipating the heat, and the concentrations of C and D increase while those of A and B decrease. For an exothermic reaction, by which heat is released,

$$A + B \rightleftarrows C + D + heat \tag{6.18}$$

the effect of a temperature increase is the opposite, that is, in this case the equilibrium shifts to the side of the reactants upon increasing the temperature. Therefore, in practice, conversion of such reactions is limited at high temperature because of the low equilibrium coefficient, while at low temperature it is limited by the low reaction rate coefficient, which presents a challenge in reactor design.

Table 6.1 summarizes a number of effects of changes on a chemical equilibrium.

Table 6.1 Le Chatelier's principle and effects of changes on a gas-phase system at equilibrium for a reaction as written

Change	Equilibrium Shifts To
	Any reaction:
Addition of reactant/product	Product side/reactant side
Subtraction of reactant/product	Reactant side/product side
Pressure/volume	*Different number of moles on both sides of the reaction equation:*
Increase/decrease	Side of fewer moles
Decrease/increase	Side of more moles
	Equal number of moles on both sides of the reaction equation:
Increase or decrease	No effect
Temperature	*Exothermic/endothermic reaction*
Increase	Reactant side/product side
Decrease	Product side/reactant side

6.2.3 Relation Between Equilibrium and Optimum Regime

6.2.3.1 Equilibrium conversion

Consider the first-order reversible reaction $A \rightleftarrows B$. In a batch reactor at equilibrium conditions

$$r = 0 = k^+ c_{A,eq} - k^- c_{B,eq} = k^+ c_{A0}(1 - x_{A,eq}) - k^- c_{A0} x_{A,eq} \tag{6.19}$$

where $x_{A,eq}$ is the equilibrium conversion of A.

The kinetic coefficients k^+ and k^- are governed by the corresponding Arrhenius dependences:

$$k^+ = k_0^+ \exp\left(-\frac{E_a^+}{R_g T}\right) \tag{6.20}$$

$$k^- = k_0^- \exp\left(-\frac{E_a^-}{R_g T}\right) \tag{6.21}$$

where k_0^+ and k_0^- are the preexponential factors of the forward and reverse reactions. For an exothermic reaction, $E_a^+ < E_a^-$ whereas for an endothermic reaction $E_a^+ > E_a^-$.

Rearrangement of Eq. (6.19) yields

$$1 - x_{A,eq} = \frac{k^-}{k^+} x_{A,eq} = \frac{x_{A,eq}}{K_{eq}} \tag{6.22}$$

Eq. (6.20) can also be written as

$$x_{A,eq} = \frac{K_{eq}}{1 + K_{eq}} \tag{6.23}$$

or

$$\frac{1}{x_{A,eq}} = 1 + \frac{1}{K_{eq}} \tag{6.24}$$

The effect of a temperature increase on the equilibrium coefficient and the equilibrium conversion are shown symbolically for an exothermic and an endothermic reaction in Table 6.2 and Fig. 6.1, which is a geometric representation of Le Chatelier's principle.

Table 6.2 Effect of temperature increase on equilibrium coefficient and conversion

Reaction	K_{eq}	$1/K_{eq}$	$x_{A,eq}$
Exothermic	↓	↑	↓
Endothermic	↑	↓	↑

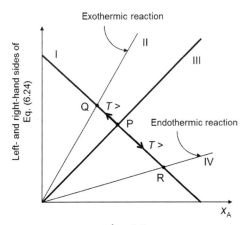

Fig. 6.1

Geometric representation of Le Chatelier's principle for exothermic and endothermic reactions at equilibrium; I: left-hand side of Eq. (6.24) II, III, IV: right-hand side of Eq. (6.24).

In Fig. 6.1, the two bold lines represent the left-hand side and right-hand side of Eq. (6.24). The intersection of these two lines (point P) represents the equilibrium conversion of A. For an exothermic reaction, the slope $(1/K_{eq})$ increases with increasing temperature and thus shifts upward, resulting in a lower value of the equilibrium conversion, denoted by point Q. Similarly, for an endothermic reaction, the equilibrium conversion increases, as shown by point R.

The dependence of the equilibrium conversion on the temperature can be characterized by the derivative of Eq. (6.21). As

$$K_{eq} = K_{eq,0} \exp\left(-\frac{\Delta_r H}{R_g T}\right) \tag{6.25}$$

where $K_{eq,0}$ is a preexponential factor and $\Delta_r H$ is the enthalpy change of reaction with $\Delta_r H = E_a^+ - E_a^-$, it follows that

$$\frac{dx_{A,eq}}{dT} = \frac{K_{eq}\left(\dfrac{\Delta_r H}{R_g T^2}\right)(1 + K_{eq}) - K_{eq}^2\left(\dfrac{\Delta_r H}{R_g T^2}\right)}{(1 + K_{eq})^2} = \frac{K_{eq}}{(1 + K_{eq})^2}\left(\frac{\Delta_r H}{R_g T^2}\right) \tag{6.26}$$

Clearly, the derivative $\dfrac{dx_{A,eq}}{dT}$ has the same sign as $\Delta_r H$. For an exothermic reaction, $\Delta_r H < 0$, so $\dfrac{dx_{A,eq}}{dT} < 0$. Therefore, $x_{A,eq}$ decreases with increasing temperature. For an endothermic reaction, $\Delta_r H > 0$ and $\dfrac{dx_{A,eq}}{dT} > 0$, so the temperature-dependence of $x_{A,eq}$ is the opposite.

6.2.3.2 Optimum conversion in a continuous plug-flow reactor

Now let us analyze the same reaction, $A \rightleftarrows B$, in a continuous plug-flow reactor at optimum conditions, which are certainly not equilibrium conditions. The overall reaction rate is given by

$$r = k^+ c_{A0}(1 - x_A) - k^- c_{A0} x_A \neq 0 \tag{6.27}$$

At the optimum point

$$\frac{dr}{dT} = k^+ c_{A0}(1 - x_{A,opt})\left(\frac{E_a^+}{R_g T^2}\right) - k^- c_{A0} x_{A,opt}\left(\frac{E_a^-}{R_g T^2}\right) = 0 \tag{6.28}$$

so,

$$1 - x_{A,opt} = \frac{1}{\alpha K_{eq}} x_{A,opt} \tag{6.29}$$

or

$$x_{A,opt} = \frac{\alpha K_{eq}}{1 + \alpha K_{eq}} \tag{6.30}$$

with $\alpha = \dfrac{E_a^+}{E_a^-}$.

Using simple algebraic manipulations, it can be shown that

$$\frac{d^2r}{dT^2} = k^+ c_{A0} \left(\frac{E_a^+}{R_g^2 T^4} \right) \frac{1}{1 + \alpha K_{eq}} \left(E_a^+ - E_a^- \right) = A \Delta_r H \tag{6.31}$$

Coefficient A is obviously positive and, therefore, the sign of the second-order derivative only depends on the sign of the enthalpy change of reaction. For an exothermic reaction $\Delta_r H < 0$, so the extremum is the maximum of the reaction rate $\left(\dfrac{d^2r}{dT^2} < 0 \right)$, whereas for an endothermic reaction $\Delta_r H > 0$ and the extremum is the minimum of the reaction rate $\left(\dfrac{d^2r}{dT^2} > 0 \right)$.

So, the conclusions for exothermic and endothermic reactions are different and thus the strategies have to be different. For an exothermic reaction, K_{eq} decreases with temperature. As $E_a^+ < E_a^-$, $\alpha < 1$, so $\alpha K_{eq} < K_{eq}$. As a result, the optimum conversion will be less than the equilibrium conversion (see Eq. 6.30). This is shown graphically in Fig. 6.2.

The decreasing bold line represents $1 - x_{A,eq}$ or $1 - x_{A,opt}$, the left-hand sides of Eqs. (6.22), (6.29), while the increasing bold line represents the right-hand side of Eq. (6.22). The intersection of these two lines corresponds to the value of the equilibrium conversion. The solid line represents the right-hand side of Eq. (6.29). For an exothermic reaction ($\alpha < 1$), this line has a steeper slope ($1/K_{eq}$), resulting in a lower value of $x_{A,opt}$ (point Q) than the equilibrium value (point P).

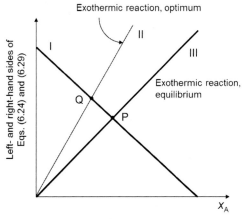

Fig. 6.2

Geometric representation of the equilibrium and optimum conversion for an exothermic reaction; I: left-hand side of Eqs. (6.24), (6.29) II: right-hand side of Eq. (6.24); III: right-hand side of Eq. (6.29).

For an endothermic reaction, the extremum is a minimum, which is not interesting from a technological point of view. Therefore, in this case, the temperature policy is to use the highest temperature possible taking into account typical limitations concerning material constraints, catalyst stability, safety, and so on.

6.2.3.3 Relationships between equilibrium and optimum conversions

Combining Eqs. (6.23), (6.30) leads to

$$\frac{x_{A,eq}}{x_{A,opt}} = \frac{\dfrac{K_{eq}}{1+K_{eq}}}{\dfrac{\alpha K_{eq}}{1+\alpha K_{eq}}} = \frac{1+\alpha K_{eq}}{\alpha(1+K_{eq})} = \frac{1}{\alpha}\left(\frac{1}{1+K_{eq}}\right) + \frac{K_{eq}}{1+K_{eq}} \tag{6.32}$$

Using Eq. (6.23) again, Eq. (6.32) can be represented as

$$\frac{X_{A,eq}}{X_{A,opt}} = \frac{1}{\alpha}\left(1 - X_{A,eq}\right) + X_{A,eq} = \frac{1}{\alpha} + \left(1 - \frac{1}{\alpha}\right)X_{A,eq} \tag{6.33}$$

Based on Eqs. (6.32), (6.33), some important conclusions can be drawn:

- Since for an exothermic reaction $\alpha < 1$, $x_{A,eq}$ is always larger than $x_{A,opt}$.
- The higher the conversion, the closer $x_{A,opt}$ is to $x_{A,eq}$; for $x_{A,eq} \approx 1$, $x_{A,opt} \approx x_{A,eq}$. At very low equilibrium conversion, $x_{A,eq} \ll 1$, Eq. (6.33) reduces to

$$\frac{x_{A,eq}}{x_{A,opt}} \approx \frac{1}{\alpha} \tag{6.34}$$

- If $E_a^+ = E_a^-$, so $\Delta_r H = 0$ and $\alpha = 1$, $x_{A,opt} = x_{A,eq}$.
- At a certain temperature, the one at which $\Delta_r G = 0$ and $K_{eq} = 1$, Eq. (6.32) reduces to

$$\frac{x_{A,eq}}{x_{A,opt}} = \frac{1+\alpha}{2\alpha} \tag{6.35}$$

All of these conclusions can help the chemical engineer with preliminary decisions on optimum technological regimes. In the optimization of an actual industrial reactor, one has to take into account many other factors, in particular, the exchange of matter and heat with the surroundings.

The effect of temperature on the equilibrium and optimum conversions of an exothermic reaction is shown qualitatively in Fig. 6.3.

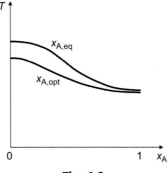

Fig. 6.3
Effect of temperature on the equilibrium and optimum conversions for an exothermic reaction.

6.2.3.4 Continuous stirred-tank reactor

The more complex case of the continuous stirred-tank reactor (CSTR) can be analyzed in a similar way as that of the plug-flow reactor (Section 6.2.3.2). Again considering the reversible exothermic reaction $A \rightleftarrows B$, the mole balance equation for component A at steady-state conditions is

$$q_{V0}c_{A0} - q_{V0}c_A - Vk^+c_A + Vk^-(c_{A0} - c_A) = 0 \tag{6.36}$$

where q_{V0} is the volumetric flow rate entering the reactor (which in this case is equal to that leaving the reactor), V is the reaction volume and $c_B = c_{A0} - c_A$. Introducing $c_{A,ss} = c_{A0}(1 - x_{A,ss})$, with $c_{A,ss}$ and $x_{A,ss}$ the steady-state concentration and conversion, and the residence time, $\tau = V/q_{V0}$, Eq. (6.36) can be rewritten as

$$c_{A0}x_{A,ss} - k^+\tau c_{A0}(1 - x_{A,ss}) + k^-\tau c_{A0}x_{A,ss} = 0 \tag{6.37}$$

from which it follows that

$$x_{A,ss} = \frac{k^+\tau}{1 + k^+\tau + k^-\tau} \tag{6.38}$$

or

$$x_{A,ss} = \frac{K_{eq}}{1 + K_{eq} + \dfrac{1}{k^-\tau}} \tag{6.39}$$

If $\tau \to \infty$, the steady-state conversion reaches the equilibrium conversion (Eq. 6.23), which is the maximum attainable conversion.

Using Eq. (6.38), with Eqs. (6.20), (6.21), it is possible to determine the conversion that relates to the optimum temperature policy:

$$\frac{dx_{A,ss}}{dT} = \frac{k^+ E_a^+ \tau (1 + k^+\tau + k^-\tau) - k^+\tau (k^+ E_a^+\tau + k^- E_a^-\tau)}{R_g T^2 (1 + k^+\tau + k^-\tau)^2}$$

$$= \frac{k^+ E_a^+\tau + k^+ k^- E_a^+\tau^2 - k^+ k^- E_a^-\tau^2}{R_g T^2 (1 + k^+\tau + k^-\tau)^2} = 0 \tag{6.40}$$

so

$$k^-\tau = \frac{E_a^+}{E_a^- - E_a^+} = \frac{\alpha}{1-\alpha} \tag{6.41}$$

with $\alpha = E_a^+/E_a^-$. Substitution of Eq. (6.41) into Eq. (6.39) yields

$$x_{A,ss,opt} = \frac{K_{eq}}{1 + K_{eq} + \dfrac{1-\alpha}{\alpha}} = \frac{\alpha K_{eq}}{1 + \alpha K_{eq}} \tag{6.42}$$

In fact, this is the same equation as that for the optimum conversion that was found using a simplified analysis (Eq. 6.30). Comparison of the two values of conversion, $x_{A,ss}$ (Eq. 6.39) and $x_{A,ss,opt}$ (Eq. 6.42 or 6.30), may yield different results, depending on the process parameters.

In order to determine the optimum temperature policy at a given residence time τ, we need to use Eq. (6.41), (6.42) together. For known value of α, the value of k^- can be determined from Eq. (6.41). Knowing the Arrhenius dependence of k^- (Eq. 6.21), the corresponding temperature can be calculated. K_{eq} at this temperature then follows from an equation like Eq. (6.17). Finally, $x_{A,ss,opt}$ follows from Eq. (6.42). Eq. (6.42) can be used directly to determine the optimum steady-state conversion if the temperature, and thus K_{eq}, is known. The residence time required for this conversion can then be calculated from Eq. (6.41).

6.2.4 Summary

The analysis presented here can be used for many reversible reactions, see Tables 6.3 and 6.4.

A number of industrial reactions of these types with corresponding equilibrium data can also be found in Fogler (2005) and Schmidt (1998).

Table 6.3 Elementary reactions

Molecularity of Forward Reaction	Elementary Step
Unimolecular	$A \rightleftarrows$ Products
Bimolecular	$2A \rightleftarrows$ Products
	$A + B \rightleftarrows$ Products
Termolecular	$3A \rightleftarrows$ Products
	$2A + B \rightleftarrows$ Products
	$A + B + C \rightleftarrows$ Products
Autocatalytic	$A + B \rightleftarrows 2B$
	$A + 2B \rightleftarrows 2B + D$

Table 6.4 Examples of processes with single overall reaction

Reaction type	Example
$A \rightleftarrows B$	$c\text{-}C_4H_8 \rightleftarrows t\text{-}C_4H_8$
$2A \rightleftarrows B$	$2NO_2 \rightleftarrows N_2O_4$
$3A \rightleftarrows B$	$3C_2H_2 \rightleftarrows C_6H_6$
$A + B \rightleftarrows C$	$NO_2 + NO \rightleftarrows N_2O_3$
$A + B \rightleftarrows C + D$	$CO + H_2O \rightleftarrows CO_2 + H_2$
$2A + B \rightleftarrows C$	$2NO + O_2 \rightleftarrows N_2O_4$

6.3 Is It Possible to Overshoot an Equilibrium?

6.3.1 Introduction

This section is based on original results obtained by Gorban (1979, 1981, 2013), Gorban et al. (2006), and Yablonskii et al. (1991). Gorban (1984) gave a detailed explanation of the theory of thermodynamic constraints and the effect on overshooting an equilibrium. His first paper on domains of attainability and thermodynamically admissible paths appeared in 1979 and a physicochemical interpretation of this phenomenon was published in 1982 (Gorban, 2013). Kaganovich et al. (2005, 2006, 2007, 2010) applied these ideas to the analysis of energy problems. Shinnar (1988), Shinnar and Feng (1985), and Vuddagiri et al. (2000) also analyzed the influence of thermodynamic constraints on the dynamic behavior of complex chemical systems. Shinnar and Feng (1985) explained a theory on thermodynamic constraints using simple examples and demonstrated how to apply this theory in the design of reactors (Shinnar, 1988). Vuddagiri et al. (2000) also attempted to apply thermodynamic constraints to simple chemical systems.

6.3.2 Mass-Action Law and Overshooting an Equilibrium

This section presents basic knowledge for understanding the problem of overshooting an equilibrium in a complex chemical system in which many reactions occur. An elementary reaction step in such a system can be described by the general equation

$$\Sigma \alpha_i \cdot A_i \underset{k^-}{\overset{k^+}{\rightleftharpoons}} \Sigma \beta_i \cdot B_i \tag{6.43}$$

where A_i and B_i are reactants and products with α_i and β_i the absolute values of their stoichiometric coefficients, which represent the number of molecules participating in the reaction. In accordance with the mass-action law, the rates of the forward and reverse reactions are expressed as

$$r^+ = k^+(T)\Pi c_{A_i}^{\alpha_i} \tag{6.44}$$

$$r^- = k^-(T)\Pi c_{B_i}^{\beta_i} \tag{6.45}$$

where c_{Ai} and c_{Bi} are the concentrations of reactants and products. The kinetic coefficients $k^+(T)$ and $k^-(T)$ are governed by Arrhenius dependences. The maximum number of different kinetic coefficients is $2N_s$, where N_s is the number of elementary steps, that is, pairs of elementary forward and corresponding reverse reactions. As described in Section 6.1.5, not all kinetic coefficients are independent. The dependences are governed by the principle of detailed balance: for any T there is a set of equilibrium concentrations c_{eq} such that for any elementary step

$$r_{eq}^+\left(c_{Ai,eq}, T\right) = r_{eq}^-\left(c_{Bi,eq}, T\right) \tag{6.46}$$

Understanding the possibility of overshooting a chemical equilibrium is based on the properties of the Gibbs free energy of a chemical system, which can be represented as

$$G = R_g T \sum_{i=1}^{N_c} c_i \left(\ln \frac{c_i}{c_{i,eq}} - 1 \right) \tag{6.47}$$

with G expressed in $J\,m^{-3}$.

During the course of a reaction at constant volume and temperature, in line with the second law of thermodynamics, the Gibbs free energy decreases monotonously.

According to the mass-action law, a complex chemical transformation can be described by a set of ordinary differential equations,

$$\frac{dc_i}{dt} = \sum_{s=1}^{N_s} (\beta_s - \alpha_s) r_s(c, T) \tag{6.48}$$

It has been shown that for such a mass-action-law system, the Gibbs free energy changes in time as follows:

$$\frac{dG}{dt} = -R_g T \sum_{s=1}^{N_s} \left(r_s^+ - r_s^-\right) \ln \frac{r_s^+}{r_s^-} \leq 0 \tag{6.49}$$

The inequality is obvious: for any positive a and b, $(a - b)(\ln a - \ln b) \geq 0$ because the natural logarithm is an increasing function. From Eqs. (6.48), (6.49), it follows that the Gibbs free energy is a Lyapunov-like function, see Gorban (1984) and Yablonskii et al. (1991).

6.3.3 Example of Equilibrium Overshoot: Isomerization

The analysis in Section 6.2 demonstrated that both the steady-state conversion and the optimum conversion in open chemical reaction systems (continuous-flow reactors) are always lower than the equilibrium conversion in closed chemical systems (batch reactors).

Now the question arises whether it is possible to overshoot the equilibrium during a nonsteady-state chemical reaction, and if so, then how far? Stated more generally, what is prohibited in the course of a chemical process?

Consider a batch reactor initially containing only component A. Then, conversion of A to B starts and the formed B is converted back to A. The temperature and volume of the reaction mixture are assumed to be constant. Let us assume that A and B are isomers and that their equilibrium concentrations are equal, that is, at equilibrium, there is 50% A and 50% B. One may wonder whether it is possible for the amount of B to become larger than 50% of the total on the way to this equilibrium. The obvious answer is, "No!" If concentrations of A and B become equal, their values do not change any more. At this point, the rate of the forward reaction is balanced by the rate of the reverse reaction and the concentrations will remain constant in time. This is an example of a one-dimensional system with two components (A and B) and one mass balance ($c_A + c_B =$ constant). One coordinate, say c_A, completely describes the state of the system at fixed temperature and volume. It is impossible to overshoot the equilibrium for a one-dimensional system, in which the concentration trajectories are just lines.

A rigorous mathematical statement can be formulated as follows: If a reaction mixture consists of only two isomers, A_1 and A_2, the system moves toward the equilibrium, never overshooting it. The concentrations $c_{A_1}(t)$ and $c_{A_2}(t)$ approach the equilibrium concentrations $c_{A_1,eq}$ and $c_{A_2,eq}$ monotonically without ever exceeding these concentrations. The type of dynamic trajectory is not arbitrary, but is governed by thermodynamic functions of the chemical composition, such as the Gibbs free energy function G, which decrease monotonously in time. Such a limitation on dynamic trajectories is eliminated if there are more than two chemical components in the system.

Imagine that an additional component, C, takes part in the reaction. In this case, the answer to the question of whether it is possible to overshoot the equilibrium is: "Yes!" However, then another question arises: "How much can this equilibrium be overshot?" Let us assume that A, B, and C are isomers again and that at equilibrium their amounts are equal (each $\sim33\%$ of the total amount). Also assume that all transformations are possible (so A can be converted to B and to C, B to A and C, and C to A and B). For this system overshooting of the equilibrium is possible, provided that there are no limitations on the detailed mechanism of chemical transformations. Then, if the reactor initially contains 100% A, the concentration of B may overshoot its equilibrium concentration, but may not exceed a certain boundary ($\sim77\%$ of the total amount).

Let us assume that there are n isomers, $A_1, \dots, A_n$, in our system, which is closed and is assumed to be ideal. At these conditions, G decreases in time (Eq. 6.47): $G(\mathbf{c}(t_1)) > G(\mathbf{c}(t_2))$ for $t_1 > t_2$, where $\mathbf{c}(t)$ is a vector of concentrations at time t. Obviously, for $t > 0$ $G(\mathbf{c}(t_0)) > G(\mathbf{c}(t))$, but this is not the only requirement on the possible values of the concentrations.

In the one-dimensional case, there are two isomers and one mass-conservation law $(c_{A_1} + c_{A_1} = \text{constant})$. The state space is divided into two parts by the equilibrium concentration c_{eq}, and $c_i(t_0)$ and $c_i(t)$ with $t > t_0$ are always on the same side of the equilibrium. If $c_{A_1}(t_0) > c_{A_1,eq}$, then $c_{A_2}(t_0) < c_{A_2,eq}$ and the nonsteady-state concentrations meet the following inequalities: $c_{A_1}(t) > c_{A_1,eq}$ and $c_{A_2}(t) < c_{A_2,eq}$, while if $c_{A_1}(t_0) < c_{A_1,eq}$, $c_{A_2}(t_0) > c_{A_2,eq}$, $c_{A_1}(t) < c_{A_1,eq}$, and $c_{A_2}(t) > c_{A_2,eq}$. If the number of different isomers in the system is larger than two, we need to analyze the thermodynamically admissible paths of the system in order to find the constraints. These paths are continuous curves $c_{A_i}(t)$ with $t \geq 0$ for which

- $c_{A_i}(t) \geq 0$ for all $t \geq 0$.

- the mass-conservation constraint is: $\displaystyle\sum_{i=1}^{N_c} c_{A_i}(t) = c_t = \text{constant}$.

- $G(c_{A_i}(t))$ is a monotonically decreasing function.

At the given conditions, a transition from point $\mathbf{c}^{(0)}$ to point $\mathbf{c}^{(1)}$ is thermodynamically admissible if there is a thermodynamically admissible reaction path for which $\mathbf{c}(t_0) = \mathbf{c}^{(0)}$ and $\mathbf{c}(t) = \mathbf{c}^{(1)}$ at some time $t > 0$. Otherwise the transition is thermodynamically prohibited. For studying the thermodynamically prohibited paths, it is useful to analyze the level sets of the function G given by equations $G(\mathbf{c}) = g = \text{constant}$. At some critical value of g, g_0, these level sets become disconnected.

We will illustrate this graphically for a two-dimensional system with three isomers, simplifying the problem as much as possible. Let us assume that all three equilibrium concentrations are equal: $c_{A_1,eq} = c_{A_2,eq} = c_{A_3,eq} = c_{eq}$. Clearly, $c_{A_1} + c_{A_2} + c_{A_3} = c_t = 3c_{eq}$. At a given value of the equilibrium concentration, the state of this system is described by a point located in an equilateral triangle, see Fig. 6.4A, for which $c_{A_1} + c_{A_2} + c_{A_3} = c_t$ and $c_{A_i} \geq 0$.

The height of the triangle is c_t. It is easy to show that the length of every side of the triangle equals $\dfrac{2}{\sqrt{3}} c_t$. Every vertex (A_1, A_2, A_3) represents a state in which only one isomer is present,

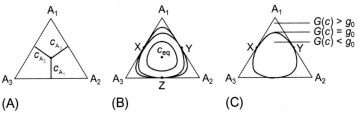

Fig. 6.4

Chemical system with three isomers: (A) coordinates in balance triangle (simplex); (B) levels of the Gibbs free energy G; (C) domain of inaccessibility in the vicinity of A_1; point X, Y, and Z denote two-component mixtures with equal concentrations ($c_t/2$) of both components;

X: $c_{A_1} = c_{A_3}$; Y: $c_{A_1} = c_{A_2}$; Z: $c_{A_2} = c_{A_3}$.

while every side represents a two-component state. The concentration of an isomer A_i is the length of the line perpendicular to the side opposite its vertex A_i, measured from a point inside the triangle. Fig. 6.4B shows the level sets of the function $G(\mathbf{c}) = g$, Eq. (6.47). Conditional minimums of $G(\mathbf{c})$, g_0, are achieved on the sides of the triangle. Since $c_{A_1,eq} = c_{A_2,eq} = c_{A_3,eq} = c_{eq} = c_t/3$, these minimums are located on the centers of the sides and are equal for all sides:

$$g_0 = R_g T c_t \left[\ln \left(\frac{3}{2} \right) - 1 \right] \tag{6.50}$$

If g is close to $G(c_{eq})$, $G(c_{eq}) < g \leq g_0$, the level set $G(\mathbf{c}) = g$ is connected. If g becomes larger than the minimum of $G(\mathbf{c})$ on the sides of the triangle, g_0, but remains less than the maximum $G(\mathbf{c})$, g_{max}, which is reached at the vertices, the corresponding level set $G(\mathbf{c}) = g$ consists of three connected parts (Fig. 6.4B). Arcs of the level sets of $G(\mathbf{c})$ that connect the centers of the sides generate impassable thermodynamic boundaries that cannot be crossed from the inside, that is, a thermodynamically admissible path cannot go from values $G(\mathbf{c}) \leq g_0$ to values $G(\mathbf{c}) > g_0$. This impossibility of crossing the thermodynamic boundary in a two-dimensional system with two independent chemical components is very similar to the impossibility of overshooting the equilibrium of a one-dimensional system with one independent chemical component.

The balance triangle can be divided into four domains (Fig. 6.4C): three curvilinear triangles A_1XY, A_2YZ, and A_3XZ, and the domain $G(\mathbf{c}) < g_0$. One side of every curvilinear triangle is part of the curve $G(\mathbf{c}) = g_0$. Any inner point of such a triangle cannot be connected with a similar point of another curvilinear triangle by a thermodynamically permissible path. This is even impossible in case $G(\mathbf{c}^{(0)}) > G(\mathbf{c}^{(1)})$, because on the way from $\mathbf{c}^{(0)}$ to $\mathbf{c}^{(1)}$, the trajectory inevitably intersects one of the arcs of the curve $G(\mathbf{c}) = g_0$. The permissibility of trajectories inside the triangles and in the domain $G(\mathbf{c}) \leq g_0$ is determined by the inequality $G(\mathbf{c}^{(0)}) > G(\mathbf{c}^{(1)})$. Assuming that one of the following conditions is fulfilled:

- both $\mathbf{c}^{(0)}$ and $\mathbf{c}^{(1)}$ are located in the same curvilinear triangle A_1XY, A_2YZ, or A_3XZ,
- $\mathbf{c}^{(0)}$ is located in one of the curvilinear triangles and $\mathbf{c}^{(1)}$ belongs to the domain $G(\mathbf{c}) < g_0$,
- $G(\mathbf{c}^{(0)}) \leq g_0$ and $G(\mathbf{c}^{(1)}) < g_0$,

a transition from $\mathbf{c}^{(0)}$ to $\mathbf{c}^{(1)}$ is thermodynamically permissible if and only if $G(\mathbf{c}^{(0)}) > G(\mathbf{c}^{(1)})$. Transitions from one curvilinear triangle to another are prohibited. Thus, it can be concluded that all permissible trajectories are covered by these three conditions.

As mentioned earlier, near every vertex, there is a domain, namely the adjacent curvilinear triangle with $G(\mathbf{c}) > g_0$, that is inaccessible from the inside, where $G(\mathbf{c}) > g_0$, see Fig. 6.4C. An interesting question then arises: how closely can this vertex, say A_1, be approached if the initial state is given and the initial concentrations are $c_{A_1}^{(0)}, c_{A_2}^{(0)}$, and $c_{A_3}^{(0)}$? In order to solve this problem, we have to find sup $\mathbf{c}^{(1)}$, the upper value of all states $\mathbf{c}^{(1)}$, where $\mathbf{c}^{(1)}$, with concentrations $c_{A_1}^{(1)}, c_{A_2}^{(1)}, c_{A_3}^{(1)}$, is a state that can be accessed from $\mathbf{c}^{(0)}$. From the geometrical point of view (Fig. 6.4B), it is clear that sup $\mathbf{c}^{(1)}$ has to be determined as follows:

- If $\mathbf{c}^{(0)}$ is located in the triangle A_1XY or in the domain $G(\mathbf{c}^{(0)}) \leq g_0$, sup $c_{A_1}^{(1)}$ is a maximum c_{A_1} on the level set $G(\mathbf{c}) = G(\mathbf{c}^{(0)})$.
- If $\mathbf{c}^{(0)}$ is located in A_2YZ or A_3XZ, sup $c_{A_1}^{(1)}$ is a maximum c_{A_1} on the level set $G(\mathbf{c}) = g_0$.

In order to determine sup $\mathbf{c}^{(1)}$ at a given value of $\mathbf{c}^{(0)}$, we have to find the indicated maximums and apply analytical criteria to determine to which domain point $\mathbf{c}^{(0)}$ belongs. These criteria are that

- $\mathbf{c}^{(0)}$ is located in the domain $G(\mathbf{c}) \leq g_0$ if and only if $G\left(\mathbf{c}^{(0)}\right) \leq g_0$.
- $\mathbf{c}^{(0)}$ is located in the triangle A_1XY if and only if $c_{A_1}^{(0)} > \dfrac{c_t}{2}, G\left(\mathbf{c}^{(0)}\right) > g_0$.
- $\mathbf{c}^{(0)}$ is located in the triangle A_2YZ if and only if $c_{A_2}^{(0)} > \dfrac{c_t}{2}, G\left(\mathbf{c}^{(0)}\right) > g_0$.
- $\mathbf{c}^{(0)}$ is located in the triangle A_3XZ if and only if $c_{A_3}^{(0)} > \dfrac{c_t}{2}, G\left(\mathbf{c}^{(0)}\right) > g_0$.

Formally, the criterion for determining in which part of the triangle $A_1A_2A_3$ a point is located, is presented as follows: the curvilinear triangle A_1XY includes points for which $G(\mathbf{c}) > g_0$. From the point of symmetry, it is obvious that max $\mathbf{c}$, on the level set of $G(\mathbf{c}) = g$, is achieved on the line $c_{A_2} = c_{A_3} = (c_t - c_{A_1})/2$. Points of intersection of this line with the level set $G(\mathbf{c}) = g$ are determined based on the equation

$$c_{A_1} \ln\left(\frac{c_{A_1}}{c_{eq}}\right) + (c_t - c_{A_1}) \ln\left(\frac{c_t - c_{A_1}}{2c_{eq}}\right) - c_t = \frac{g}{R_g T} \tag{6.51}$$

The dependence $\dfrac{c_{A_1}}{c_{eq}}$ on $\dfrac{g}{c_{eq}}$ based on Eq. (6.51) is presented in Fig. 6.5A.

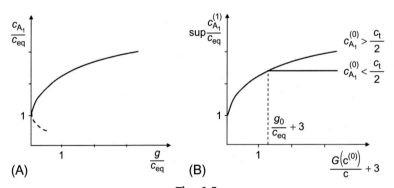

Fig. 6.5

(A) Solutions of Eq. (6.51); (B) two branches of the dependence of sup $c_{A_1}^{(1)}$ on $G(\mathbf{c}^{(0)})$.

Using Eq. (6.51) it is possible to find the dependence of $\sup\dfrac{c_{A_1}^{(1)}}{c_{eq}}$ on $\dfrac{G(\mathbf{c}^{(0)})}{\mathbf{c}} + 3$, see Fig. 6.5B.

For $G\left(\mathbf{c}^{(0)}\right) \leq g_0$, this dependence is determined uniquely. If $G\left(\mathbf{c}^{(0)}\right) > g_0$, two scenarios are possible:

- If $c_{A_1}^{(0)} < \dfrac{c_t}{2}$, that is, $\mathbf{c}^{(0)}$ is located in A_2YZ or A_3XZ, $\sup c_{A_1}^{(1)}$ is equal to the maximum of c_{A_1} on the line $G(\mathbf{c}) = g_0$, see the lower branch of the dependence in Fig. 6.5B.

- If $c_{A_1}^{(0)} > \dfrac{c_t}{2}$, the dependence of $\sup c_{A_1}^{(1)}$ on $G(\mathbf{c}^0)$ is determined using Eq. (6.51), see the upper branch of the dependence in Fig. 6.5B.

In a one-dimensional chemical system with two isomers and equilibrium concentrations $c_{A_1,eq} = c_{A_2,eq}$ (so the equilibrium coefficient equals one), the impossibility of overshooting the equilibrium can be formulated as follows: if $c_{A_1}^{(0)} \leq \dfrac{c_t}{2}$, $c_{A_1}^{(1)} < \dfrac{c_t}{2}$. For the two-dimensional chemical system with three isomers, which we analyzed in detail, a similar statement can be made, although the numerical values are different: if $c_{A_1}^{(0)} \leq \dfrac{c_t}{2}$, $c_{A_1}^{(1)} \leq 0.773 c_t$. The value $0.773\,c_t$ is an approximate solution of Eq. (6.51) at $g = g_0$.

6.3.4 Requirements Related to the Mechanism of a Complex Chemical Reaction

Thermodynamic limitations on kinetic paths are determined only by the list of reactants and products, the equilibrium composition, and the function G, see Eq. (6.47). These limitations are valid even if the assumed reaction steps have no physicochemical meaning, for example, in the reaction $11A_1 + 12A_2 \rightleftarrows 23A_3$. Reducing the list of possible elementary steps to a rational one, the number of possible paths may be dramatically reduced in comparison with the number of thermodynamically admissible paths.

Let us start our analysis of a chemical system with three isomers by considering three possible mechanisms (Fig. 6.6):

(1) $A_1 \rightleftarrows A_2$, $A_2 \rightleftarrows A_3$ (Fig. 6.6A)
(2) $A_1 \rightleftarrows A_2$, $A_1 \rightleftarrows A_3$ (Fig. 6.6B)
(3) $A_1 \rightleftarrows A_2$, $A_2 \rightleftarrows A_3$, $A_3 \rightleftarrows A_1$ (Fig. 6.6C)

Mechanism 3 is a combination of mechanisms 1 and 2. As before, we assume the equilibrium concentrations to be equal: $c_{A_1,eq} = c_{A_2,eq} = c_{A_3,eq} = c_{eq} = c_t/3$. Each step is characterized by a specific geometrical surface along which the forward reaction rate is balanced by the reverse reaction rate. In our example, this surface is reduced to a line. Equilibrium lines divide the triangle $A_1A_2A_3$ into sections in which the direction of each reaction—forward or reverse—is determined uniquely (see Fig. 6.6). Step $A_1 \rightleftarrows A_2$ shifts along the line $A_3 = $ constant toward the equilibrium line $A_1 \rightleftarrows A_2$. The other steps shift toward their own equilibria in a similar way. For a reaction mechanism consisting of only one step, the system would move along the equilibrium line for that step. For multistep reactions, a possible direction of movement is a linear combination with positive coefficients of leading vectors of the corresponding steps. A set of

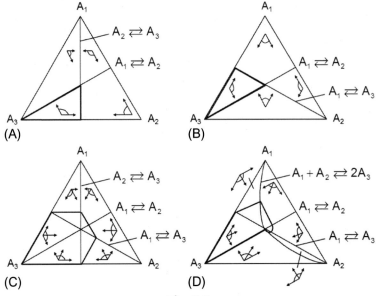

Fig. 6.6

Compositions accessible from A_3. Boundaries of domains M are represented by fat lines and equilibrium lines are represented by thin lines. The angles of possible directions are also shown. Mechanisms of the reactions are (A) $A_1 \rightleftarrows A_2$, $A_2 \rightleftarrows A_3$; (B) $A_1 \rightleftarrows A_2$, $A_1 \rightleftarrows A_3$; (C) $A_1 \rightleftarrows A_2$, $A_2 \rightleftarrows A_3$, $A_3 \rightleftarrows A_1$; (D) $A_1 \rightleftarrows A_2$, $A_1 \rightleftarrows A_3$, $A_1 + A_2 \rightleftarrows 2A_3$.

such combinations inside each section forms an angle of possible directions of movement, as shown in Fig. 6.6 for different mechanisms of complex reactions.

A smooth curve $c(t)$ is a thermodynamically admissible path of the reaction if

- $c_{A_i}(t) \geq 0$ for any $t \geq 0$, $i = 1,2,3$,
- $c_{A_1}(t) + c_{A_2}(t) + c_{A_3}(t) = c_t = \text{constant}$,
- a tangent vector $\left(\dfrac{dc_{A_1}}{dt}, \dfrac{dc_{A_2}}{dt}, \dfrac{dc_{A_3}}{dt} \right)$ belongs to the angle of possible directions that

corresponds to the point $(c_{A_1}, c_{A_2}, c_{A_3})$.

The function $G(\mathbf{c}(t))$ decreases monotonically along these paths.

In the course of a reaction, the transition from composition $\mathbf{c}^{(0)}$ to composition $\mathbf{c}^{(1)}$ is allowed if and only if an admissible path exists for which $\mathbf{c}(t_0) = \mathbf{c}^{(0)}$ and $\mathbf{c}(t) = \mathbf{c}^{(1)}$ for some $t > 0$. Generally, for a given initial state $\mathbf{c}^{(0)}$, a set of compositions $\mathbf{c}^{(1)}$ can be constructed for which the transition from $\mathbf{c}^{(0)}$ to $\mathbf{c}^{(1)}$ is admissible.

Fig. 6.6A–C shows sets of admissible states M starting from initial state $c_{A_1}^{(0)} = c_{A_2}^{(0)} = 0$, $c_{A_3}^{(0)} = c_t$ for different reaction mechanisms. For the two-step mechanisms $A_1 \rightleftarrows A_2, A_2 \rightleftarrows A_3$ and $A_1 \rightleftarrows A_2, A_1 \rightleftarrows A_3$ (Fig. 6.6A and B), the set M coincides with M_0, while for the three-step mechanism $A_1 \rightleftarrows A_2$, $A_2 \rightleftarrows A_3$, $A_3 \rightleftarrows A_1$ (Fig. 6.6C) the set M is considerably larger. For the two-step mechanism $A_1 \rightleftarrows A_2$, $A_2 \rightleftarrows A_3$ the value of $c_{A_1}^{(1)}$ does not exceed $c_{eq} = c_t/3$ (Fig. 6.6A). For mechanisms $A_1 \rightleftarrows A_2$, $A_1 \rightleftarrows A_3$ (Fig. 6.6B) and $A_1 \rightleftarrows A_2$, $A_2 \rightleftarrows A_3$, $A_3 \rightleftarrows A_1$ (Fig. 6.6C) the value of $c_{A_1}^{(1)}$ can become larger than c_{eq}, but does not exceed $c_t/2$, which is a situation similar to the one-dimensional case.

It can be demonstrated how it is possible to overshoot the equilibrium for the system with three isomers by modifying the reaction mechanism so as to eliminate the prohibition of overshooting the equilibrium. This can be done by adding the nonlinear step $A_1 + A_2 \rightleftarrows 2A_3$ to the two-step mechanism $A_1 \rightleftarrows A_2$, $A_1 \rightleftarrows A_3$. Then, the equilibrium curve is given by the equation $c_{A_1,eq} c_{A_2,eq} = c_{A_3,eq}^2$. Because of the stoichiometry of the step $A_1 + A_2 \rightleftarrows 2A_3$, the composition moves along a line parallel to the bisector of the angle A_3, see Fig. 6.6D. The leading vector has components $(1, 1, -2)$. The maximum value of $c_{A_1}^{(1)}$ for the initial conditions $c_{A_1}^{(0)} = c_{A_2}^{(0)} = 0$ and $c_{A_3}^{(0)} = c_t$ is determined by the admissible path that consists of two rectilinear parts:

(1) movement from $\mathbf{c}^{(0)}$ to the point $c_{A_1} = c_{A_3} = c_t/2$, $c_{A_2} = 0$ along the line $c_{A_2} = 0$
(2) movement from the point $c_{A_1} = c_{A_3} = c_t/2$, $c_{A_2} = 0$ along the line parallel to the bisector of angle A_3 to the equilibrium line of the step $A_1 + A_2 \rightleftarrows 2A_3$

This path can be interpreted as follows: steps $A_1 \rightleftarrows A_3$ and $A_1 + A_2 \rightleftarrows 2A_3$ become equilibrated. At the final point, $c_{A_1} = c_t \left(\dfrac{11 - \sqrt{13}}{12} \right) = 0.616 c_t$ and this value is larger than that at the equilibrium point, $c_{eq} = c_t/3$. Therefore, overshooting the equilibrium is possible for a mechanism including a nonlinear step, in this case $A_1 + A_2 \rightleftarrows 2A_3$, in which interaction between different components takes place.

In conclusion, for every specific mechanism of a complex chemical reaction, the sets of admissible paths and the sets of admissible states are smaller than the corresponding sets of thermodynamically admissible paths and states. According to the second law of thermodynamics, a perpetual motion machine is impossible. However, overshooting of the chemical equilibrium during the course of a reaction may be possible for some reaction mechanisms starting from certain initial conditions. The extent of this overshoot can be estimated based on the Gibbs free energy levels and known values of mole balances and can be used for distinguishing mechanisms. Kaganovich et al. (2006) have used this approach for industrial applications in energetic processes. For more detailed information on the theory and practice of overshooting an equilibrium, see Gorban (1984, 2013) and Gorban et al. (2006).

6.4 Equilibrium Relationships for Nonequilibrium Chemical Dependences

6.4.1 Introduction

In presenting the foundations of physical chemistry, the basic difference between equilibrium chemical thermodynamics and chemical kinetics is always stressed. A typical problem in equilibrium thermodynamics is calculating the composition of a chemical mixture reacting in a closed system for an infinitely long time. It does not consider time. In contrast, chemical kinetics is the science of the evolution of a chemical mixture with time. Some results of theoretical chemical kinetics are related to features of nonsteady-state kinetic behavior. These results have been obtained from thermodynamic principles, especially from the principle of detailed balance:

- uniqueness and stability of the equilibrium in any closed system, see Zel'dovich (1938) and the analysis by Yablonskii et al. (1991)
- absence of damped oscillations near the point of detailed balance (Wei and Prater, 1962; Yablonskii et al., 1991)
- limitations on the kinetic relaxation from the initial conditions. For example, based on a known set of equilibrium coefficients that determine the equilibrium composition, one can find a forbidden domain of compositions that is impossible to reach from the initial conditions (Gorban, 1984; Gorban et al., 1982, 2006).

The present dogma of physical chemistry holds that it is impossible to find an expression for any nonsteady-state chemical system based on its description at equilibrium conditions, except for some relations describing the behavior in the linear vicinity of equilibrium, see for example, Boudart (1968). Recently, however, Yablonsky et al. (2011a, 2011b) and Constales et al. (2011, 2012, 2015) have obtained new equilibrium-type relations for certain nonsteady-state chemical systems. These results, which we will discuss in this and subsequent sections, have been achieved for linear reaction systems, with a general proof, and for some nonlinear ones. Based on these results, equilibrium thermodynamic relationships can be considered not only as a description of the final point of the temporal evolution, but also as inherent characteristics of the dynamic picture. In this section, we present an analysis of a number of linear examples, using models of chemical kinetics based on the mass-action law. Processes described by these models occur in closed nonsteady-state chemical systems with perfect mixing (ideal batch reactors) and in open steady-state chemical systems without radial gradients (ideal plug-flow reactors (PFRs)). The model description of the ideal PFR is identical to that of the batch reactor, but with the astronomic time t replaced by the residence time τ.

6.4.2 Dual Experiments

In this and subsequent sections, a type of thought experiment is described that leads to a new type of chemical time invariance. For the single reversible reaction $A \rightleftarrows B$, the value of this invariance, $B_A(t)/A_B(t)$, is the equilibrium coefficient. It is constructed based on data of nonequilibrium experiments performed starting from different extreme initial conditions of the reacting mixture, called "dual experiments." In the case of the reaction $A \rightleftarrows B$, this involves two separate experiments, one taking place in a reactor primed with substance A only, and one in a reactor primed with substance B only. In both cases, the concentrations of A and B are monitored and particular attention is paid to the dependencies of "B produced from A" and "A produced from B." The following notation is used: $A_A(t)$ for the temporal normalized concentration (c_A/c_t) dependence of A given the initial condition $(A, B) = (1, 0)$, that is, initially only A is present, with normalized concentration 1. Similarly, $B_A(t)$ is the concentration of B for the same initial condition and $A_B(t)$ and $B_B(t)$ are the normalized concentrations of A and B for the initial condition $(A, B) = (0, 1)$. Some relatively simple linear examples are presented here.

Example 6.1

For a single first-order reversible reaction,

$$A \underset{k^-}{\overset{k^+}{\rightleftharpoons}} B, \tag{6.52}$$

the kinetic model is given by

$$\frac{dc_A}{dt} = -\frac{dc_B}{dt} = -k^+ c_A + k^- c_B \tag{6.53}$$

$$c_A + c_B = c_t \Leftrightarrow \frac{c_A}{c_t} + \frac{c_B}{c_t} = A + B = 1 \tag{6.54}$$

Then for $t \geq 0$, we can write:

$$A_A(t) = \frac{k^- + k^+ \exp\left(-(k^+ + k^-)t\right)}{k^+ + k^-} \tag{6.55}$$

$$B_A(t) = \frac{k^+ \left(1 - \exp\left(-(k^+ + k^-)t\right)\right)}{k^+ + k^-} \tag{6.56}$$

$$A_B(t) = \frac{k^- \left(1 - \exp\left(-(k^+ + k^-)t\right)\right)}{k^+ + k^-} \tag{6.57}$$

$$B_B(t) = \frac{k^+ + k^- \exp\left(-(k^+ + k^-)t\right)}{k^+ + k^-} \tag{6.58}$$

Comparing $B_A(t)$ and $A_B(t)$, it is clear that their ratio is constant:

$$\frac{B_A(t)}{A_B(t)} = \frac{k^+}{k^-} = K_{eq} \tag{6.59}$$

Remarkably, this ratio holds at any moment in time $t \geq 0$ and not merely in the limit $t \to \infty$. Other simple relationships for this reaction are

$$A_A(t) + B_B(t) = 1 + \exp\left(-(k^+ + k^-)t\right) \tag{6.60}$$

and

$$\frac{A_A(t) - A_{eq}}{B_B(t) - B_{eq}} = K_{eq} \tag{6.61}$$

where the normalized equilibrium concentrations of A and B are given by

$$A_{eq} = \frac{k^-}{k^+ + k^-} = \frac{1}{1 + K_{eq}} \tag{6.62}$$

and

$$B_{eq} = \frac{k^+}{k^+ + k^-} = \frac{K_{eq}}{1 + K_{eq}} \tag{6.63}$$

Example 6.2

In the case of two consecutive first-order reactions, the first being reversible and the second irreversible,

$$A \underset{k_1^-}{\overset{k_1^+}{\rightleftharpoons}} B \xrightarrow{k_2^+} C \tag{6.64}$$

using the Laplace domain techniques, we can define the following analytical expressions for the roots of the characteristic equation:

$$\lambda_{p,m} = \frac{k_1^+ + k_1^- + k_2^+ \pm \sqrt{(k_1^+ + k_1^- + k_2^+)^2 - 4k_1^+ k_2^+}}{2} \tag{6.65}$$

where subscripts p and m denote the roots with respectively the plus and minus sign, and verify that (see Appendix) $\lambda_p > k_2^+ > \lambda_m > 0$ and $\lambda_p > k_1^+ > \lambda_m > 0$. We then can write

$$A_A(t) = \frac{\lambda_p \left(k_2^+ - \lambda_m\right) \exp\left(-\lambda_m t\right) + \lambda_m \left(\lambda_p - k_2^+\right) \exp\left(-\lambda_p t\right)}{k_2^+ \left(\lambda_p - \lambda_m\right)} \tag{6.66}$$

$$B_A(t) = \frac{k_1^+ \left(\exp\left(-\lambda_m t\right) - \exp\left(-\lambda_p t\right)\right)}{\lambda_p - \lambda_m} \tag{6.67}$$

$$C_A(t) = 1 - \frac{\lambda_p \left(\exp\left(-\lambda_m t\right) - \lambda_m \exp\left(-\lambda_p t\right)\right)}{\lambda_p - \lambda_m} \tag{6.68}$$

$$A_B(t) = \frac{k_1^- \left(\exp\left(-\lambda_m t\right) - \exp\left(-\lambda_p t\right)\right)}{\lambda_p - \lambda_m} \tag{6.69}$$

$$B_B(t) = \frac{\lambda_m \left(\lambda_p - k_2^+\right) \exp\left(-\lambda_m t\right) + \lambda_p \left(k_2^+ - \lambda_m\right) \exp\left(-\lambda_p t\right)}{k_2^+ \left(\lambda_p - \lambda_m\right)} \tag{6.70}$$

$$C_B(t) = 1 - \frac{\left(\lambda_p - k_2^+\right)\left(\exp\left(-\lambda_m t\right) + \left(k_2^+ - \lambda_m\right) \exp\left(-\lambda_p t\right)\right)}{\lambda_p - \lambda_m} \tag{6.71}$$

The ratio $B_A(t)/A_B(t)$ again equals the equilibrium coefficient, $K_{eq,1} = k_1^+/k_1^-$. Also, it is easy show that

$$A_A(t) + B_B(t) = \exp\left(-\lambda_p t\right) + \exp\left(-\lambda_m t\right) \tag{6.72}$$

Example 6.3

For the cycle of three reversible first-order reactions,

$$A \underset{k_1^-}{\overset{k_1^+}{\rightleftharpoons}} B \underset{k_2^-}{\overset{k_2^+}{\rightleftharpoons}} C \underset{k_3^-}{\overset{k_3^+}{\rightleftharpoons}} A, \tag{6.73}$$

in the Laplace domain we can define the symbols

$$\sigma_1 = k_1^+ + k_1^- + k_2^+ + k_2^- + k_3^+ + k_3^- \tag{6.74}$$

$$\sigma_2 = k_1^+ k_2^+ + k_2^+ k_3^+ + k_3^+ k_1^+ + k_1^+ k_2^- + k_2^+ k_3^- + k_3^+ k_1^- + k_1^- k_3^- + k_2^- k_1^- + k_3^- k_2^- \tag{6.75}$$

and

$$\Delta(s) = s\left(s^2 + \sigma_1 s + \sigma_2\right) \tag{6.76}$$

Then the transformed normalized concentrations $A_A(t)$, $B_A(t)$ and $A_B(t)$ are given by

$$\mathcal{L}\{A_A\}(s) = \frac{s^2 + \left(k_1^- + k_3^+ + k_2^+ + k_2^-\right) + s\left(k_1^- k_3^+ + k_1^- k_2^- + k_2^+ k_3^+\right)}{\Delta(s)} \tag{6.77}$$

$$\mathcal{L}\{B_A\}(s) = \frac{k_1^+ s + \left(k_1^+ k_3^+ + k_1^+ k_2^- + k_2^- k_3^-\right)}{\Delta(s)} \tag{6.78}$$

$$\mathcal{L}\{A_B\}(s) = \frac{k_1^- s + \left(k_1^- k_3^+ + k_1^- k_2^- + k_2^+ k_3^+\right)}{\Delta(s)} \tag{6.79}$$

and the ratio of the latter two is

$$\frac{\mathcal{L}\{B_A\}(s)}{\mathcal{L}\{A_B\}(s)} = \frac{k_1^+ s + \left(k_1^+ k_3^+ + k_1^+ k_2^- + k_2^- k_3^-\right)}{k_1^- s + \left(k_1^- k_3^+ + k_1^- k_2^- + k_2^+ k_3^+\right)}$$

$$= \frac{k_1^+}{k_1^-}\left(1 - \frac{1}{1 + k_1^- \dfrac{k_1^+ k_3^+ + k_1^+ k_2^- + k_2^- k_3^- + s k_1^+}{k_1^+ k_2^+ k_3^+ - k_1^- k_2^- k_3^-}}\right) = \frac{k_1^+}{k_1^-} = K_{eq,1} \tag{6.80}$$

where the Onsager relationship, $k_1^+ k_2^+ k_3^+ = k_1^- k_2^- k_3^-$, was used in the final step. Thus, for all s, the ratio of $\mathcal{L}\{B_A\}(s)$ to $\mathcal{L}\{A_B\}(s)$ is fixed and equal to $K_{eq,1}$. Since the inverse Laplace transform is linear, the same ratio holds in the time domain:

$$\frac{B_A(t)}{A_B(t)} = \frac{k_1^+}{k_1^-} = K_{eq,1} \tag{6.81}$$

Similarly,

$$\frac{C_B(t)}{B_C(t)} = \frac{k_2^+}{k_2^-} = K_{eq,2} \tag{6.82}$$

and

$$\frac{A_C(t)}{C_A(t)} = \frac{k_3^+}{k_3^-} = K_{eq,3} \tag{6.83}$$

It is also possible to show that

$$A_A(t) + B_B(t) + C_C(t) = 1 + \exp\left(-\lambda_p t\right) + \exp\left(-\lambda_m t\right) \tag{6.84}$$

As an example, Fig. 6.7 shows the time dependence of the concentration ratios $B_A(t)/A_A(t)$ and $B_B(t)/A_B(t)$ and the time-invariant ratio $B_A(t)/A_B(t)$ for the isomerization of butenes analyzed by Wei and Prater (1962), which can be represented by Eq. (6.73) where A = *cis*-2-butene, B = 1-butene, and C = *trans*-2-butene.

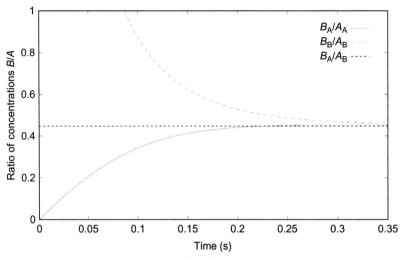

Fig. 6.7

Time dependence of B_A/A_A and B_B/A_B and the time-invariant B_A/A_B for the isomerization of butenes analyzed by Wei and Prater (1962, Eq. 129); A = cis-2-butene, B = 1-butene; the relative values of the rate coefficients are $k_1^+ = 4.623$; $k_2^+ = 3.724$; $k_3^+ = 3.371$; $k_1^- = 10.344$; $k_2^- = 1.000$; $k_3^- = 5.616\,s^{-1}$. Reprinted from Yablonsky, G.S., Constales, D., Marin, G.B., 2011a. Equilibrium relationships for nonequilibrium chemical dependences. Chem. Eng. Sci. 66, 111–114. Copyright (2011), with permission from Elsevier.

This behavior is typical of all examples given here: if the initial normalized concentrations of the system are $(A, B, C) = (1, 0, 0)$, the ratio B_A/A_A first increases from zero, eventually reaching the equilibrium value. Note that for these parameters B_A/A_A slightly overshoots the limit value. Similarly, when starting from $(A, B, C) = (0, 1, 0)$, the corresponding ratio B_B/A_B decreases from $+\infty$ and eventually reaches the same limit value, namely the equilibrium ratio. Surprisingly though, the ratio B_A/A_B is equal to the equilibrium value for all times $t \geq 0$.

Example 6.4

For the cycle of four reversible first-order reactions,

$$A \underset{k_1^-}{\overset{k_1^+}{\rightleftarrows}} B \underset{k_2^-}{\overset{k_2^+}{\rightleftarrows}} C \underset{k_3^-}{\overset{k_3^+}{\rightleftarrows}} D \underset{k_4^-}{\overset{k_4^+}{\rightleftarrows}} A, \qquad (6.85)$$

the expressions become more complex, but verifying that in this case the fixed equilibrium ratios also hold is straightforward given the Onsager relation, $k_1^+ k_2^+ k_3^+ k_4^+ = k_1^- k_2^- k_3^- k_4^-$.

Again Eq. (6.81) holds and similar equations for C_B/B_C, D_C/C_D, and A_D/D_A. Furthermore,

$$\frac{C_A(t)}{A_C(t)} = \frac{k_1^+ k_2^+}{k_1^- k_2^-} = K_{eq,1} K_{eq,2} \qquad (6.86)$$

and

$$\frac{D_B(t)}{B_D(t)} = K_{eq,2}K_{eq,3} \qquad (6.87)$$

In addition, similar to Eq. (6.83), we now have

$$A_A(t) + B_B(t) + C_C(t) + D_D(t) = 1 + \exp(-\lambda_1 t) + \exp(-\lambda_2 t) + \exp(-\lambda_3 t) \qquad (6.88)$$

6.5 Generalization. Symmetry Relations and Principle of Detailed Balance

6.5.1 Introduction

The equilibrium relationships for nonequilibrium chemical dependencies described in Section 6.4 are directly related to Onsager's famous reciprocal relations. In fact, such relations already had been introduced in the 19th century by Thomson (now Lord Kelvin) and Helmholtz, but Onsager (1931a,b) developed the background and generalizations to these reciprocal relations. In his influential papers, Onsager also mentioned the close connection between these relations and the detailed balancing of elementary processes: at equilibrium, each elementary process should be balanced by its reverse process.

According to Onsager's work, the fluxes near chemical equilibrium are linear functions of potentials and the reciprocal relations state that the matrix of coefficients of these functions is symmetric. It is impossible to measure these coefficients directly. In order to find them, the inverse problem of chemical kinetics, which is often ill posed, needs to be solved. Sometimes it is possible though to find them directly in what we call dual experiments.

Here we will demonstrate how general reciprocal relations between measurable quantities can be formulated. These relations between kinetic curves use the symmetry of the so-called propagator in the entropic scalar product. A dual experiment is defined for each ideal kinetic experiment. For this dual experiment, the initial data and the observables are different (their positions are exchanged), but the result of the measurement is essentially the same function of time.

6.5.2 Symmetry Between Observables and Initial Data

In the original form of Onsager's relations, the vector of thermodynamic fluxes $\mathbf{J}$ and the vector of thermodynamic forces $\mathbf{F}$ are connected by a matrix $\mathbf{L}$ that is symmetric in the standard scalar product:

$$\mathbf{J} = \mathbf{LF}; \quad L_{ij} = L_{ji} \qquad (6.89)$$

The vector **F** is the gradient of the corresponding thermodynamic potential:

$$F_i = \frac{\partial \Phi}{\partial x_i} \tag{6.90}$$

where F_i is the ith thermodynamic force and x_i is any variable with respect to component i. For isolated systems, Φ is the entropy.

For finite-dimensional systems, such as in chemical kinetics, the dynamics satisfy linear kinetic equations or kinetic equations linearized near the equilibrium:

$$\frac{d\mathbf{c}}{dt} = \mathbf{kc} \tag{6.91}$$

where **c** is a column vector containing the concentrations of N_c reacting chemical components and **k** is a matrix of kinetic coefficients with

$$k_{ij} = \sum_l L_{il} \frac{\partial^2 \Phi}{\partial c_i c_j}\bigg|_{c_{eq}} \tag{6.92}$$

that is,

$$\mathbf{k} = \mathbf{L}\left(\mathbf{D}^2 \Phi\right)_{c_{eq}} \tag{6.93}$$

This matrix is not symmetric but the product $\left(\mathbf{D}^2 \Phi\right)_{c_{eq}} \mathbf{k} = \left(\mathbf{D}^2 \Phi\right)_{c_{eq}} \mathbf{L}\left(\mathbf{D}^2 \Phi\right)_{c_{eq}}$ is symmetric and thus **k** is symmetric in the entropic scalar product

$$\langle \mathbf{a}|\mathbf{kb}\rangle_\Phi \equiv \langle \mathbf{ka}|\mathbf{b}\rangle_\Phi \tag{6.94}$$

where

$$\langle \mathbf{a}|\mathbf{b}\rangle_\Phi = -\sum_{i,j} a_{ij} \frac{\partial^2 \Phi}{\partial c_i c_j}\bigg|_{c_{eq}} b_j \tag{6.95}$$

Generalizations to spatially distributed systems in which transport processes occur, inhomogeneous equilibria and nonisotropic and non-Euclidian spaces can also be made (see Yablonsky et al. 2011b).

Real functions of symmetric operators are also symmetric. In particular, the propagator $\exp(\mathbf{k}t)$ is symmetric. Therefore, we can formulate reciprocal relations between kinetic curves as symmetry relations between observables and initial data (the observables-initial data symmetry). Because these relations do not include fluxes and time derivatives, they are more robust.

As a result of the function $\exp(\mathbf{k}t)$ being symmetric, Onsager's relations, Eq. (6.94), directly imply that

$$\langle \mathbf{a} | \exp(\mathbf{k}t)\mathbf{b} \rangle_\Phi \equiv \langle \exp(\mathbf{k}t)\mathbf{a} | \mathbf{b} \rangle_\Phi \tag{6.96}$$

The expression

$$\mathbf{c}(t) = \exp(\mathbf{k}t)\mathbf{b} \tag{6.97}$$

is a solution to the kinetic equations (6.91) with initial conditions $\mathbf{c}(0) = \mathbf{b}$. The left-hand side of Eq. (6.96) represents the result of an experiment in which we prepare an initial state $\mathbf{c}(0) = \mathbf{b}$, start the process from this state, and measure $\langle \mathbf{a} | \mathbf{c}(t) \rangle$, the scalar product of vector $\mathbf{a}$ on a state $\mathbf{c}$. On the right-hand side of Eq. (6.96), the positions and roles of $\mathbf{b}$ and $\mathbf{a}$ have been exchanged: we start from the initial condition $\mathbf{c}(0) = \mathbf{a}$ and measure $\langle \mathbf{b} | \mathbf{c}(t) \rangle$. The result is the same function of time. This exchange of the initial state and the observed state transforms an ideal experiment into another ideal experiment. The left- and the right-hand sides of Eq. (6.96) represent different experimental situations with identical results.

We consider a general network of linear monomolecular first-order reversible reactions, $A_i \rightleftarrows A_{i+1}$ with $i = 1, 2, \ldots, N_c$. This network is represented as a directed graph (Yablonskii et al., 1991) with nodes corresponding to components A_i ($i = 1, 2, \ldots, N_c$). The kinetic equations have the standard form

$$\frac{dc_{A_i}}{dt} = \sum_{i=1}^{N_c} \left(k_{i+1 \to i} c_{A_{i+1}} - k_{i \to i+1} c_{A_i} \right) \tag{6.98}$$

where c_{A_i} are the concentrations of components A_i and $k_{i \to i+1}$ and $k_{i \to i+1} > 0$ are the reaction rate coefficients for respectively the reactions $A_i \to A_{i+1}$ and $A_{i+1} \to A_i$.

The principle of detailed balance (time-reversibility, see Section 6.1.4) means that there exists a positive vector of equilibrium concentrations, $\mathbf{c}_{A_i,eq} > 0$, such that for all i

$$k_{i+1 \to i} c_{A_{i+1},eq} = k_{i \to i+1} c_{A_i,eq} \tag{6.99}$$

The following conditions are necessary and sufficient for the existence of such an equilibrium:

- Reversibility (in the sense that if transition $A \to B$ exists, then transition $B \to A$ also exists): if $k_{i \to i+1} > 0$ then $k_{i \to i+1} > 0$;
- For any cycle $A_1 \rightleftarrows A_2 \rightleftarrows \cdots \rightleftarrows A_{N_c} \rightleftarrows A_1$, the product of the rate coefficients of the forward reactions is equal to the product of the rate coefficients of the reverse reactions:

$$\prod_{i=1}^{N_c} k_{i \to i+1} = \prod_{i=1}^{N_c} k_{i+1 \to i} \tag{6.100}$$

where $k_{N_c+1 \to N_c} = k_{1 \to N_c}$. This is the Wegscheider relation (Wegscheider, 1901).

It is sufficient to consider a finite number of basic cycles in Eq. (6.100) (Yablonskii et al., 1991).

The entropic scalar product for the free entropy is

$$\langle \mathbf{a} | \mathbf{b} \rangle_\Phi = \sum_i \frac{a_{A_i} b_{A_i}}{c_{A_i,eq}} \tag{6.101}$$

Let $\mathbf{c}^a(t)$ be a solution of Eq. (6.98) with initial condition $\mathbf{c}^a(0) = \mathbf{a}$. Then the reciprocal relations, Eq. (6.96), for linear systems with detailed balance take the form

$$\sum_i \frac{b_{A_i} c_{A_i}^a(t)}{c_{A_i,eq}} = \sum_i \frac{a_{A_i} c_{A_i}^b(t)}{c_{A_i,eq}} \tag{6.102}$$

We now compare two experimental situations:

(1) A process starts with only A_q and we measure the concentration of A_r. The result is $c_{A_r}^a(t)$ (How much A_r is produced from the initial A_q?).
(2) A process starts with only A_r and we measure the concentration of A_q. The result is $c_{A_q}^b(t)$ (How much A_q is produced from the initial A_r?).

The results of this dual experiment are connected by the equality

$$\frac{c_{A_r}^a(t)}{c_{A_r,eq}} = \frac{c_{A_q}^b(t)}{c_{A_q,eq}} \tag{6.103}$$

It is much more straightforward to experimentally check these relations between kinetic curves than it is to check the original Onsager relations between kinetic coefficients. Specific examples of such relations have been presented in the Section 6.4 and in Yablonsky et al. (2011a).

For many real chemical and biochemical processes some of the elementary reaction steps are virtually irreversible. For these processes, the microreversibility conditions (see Section 6.1.4) are not applicable directly. Nevertheless, such systems may be considered as limits of systems with detailed balance with some of the kinetic coefficients tending to zero if and only if (1) the reversible part of the system satisfies the principle of detailed balance and (2) the irreversible reactions are not included in oriented cyclic pathways. This is a weak form of detailed balance without the necessary existence of a positive equilibrium. See Yablonsky et al. (2011b) for more detailed information.

6.5.3 Experimental Evidence

The validity of the reciprocal relations was shown by Yablonsky et al. (2011b) using the technique of Temporal Analysis of Products (TAP) developed by Gleaves in 1988 (Gleaves et al., 1988; see Chapter 5). The reactor used was a so-called thin-zone TAP reactor (TZTR), in which a thin catalyst zone is packed between two inert zones. The reaction studied is part of the reversible water-gas shift reaction, $H_2O + CO \rightleftarrows H_2 + CO_2$, taking place over an iron oxide catalyst.

In a TZTR, diffusion in the inert zones on both sides of the thin catalyst zone must be accounted for. The Knudsen regime in these zones guarantees linear behavior, so that the resulting outlet fluxes can be expressed in terms of convolutions. Switching to the Laplace domain greatly facilitates the analysis, and it can be proven that the following equality holds in terms of the exit flux of B given a unit inlet pulse of A, J_{B_A}, and the exit flux of A given a unit inlet pulse of B, J_{A_B}:

$$K_{eq} = \frac{\left(\cosh \sqrt{s\tau_{1,A}}\right)\left(\sqrt{\tau_{3,B}} \sinh \sqrt{s\tau_{3,B}}\right)\mathcal{L}\{J_{B_A}\}}{\left(\cosh \sqrt{s\tau_{1,B}}\right)\left(\sqrt{\tau_{3,A}} \sinh \sqrt{s\tau_{3,A}}\right)\mathcal{L}\{J_{A_B}\}} \qquad (6.104)$$

where $\tau_{i,G} = \dfrac{\varepsilon_i L_i^2}{D_{\text{eff},G}}$ with ε_i the void fraction of zone i, L_i its length, and $D_{\text{eff},G}$ the effective diffusivity of gas G (A or B) with A denoting CO and B CO_2. In the Fourier domain, the results of Fig. 6.8 are obtained. The real and imaginary parts of the right-hand side of Eq. (6.104) are plotted, with error bars corresponding to three times the standard deviation estimated from resampling the exit flux measurements 10,000 times using their principal error components. Ideally, all imaginary values should be zero and zero indeed is within all the confidence intervals. Furthermore, the smallest error in the real parts occurs for the second frequency,

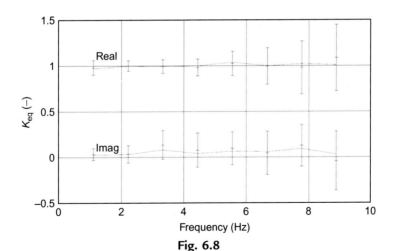

Fig. 6.8

Real and imaginary parts of Fourier domain result values for the "B from A to A from B" ratio (Eq. 6.104), as a function of the frequency f in Hz (so that $\omega = 2\pi f$, $s = i\omega$). The error bars were obtained from 10,000 resampled measurements. *From Yablonsky, G.S., Gorban, A.N., Constales, D., Galvita, V.V., Marin, G.B., 2011b. Reciprocal relations between kinetic curves. Europhys. Lett. 93, 20004–20007. Copyright (2011), with permission.*

2.2 Hz. The confidence interval at this frequency lies well within the others, providing confirmation that the same value for all frequencies is obtained (within experimental error). Summarizing, this experimental result can be considered as one of few straightforward proofs of the validity of Onsager's reciprocal relations for chemical reactions.

6.5.4 Concluding Remarks

The operator, $\exp(\mathbf{k}t)$, is symmetric in the entropic scalar product. This enables the formulation of symmetry relations between observables and initial data, which can be validated without differentiation of empirical curves and are, in that sense, more robust and closer to direct measurements than the classical Onsager relations. In chemical kinetics, there is an elegant form of symmetry between "A produced from B" and "B produced from A": their ratio is equal to the equilibrium coefficient of the reaction A $\rightleftarrows$ B and does not change in time. The symmetry relations between observables and initial data have a rich variety of realizations, which makes direct experimental verification possible. This symmetry also provides the possibility of extracting additional experimental information about the detailed reaction mechanism through dual experiments. The symmetry relations are applicable to all systems with microreversibility.

6.6 Predicting Kinetic Dependences Based on Symmetry and Balance

6.6.1 Introduction

In this section a strategy is described for predicting the temporal evolution of a complex chemical reaction that is described by a linear model, based on known equilibrium coefficients and selected known temporal concentration dependences. Such a model may relate to a linear mechanism, for example, a set of monomolecular reactions.

Sections 6.4 and 6.5 were devoted to a new type of relations between kinetic dependences that are obtained from the symmetrical initial conditions during the course of a so-called dual experiment. This result is the consequence of Onsager's reciprocity for linear (or linearized) kinetics of the type $d\mathbf{c}/dt = \mathbf{k}\mathbf{c}$ (Eq. 6.91) with $\mathbf{k}$ symmetric in the entropic scalar product given by $\langle \mathbf{a}|\mathbf{k}\mathbf{b}\rangle_\Phi \equiv \langle \mathbf{k}\mathbf{a}|\mathbf{b}\rangle_\Phi$ (Eq. 6.94) if isothermal and isobaric conditions are assumed.

This form of Onsager's reciprocity implies that the time shift operator, $\exp(\mathbf{k}t)$, which generates reciprocal relations between kinetic curves, is also symmetric. Obviously, the next step is to address the following questions (Constales et al., 2015):

- Is it possible to find all transient dependences based on a known equilibrium composition or on known equilibrium coefficients using this approach?
- How many kinetic dependences do we have to determine for this purpose in addition to the equilibrium coefficients?

- How can the unknown kinetic dependences be determined based on the ones already known?
- Is this procedure always successful?

In this section, some of these questions are answered.

6.6.2 Theoretical Model

We start from the linear kinetic model of Eq. (6.91):

$$\frac{d\mathbf{c}}{dt} = \mathbf{kc} \tag{6.105}$$

which is related to a system of monomolecular reversible first-order reactions, $A_i \rightleftarrows A_{i+1}$ with $i = 1, 2, \cdots, N_c$. Every nondiagonal element of the kinetic matrix is the rate coefficient of a reaction. For example, the element $(1,2)$ represents the rate coefficient k_{12} of the reaction $A_1 \rightarrow A_2$. As a result of mass conservation, each column of the matrix must sum to zero, so that the diagonal element in each column contains minus the sum of all rate coefficients that are not on the diagonal. For example, the element $(2,2)$ contains the value $-(k_{2\rightarrow1} + k_{2\rightarrow3} + \cdots k_{2\rightarrow N_c})$.

The two main properties of the kinetic model are the mass balance for the concentrations,

$$\sum_{i=1}^{N_c} c_{A_i} = c_t \quad \text{or} \quad \sum_{i=1}^{N_c} A_i = 1 \tag{6.106}$$

and the symmetry of the kinetic operator $\mathbf{k}$,

$$\prod_{i=1}^{N_c} k_{i \rightarrow i+1} = \prod_{i=1}^{N_c} k_{i+1 \rightarrow i} \quad \text{or} \quad \prod_{i=1}^{N_c} K_{\text{eq}, i \rightarrow i+1} = 1 \tag{6.107}$$

The mass balance can be used to verify that all participating components had been identified and accounted for in the model.

6.6.3 Matrix of Experimental Curves

Mathematically, the fact that the symmetry of the kinetic operator $\mathbf{k}$ determines symmetry relationships between observed experimental dependences is based on the fact that entire functions of symmetric operators are also symmetric, which means that the characteristic $\exp(\mathbf{k}t)$, the propagator, is also symmetric. The propagator coincides with the matrix of observed experimental curves. For example, for a system consisting of three components ($N_c = 3$), A, B, and C, with reactions $A \rightleftarrows B \rightleftarrows C \rightleftarrows A$, this matrix is the 3×3 matrix

$$
\begin{bmatrix}
A_A(t) & \boxed{A_B(t)} & \boxed{A_C(t)} \\
B_A(t) & B_B(t) & \boxed{B_C(t)} \\
C_A(t) & C_B(t) & C_C(t)
\end{bmatrix}
\tag{6.108}
$$

with the same initial components in the columns and the same products in the rows. For example, $A_A(t)$ denotes the dependence of the normalized concentration of component A, starting from pure species A, $B_A(t)$ that of B starting from A, and so on. In the general case of N_c components, the propagator matrix is an $N_c \times N_c$ matrix of similar form. The symmetry of this matrix for the entropic scalar product $\langle . | . \rangle$ means that, for example,

$$
B_A(t) = A_B(t) \frac{B_{eq}}{A_{eq}} = A_B(t) K_{eq, A \rightleftarrows B}
\tag{6.109}
$$

Together with the balance requirement

$$
A(t) + B(t) + C(t) = A_{eq} + B_{eq} + C_{eq} = 1
\tag{6.110}
$$

this can be directly applied to the problem of predicting the transient regime from arbitrary initial conditions based on the known equilibrium composition and a limited amount of transient information. If, for example, in this case $A_B(t)$, $A_C(t)$, and $B_C(t)$ (shown framed in Eq. 6.108) are known, we can first use symmetry to determine their counterparts, $B_A(t), C_A(t)$, and $C_B(t)$ using Eq. (6.109) and equivalents thereof, and then the remaining three concentration dependences can be found from the balance equations

$$
A_A(t) = 1 - B_A(t) - C_A(t)
\tag{6.111}
$$

$$
B_B(t) = 1 - A_B(t) - C_B(t)
\tag{6.112}
$$

$$
c_C(t) = 1 - A_C(t) - B_C(t)
\tag{6.113}
$$

To illustrate this approach for the prediction of kinetic dependences, we will use the mechanism of the isomerization of butenes over a pure alumina catalyst analyzed by Wei and Prater (1962, p. 257) , see Example 6.3 in Section 6.4. Let us assume that the three equilibrium coefficients are known, together with the three dependences $A_B(t)$, $A_C(t)$, and $B_C(t)$ (framed in Eq. 6.108). Fig. 6.9 shows how symmetry is used to determine $B_A(t)$ from $A_B(t)$. Similarly, $C_A(t)$ can be obtained from $A_C(t)$ using symmetry. Then $A_A(t)$ follows from $B_A(t)$ and $C_A(t)$ using the balance equation (6.111).

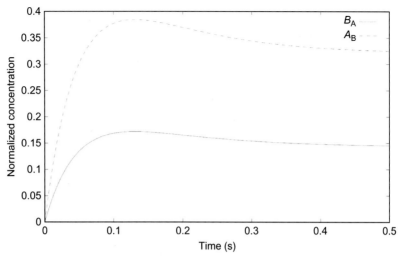

Fig. 6.9

Dependences $A_B(t)$ assumed known and $B_A(t) = A_B(t)K_{eq,A \rightleftarrows B}$ obtained from symmetry; $K_{eq,A \rightleftarrows B} = 0.4469$. *Reprinted from Constales, D., Yablonsky, G.S., Marin, G.B., 2015. Predicting kinetic dependences and closing the balance: Wei and Prater revisited. Chem. Eng. Sci. 123, 328–333. Copyright (2015), with permission from Elsevier.*

A similar procedure is used to determine the two remaining dependences, $B_B(t)$ and $C_C(t)$. All dependences have been plotted in Fig. 6.10. The concentration dependences $A_A(t)$, $A_B(t)$, and $A_C(t)$, all tend to A_{eq} and similar statements can be made for the concentrations of B and C.

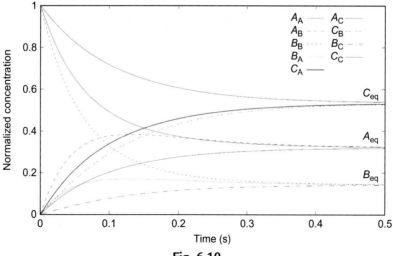

Fig. 6.10

Dependences $A_B(t)$, $A_C(t)$, and $B_C(t)$ that are assumed known; their counterparts $B_A(t)$, $C_A(t)$, and $C_B(t)$ that are obtained by symmetry; and the diagonal element dependences $A_A(t)$, $B_B(t)$, and $C_C(t)$ that are obtained from the balance equation. *Reprinted from Constales, D., Yablonsky, G.S., Marin, G. B., 2015. Predicting kinetic dependences and closing the balance: Wei and Prater revisited. Chem. Eng. Sci. 123, 328–333. Copyright (2015), with permission from Elsevier.*

6.6.4 General Considerations

Now we are ready to answer the questions posed in the beginning of this section. In general, when N_c components are involved, the upper triangle of concentration dependences in the matrix of experimental curves contains all $A_{iA_{i+1}}(t)$ with $1 \leq i \leq N_c$, which means there are $N_c(N_c - 1)/2$ elements in this upper triangle. For instance, for $N_c = 3$, there are three elements in the upper triangle: A_{1A_2}, A_{1A_3}, and A_{2A_3} (or A_B, A_C, and B_C in our example). From these elements, their counterparts (B_A, C_A, and C_B) in the lower triangle of the matrix can be uniquely determined using symmetries. Then in each column, all elements except the diagonal ones (A_A, B_B, and C_C) are known. These can now be determined using balance. Consequently, the concentration dependences in the upper triangle are a basis for all matrix elements.

For other sets of $N_c(N_c - 1)/2$ elements, we need to investigate whether these too can be used as a basis. Clearly, two symmetric elements cannot occur in such a set simultaneously, nor can all elements in one column or in one row be part of the same basis. For $N_c = 3$, the following sets of three elements with none in the lower triangle, fail to be a basis:

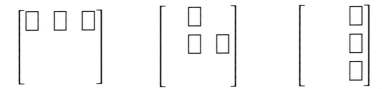

For the second selection, this is due to the fact that the (3,2) entry is determined from the (2,3) element through symmetry and also independently by the balance of the second column. Therefore, the selected elements are not independent and thus do not form a basis. The other nonlower-triangle sets are bases, and in most cases this is easy to prove by applying symmetry and balance; such cases are trivial. However, the set of diagonal elements is also a basis, but a nontrivial one. If we use this basis to express, for example, the elements $A_B(t)$ and $B_A(t)$, we obtain

$$2B_{eq}\left(A_B(t) - A_{eq}\right) = 2A_{eq}\left(B_A(t) - B_{eq}\right)$$
$$= C_{eq}\left(C_C(t) - C_{eq}\right) - A_{eq}\left(A_B(t) - A_{eq}\right) - B_{eq}\left(B_A(t) - B_{eq}\right) \quad (6.114)$$

Constales et al. (2015) have analyzed each case for $N_c = 3$ using Maple to solve the system of linear equations expressing symmetry and balance. Finding a solution means that the chosen set is a basis. Among the $20 = \binom{6}{3}$ possible choices, there are three nonbases (listed previously) and thus 17 bases. A similar exhaustive search for $N_c = 4$ involves $210 = \binom{10}{6}$ candidates. Table 6.5 summarizes results for a system containing from two to six components.

Table 6.5 Bases for concentration dependences in systems containing N_c components (Constales et al., 2015)

N_c	Candidates	Bases	Nontrivial
2	3	3	0
3	20	17	1
4	210	141	16
5	3003	1548	252
6	54,264	21,169	4362

6.6.5 Checking Model Completeness

6.6.5.1 Closing the mass balance

In many cases, closing the mass balance of a multicomponent chemical system is a challenge for the experimental chemist. This is caused by the difficulty that the measured concentrations of some reaction components, typically those present in small amounts, are rather inaccurate and often irreproducible being overloaded with experimental errors. The results of the present analysis provide an interesting method of closing the mass balance, based on measurements of only one or a few reaction components in several experiments. We illustrate this procedure for a mixture of three components ($N_c = 3$) using the data of Wei and Prater (1962) as previously, and adding a 2% normally distributed experimental error to the calculated concentration dependences, sampled at 0.01 conventional time intervals. The goal then is to obtain a linear equation,

$$A_A + \alpha A_B + \beta A_C = \gamma \tag{6.115}$$

where α and β are weights of experimental dependences and γ is the total normalized concentration (a balance constant) to be determined. The true values of these parameters are $\alpha = A_{eq}/B_{eq} = K_{eq,A \rightleftarrows B}$, $\beta = A_{eq}/C_{eq} = K_{eq,A \rightleftarrows C}$, and $\gamma = 1$. These parameters can easily be calculated from the values of $A_A(t), A_B(t)$, and $A_C(t)$ at three or more different moments in time, applying a least-squares method to a system of linear equations. For the example plotted in Fig. 6.11, the results obtained are $\alpha = 0.4135$, $\beta = 1.703$, and $\gamma = 0.9980$. Comparing these values with the true values: $\alpha = 0.4469$, $\beta = 1.666$, and $\gamma = 1.000$, the usefulness of the procedure is clearly established, proving that the mass balance can be closed effectively based on the measurement of the concentration of a single component during a small number of experiments.

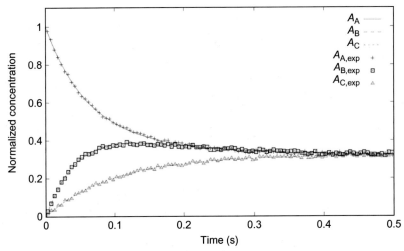

Fig. 6.11

Example of closing the balance: simulated data of observed values of $A_A(t)$, $A_B(t)$, and $A_C(t)$ with normally distributed errors. *Reprinted from Constales, D., Yablonsky, G.S., Marin, G.B., 2015. Predicting kinetic dependences and closing the balance: Wei and Prater revisited. Chem. Eng. Sci. 123, 328–333, Copyright (2015), with permission from Elsevier.*

6.6.5.2 Completeness test

Closing of mass balances is one way of checking the completeness of a model. However, the following independent test can also be used for linear models. It is based on the fact that the trace of an entire function of a matrix is equal to the sum of its eigenvalues, $f(\lambda_1) + \cdots f(\lambda_{N_c})$. In our case, the trace $\mathrm{tr}(t)$ of the matrix $\exp(\mathbf{k}t)$ is then necessarily equal to the sum of all exponential decays corresponding to the eigenvalues of $\mathbf{k}$ with coefficients 1, that is,

$$\mathrm{tr}(t) = \sum_{i=1}^{N_c} \exp(\lambda_i t) \tag{6.116}$$

For example, for $N_c = 3$,

$$\mathrm{tr}(t) = \underbrace{c_{A_A}(t) + c_{B_B}(t) + c_{C_C}(t)}_{T_1} = \underbrace{1 + e^{\lambda_1 t} + e^{\lambda_2 t}}_{T_2} \tag{6.117}$$

We try and fit $\gamma\left(1 + e^{\lambda_1 t} + e^{\lambda_2 t}\right)$ to experimental data obtained for the sum $c_{A_A}(t) + c_{B_B}(t) + c_{C_C}(t)$ through the parameters γ, λ_1, and λ_2. A successful fit for the example from Wei and Prater (1962) with $N_c = 3$ is shown in Fig. 6.12. If the normalized concentrations do not add up to the balance value of one (see Eq. 6.106), this means that one or more components have not been accounted for.

We can thus distinguish two tests involving the trace. The preliminary test, T_1, is used to confirm that the ratio $\mathrm{tr}(0)/\mathrm{tr}(+\infty)$ equals N_c, so that there is no visible deficit. The stronger

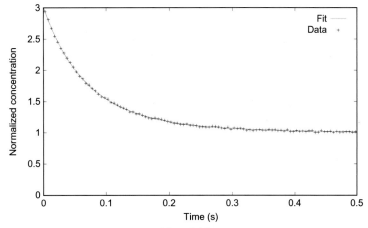

Fig. 6.12
Successful completeness test for $N_c = 3$. Reprinted from Constales, D., Yablonsky, G.S., Marin, G.B., 2015. *Predicting kinetic dependences and closing the balance: Wei and Prater revisited. Chem. Eng. Sci. 123, 328–333, Copyright (2015), with permission from Elsevier.*

test, T_2, is then used to determine whether the trace can be fitted sufficiently well by a sum of exponentials with real rate coefficients. The following can be stated about these tests:

- If the system is complete, there can be no balance deficit that can be attributed to unobserved components, hence T_1 must be fulfilled, whether the system is linear or not.
- If the system is linear and there is detailed balance, T_2 must be fulfilled since the principle of detailed balance implies that all eigenvalues are real.
- If the system is linear, but there is no detailed balance, some eigenvalues may be nonreal complex and then T_2 fails, but this is not necessarily the case.
- If the system is linear and incomplete, T_1 will certainly fail, as will the stronger test T_2.
- If the system is nonlinear and complete, T_1 will hold, but it is uncertain whether T_2 will hold.
- If the system is nonlinear and incomplete, the validity of both T_1 and T_2 is questionable.

Table 6.6 summarizes the interpretation of tests T_1 and T_2.

Table 6.6 Summary of the validity of the tests T_1 and T_2

System	T_1 $$\frac{tr(0)}{tr(+\infty)} = N_c$$	T_2 $$tr(t) = \sum_{i=1}^{N_c} \exp(\lambda_i t), \ \lambda_i \leq 0$$
Linear, complete, detailed balance	+	+
Linear, complete, no detailed balance	+	?
Linear, incomplete	−	−
Nonlinear, complete	+	?
Nonlinear, incomplete	?	?

Note: +, pass; −, fail; ?, pass or fail.

6.6.6 Concluding Remarks

The strategy for predicting the temporal evolution of a complex chemical reaction described in this section is based on the application of mass balances and symmetry relations between concentration dependences, starting from extreme initial values of the concentrations. The results obtained may be very useful for advanced analysis of complex chemical reactions and can be applied to the analysis of linear models of reversible reactions in plug-flow reactors and in the linear vicinity of nonlinear complex reversible reactions both in batch reactors (closed systems) and in plug-flow reactors. They can also be applied to the analysis of pseudomonomolecular models of the Langmuir-Hinshelwood-Hougen-Watson type for reversible reactions.

6.7 Symmetry Relations for Nonlinear Reactions

6.7.1 Introduction

The previous sections were devoted to linear reaction models. In this section, we present an analysis of nonlinear reactions that was performed by Constales et al. (2012).

For the general single reversible reaction

$$\sum_i \alpha_i A_i \underset{\longleftarrow}{\overset{\longrightarrow}{\rightleftharpoons}} \sum_i \beta_i B_i \tag{6.118}$$

where A_i are reactants, B_i are products, and α_i and β_i are the absolute values of their stoichiometric coefficients, the following well-known stoichiometric relationship holds:

$$-\frac{1}{\alpha_1}\frac{dc_{A_1}}{dt} = -\frac{1}{\alpha_2}\frac{dc_{A_2}}{dt} = \cdots = \frac{1}{\beta_1}\frac{dc_{B_1}}{dt} = \frac{1}{\beta_2}\frac{dc_{B_2}}{dt} = \cdots \tag{6.119}$$

There are only six possible single reaction types with first- or second-order kinetics and these are listed below together with their so-called kinetic balances, which readily follow from Eq. (6.119):

- First-first order:

$$A \rightleftharpoons B \quad A + B = 1$$

- Second-first order:

$$2A \underset{\longleftarrow}{\overset{\longrightarrow}{\rightleftharpoons}} B \quad A + 2B = 1$$
$$A + B \underset{\longleftarrow}{\overset{\longrightarrow}{\rightleftharpoons}} C \quad A + B + 2C = 1 \quad B - A = \Delta$$

- Second-second order:

$$2A \rightleftharpoons 2B \qquad A + B = 1$$
$$A + B \rightleftharpoons 2C \qquad A + B + C = 1 \qquad B - A = \Delta$$
$$A + B \rightleftharpoons C + D \quad A + B + C + D = 1 \quad B - A = \Delta^{\text{left}} \quad D - C = \Delta^{\text{right}}$$

In this list, reactions that can be obtained by renaming the components participating in the reactions or by reversing the reactions are omitted from the list. For example, there is no reaction $A \rightleftharpoons B + C$ in the list because this is a variation of $A + B \rightleftharpoons C$. A, B, C, and D represent the normalized concentrations of components A, B, C, and D. The symbol Δ denotes the constant difference between two normalized concentrations.

6.7.2 Single Reversible Reactions

6.7.2.1 First-first order: $A \rightleftharpoons B$

This type of reaction, a simple example of which is the isomerization reaction $cis\text{-}C_4H_8 \rightleftharpoons trans\text{-}C_4H_8$, was analyzed in Section 6.4.2, Example 6.1, Eqs. (6.55)–(6.63).

Using $A_A = 1 - B_A$ and $A_{eq} = 1 - B_{eq}$, Eq. (6.61) can be rewritten as

$$B_A(t) - B_{eq} = -K_{eq}\left(B_B(t) - B_{eq}\right) \tag{6.120}$$

with

$$K_{eq} = \frac{k^+}{k^-} = \frac{B_A(t)}{A_B(t)} \tag{6.121}$$

Thus, plotting $B_A(t)$ versus $B_B(t)$ yields a straight line with slope $-K_{eq}$ from $(B_B, B_A) = (1, 0)$ to $(B_B, B_A) = (B_{eq}, B_{eq})$. In particular, at all times the lower deviation, $(B_{eq} - B_A(t))$ is K_{eq} times the upper deviation, $(B_B(t) - B_{eq})$ (see Fig. 6.13). This and subsequent plots can be interpreted as special phase trajectories for dual experiments.

6.7.2.2 Second-first order: $2A \rightleftharpoons B$

Examples of this type of reaction are the dimerization reaction $2CH_3 \rightleftharpoons C_2H_6$ and association reactions $2H \rightleftharpoons H_2$, $2O \rightleftharpoons O_2$ and $2OH \rightleftharpoons H_2O_2$. The kinetic balance for this type of reaction is $A(t) + 2B(t) = 1$. In order to reach the equilibrium concentrations, both dual experiments must start from the same absolute amount of matter. Hence, the first experiment starts from reactant A: (A,B)=(1,0) and the second from the product B: (A,B)=(0,1/2). The concentration of the product will be diluted. However, the total mass will be the same because the molecular mass of B is twice that of A. In both cases, the kinetic balance is fulfilled.

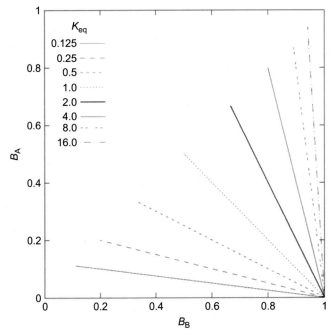

Fig. 6.13

Trajectories B_A versus B_B for the reaction $A \rightleftarrows B$ for different values of K_{eq}. *Reprinted from Constales, D., Yablonsky, G.S., Marin, G.B., 2012. Thermodynamic time invariants for dual kinetic experiments: nonlinear single reactions and more. Chem. Eng. Sci. 73, 20–29, Copyright (2012), with permission from Elsevier.*

The nonlinear differential equation for component A in normalized form is

$$\frac{dA}{dt} = -2k^+ A^2(t) + k^-(1 - A(t)) \tag{6.122}$$

which can be solved by quadratures (see Constales et al., 2012). The solutions for the normalized concentrations as a function of time then are

$$A_A(t) = \frac{\sqrt{8K_{eq}+1} + \tanh\left(\frac{k^+ t}{2}\sqrt{8K_{eq}+1}\right)}{\sqrt{8K_{eq}+1} + (4K_{eq}+1)\tanh\left(\frac{k^+ t}{2}\sqrt{8K_{eq}+1}\right)} \tag{6.123}$$

$$B_A(t) = \frac{2K_{eq}\tanh\left(\frac{k^+ t}{2}\sqrt{8K_{eq}+1}\right)}{\sqrt{8K_{eq}+1} + (4K_{eq}+1)\tanh\left(\frac{k^+ t}{2}\sqrt{8K_{eq}+1}\right)} \tag{6.124}$$

$$A_B(t) = \frac{2\tanh\left(\frac{k^+ t}{2}\sqrt{8K_{eq}+1}\right)}{\sqrt{8K_{eq}+1} + \tanh\left(\frac{k^+ t}{2}\sqrt{8K_{eq}+1}\right)} \tag{6.125}$$

The equilibrium coefficient can then be expressed as

$$K_{eq} = \frac{B_A(t)}{A_A(t)A_B(t)} \tag{6.126}$$

Furthermore,

$$B_{eq} = \frac{1}{2}\frac{\sqrt{8K_{eq}+1}-1}{\sqrt{8K_{eq}+1}+1} \tag{6.127}$$

and if we plot $B_A(t)$ versus $B_B(t) = (1-A_B(t))/2$ the curves lie on a branch of a rectangular hyperbola with axes parallel to the coordinate axes, the horizontal one being $B_A = \frac{1}{2}$ with the center at $(B_B, B_A) = \left(B_{eq} + \frac{1}{\sqrt{8K_{eq}+1}-1}, B_{eq} + \frac{1}{\sqrt{8K_{eq}+1}+1}\right)$, from the point $(1,0)$ to (B_{eq}, B_{eq}). In particular, for large t the lower deviation $(B_{eq}-B_A(t))$ is $2B_{eq}$ times the upper deviation, $(B_B - B_{eq})$ (see Fig. 6.14).

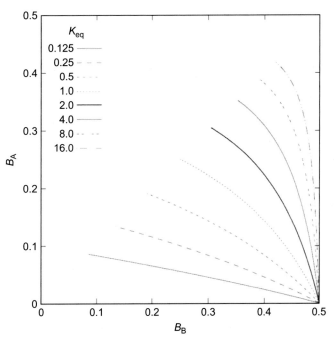

Fig. 6.14

Trajectories B_A versus B_B for the reaction $2A \rightleftarrows B$ for different values of K_{eq}. *Reprinted from Constales, D., Yablonsky, G.S., Marin, G.B., 2012. Thermodynamic time invariants for dual kinetic experiments: nonlinear single reactions and more. Chem. Eng. Sci. 73, 20–29, Copyright (2012), with permission from Elsevier.*

6.7.2.3 Second-second order: 2A ⇌ 2B

Examples of this type of reaction are second-order isomerization reactions. The kinetic balance is $A(t) + B(t) = 1$, which is the same balance as for the first-order reaction $A \rightleftarrows B$ and offers no difficulties regarding the initial values: (1,0) and (0,1). The differential equation for component A in normalized form,

$$\frac{dA}{dt} = -2k^+A^2(t) + 2k^-(1-A(t))^2 \tag{6.128}$$

can be reduced in quadratures, so that

$$A_A(t) = \frac{1}{1 + \sqrt{K_{eq}}\tanh\left(2t\sqrt{k^+k^-}\right)} \tag{6.129}$$

$$B_A(t) = \frac{\sqrt{K_{eq}}\tanh\left(2t\sqrt{k^+k^-}\right)}{1 + \sqrt{K_{eq}}\tanh\left(2t\sqrt{k^+k^-}\right)} \tag{6.130}$$

$$A_B(t) = \frac{\sqrt{\frac{1}{K_{eq}}}\tanh\left(2t\sqrt{k^+k^-}\right)}{1 + \sqrt{\frac{1}{K_{eq}}}\tanh\left(2t\sqrt{k^+k^-}\right)} \tag{6.131}$$

$$B_B(t) = \frac{1}{1 + \sqrt{\frac{1}{K_{eq}}}\tanh\left(2t\sqrt{k^+k^-}\right)} \tag{6.132}$$

Consequently,

$$K_{eq} = \frac{B_A(t)B_B(t)}{A_A(t)A_B(t)} \tag{6.133}$$

With

$$B_{eq} = \frac{\sqrt{K_{eq}}}{1 + \sqrt{K_{eq}}} \tag{6.134}$$

and plotting $B_A(t)$ versus $B_B(t)$, these curves lie on a branch of a rectangular hyperbola with axes parallel to the coordinate axes and center at $(B_B, B_A) = \left(\frac{K_{eq}}{K_{eq}-1}, \frac{K_{eq}}{K_{eq}-1}\right)$,

$$\left(B_A(t) - \frac{K_{eq}}{K_{eq}-1}\right)\left(B_B(t) - \frac{K_{eq}}{K_{eq}-1}\right) = \frac{K_{eq}}{(K_{eq}-1)^2} \tag{6.135}$$

from $(B_B, B_A) = (1, 0)$ to (B_{eq}, B_{eq}). In particular, for large t, the lower deviation $\left(B_{eq} - B_A(t)\right)$ equals the upper deviation $\left(B_B(t) - B_{eq}\right)$ (see Fig. 6.15).

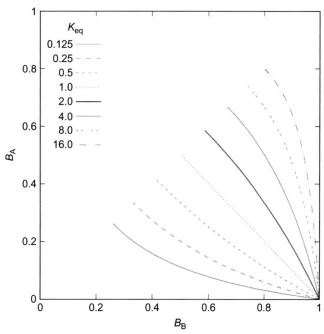

Fig. 6.15

Trajectories B_A versus B_B for the reaction $2A \rightleftarrows 2B$ for different values of K_{eq}. *Reprinted from Constales, D., Yablonsky, G.S., Marin, G.B., 2012. Thermodynamic time invariants for dual kinetic experiments: nonlinear single reactions and more. Chem. Eng. Sci. 73, 20–29, Copyright (2012), with permission from Elsevier.*

6.7.2.4 Second-first order: $A + B \rightleftarrows C$

An example of this type of reaction is $O + H \rightleftarrows OH$. In this case, two balances must be satisfied: the kinetic balance, $A(t) + B(t) + 2C(t) = 1$, as before, but also the balance expressing the difference $B(t) - A(t)$, which is constant during the reaction. Without loss of generality, we assume that $A \leq B$ (A is the reactant with the smallest concentration) and we can write $B(t) = A(t) + \Delta$.

The essential difference of this reaction compared to the ones described earlier, is that now the forward reaction involves two different components, resulting in a specific feature of the dual experiments, namely that the same reactant has to be chosen as the one with the smallest concentration in both experiments. To determine the trajectories of the experimental curves

- both experiments must involve the same total amount of A, B, and C;
- each of the dual experiments must start from the maximum possible value of A in case of the trajectory starting from A (and B), and the maximum possible value of C when starting from C;
- the Δ values of both experiments must be the same.

These requirements together are essential for tuning the dual experiments to obtain an elegant result.

Thus, for the trajectory starting from A, the maximum initial concentration of A that satisfies $B - A = \Delta$ is chosen. Together with the kinetic balance, this concentration can be determined to be $(1 - \Delta)/2$. The maximum initial concentration of C that satisfies both balances is also $(1 - \Delta)/2$. Thus the initial compositions are,

- starting from A: $(A_A(0), B_A(0), C_A(0)) = ((1-\Delta)/2, (1+\Delta)/2, 0)$
- starting from C: $(A_C(0), B_C(0), C_C(0)) = (0, \Delta, (1-\Delta)/2)$

In both experiments the value of Δ is the same, $B_A - A_A = B_C - A_C = \Delta$. Furthermore, in both experiments the total concentration is the same, $A + B + 2C = 1$, so that there is full consistency between the dual experiments. In the special case that Δ is zero, the reactant composition is stoichiometric, $A = B$, which leads to an elegant time invariance in the form

$$K_{eq} = \frac{C_A(t)}{A_C(t)B_A(t)} \tag{6.136}$$

6.7.2.5 Second-second order: $A + B \rightleftarrows 2C$

An example of this type of reaction is $H + HO_2 \rightleftarrows 2OH$. In this case again, two balances must be satisfied: the kinetic balance, $A(t) + B(t) + C(t) = 1$ and $B(t) - A(t) = \Delta$. Again we assume that A is the reactant with the smallest concentration, resulting in the following extreme initial conditions for the dual experiments:

- starting from A: $(A_A(0), B_A(0), C_A(0)) = ((1-\Delta)/2, (1+\Delta)/2, 0)$
- starting from C: $(A_C(0), B_C(0), C_C(0)) = (0, \Delta, (1-\Delta))$

Reduction to quadratures leads to the time invariance

$$K_{eq} = \frac{C_A(t)C_C(t)}{A_C(t)B_A(t)} \tag{6.137}$$

6.7.2.6 Second-second order: $A + B \rightleftarrows C + D$

An example of this type of reaction is $O_3 + OH \rightleftarrows HO_2 + O_2$. Besides the kinetic balance, $A(t) + B(t) + C(t) + D(t) = 1$, there are now two supplementary balances, in the form of constant concentration differences, one on the left side of the reaction equation and one on the right side. Without loss of generality, A and C are assumed to be the reactants with the smallest concentration on the left and the right, so $A \leq B$ and $C \leq D$. Then, with Δ^{left} and Δ^{right} the constant concentration differences, $B(t) = A(t) + \Delta^{\text{left}}$ and $D(t) = C(t) + \Delta^{\text{right}}$, with the special stoichiometric case for $\Delta^{\text{left}} = \Delta^{\text{right}} = 0$. In this case, the extreme initial conditions are

- $(A_A(0), B_A(0), C_A(0), D_A(0)) = \left(\left(1 - \Delta^{\text{left}} - \Delta^{\text{right}} \right)/2, \left(1 + \Delta^{\text{left}} - \Delta^{\text{right}} \right)/2, 0, \Delta^{\text{right}} \right)$, starting without C present.
- $(A_C(0), B_C(0), C_C(0), D_C(0)) = \left(0, \Delta^{\text{left}}, \left(1 - \Delta^{\text{left}} - \Delta^{\text{right}} \right)/2, \left(1 - \Delta^{\text{left}} + \Delta^{\text{right}} \right)/2 \right)$, starting without A present.

The time invariance obtained by reduction to quadratures is

$$K_{\text{eq}} = \frac{C_A(t) D_C(t)}{A_C(t) B_A(t)} \tag{6.138}$$

Fig. 6.16 illustrates the validity of this result showing plots of all concentration curves present in Eq. (6.152). All these trajectories change with time, but Eq. (6.138) is invariant. Comparing this equation with Eq. (6.121), we can interpret $\dfrac{D_C(t)}{B_A(t)}$ as a correction factor.

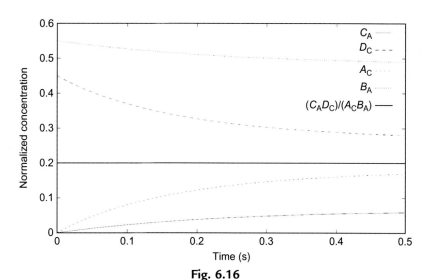

Fig. 6.16

Trajectories C_A, D_C, A_C, and B_A versus time and the invariance $(C_A D_C)/(A_C B_A)$ for the reaction $A + B \rightleftarrows C + D$; $k^+ = 2\,\text{s}^{-1}$; $k^- = 10\,\text{s}^{-1}$; $\Delta^{\text{left}} = 0.3$; $\Delta^{\text{right}} = 0.2$. *Reprinted from Constales, D., Yablonsky, G.S., Marin, G.B., 2012. Thermodynamic time invariants for dual kinetic experiments: nonlinear single reactions and more. Chem. Eng. Sci. 73, 20–29, Copyright (2012), with permission from Elsevier.*

6.7.3 Two-Step Reaction Mechanisms

6.7.3.1 Catalytic reaction mechanism

The simplest catalytic reaction mechanism is the Temkin-Boudart mechanism, which is a single-route mechanism consisting of two reversible steps

$$A + Z \underset{k_1^-}{\overset{k_1^+}{\rightleftharpoons}} AZ \tag{6.139}$$

$$AZ \underset{k_2^-}{\overset{k_2^+}{\rightleftharpoons}} B + Z \tag{6.140}$$

and can be used to represent the overall reaction $A \rightleftharpoons B$ (eg, an isomerization reaction). Here A and B are the gaseous reactant and product of the reaction, and Z and AZ are catalytic surface intermediates. Z represents a free active site and AZ represents adsorbed A. This mechanism is linear, that is, in any reaction only one surface intermediate participates. For the concentration of the surface intermediates, the site balance holds: $\Gamma_Z + \Gamma_{AZ} = \Gamma_t$.

Under the assumption of quasi steady state (see Chapter 4), the rate of generation of every surface intermediate is approximately equal to its rate of consumption, and the concentrations of the intermediates can be eliminated from the kinetic equations describing the temporal evolution of this reaction system, which can then be written in normalized form as (see Section 3.6)

$$-\frac{dA}{dt} = \frac{dB}{dt} = \frac{k_1^+ k_2^+ A(t) - k_1^- k_2^- B(t)}{k_1^+ A(t) + k_2^+ + k_1^- + k_2^- B(t)} \tag{6.141}$$

with kinetic balance $A(t) + B(t) = 1$, $\dfrac{k_1^+}{k_1^-} = K_{eq,1}$, $\dfrac{k_2^+}{k_2^-} = K_{eq,2}$, $K_{eq,1}K_{eq,2} = K_{eq}$ and $L^- = \dfrac{k_1^-}{k_2^-}$.

The following invariant equation can be obtained (see Constales et al., 2012):

$$\frac{1}{A_B(t) + B_A(t)} \ln \left(\frac{1 - \dfrac{B_A(t)}{B_{eq}}}{1 - \dfrac{A_B(t)}{A_{eq}}} \right) = \frac{(1 + K_{eq})(K_{eq,1}L^- - 1)}{(K_{eq,2} + L^-)(1 + K_{eq,1} + K_{eq})} \tag{6.142}$$

where $A_{eq} = \dfrac{1}{1 + K_{eq}}$ and $B_{eq} = \dfrac{K_{eq}}{1 + K_{eq}}$.

Fig. 6.17 illustrates this result. The function on the left-hand side of Eq. (6.142) is constant in time even though the concentrations vary. As a special case, when $K_{eq,1}L^- = 1$, that is, when $k_1^+ = k_2^-$, the right-hand side vanishes and the linear case is recovered. Using Eq. (6.142), the dependence $B_A(t)$ can be calculated knowing the dependence $A_B(t)$.

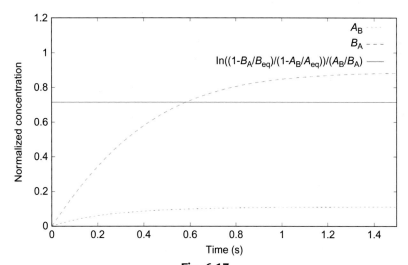

Fig. 6.17

Trajectories A_B and B_A for the reaction mechanism $A + Z \underset{k_1^-}{\overset{k_1^+}{\rightleftharpoons}} AZ, AZ \underset{k_2^-}{\overset{k_2^+}{\rightleftharpoons}} B + Z$ along with the

left-hand side of Eq. (6.142); $k_1^+ = 8\,s^{-1}$; $k_1^- = 4\,s^{-1}$; $k_2^+ = 4\,s^{-1}$; $k_2^- = 1\,s^{-1}$. *Reprinted from Constales, D., Yablonsky, G.S., Marin, G.B., 2012. Thermodynamic time invariants for dual kinetic experiments: nonlinear single reactions and more. Chem. Eng. Sci. 73, 20–29, Copyright (2012), with permission from Elsevier.*

6.7.3.2 Two-step mechanism, triple experiment

For the two-step mechanism $A + B \underset{k_1^-}{\overset{k_1^+}{\rightleftharpoons}} C \underset{k_2^-}{\overset{k_2^+}{\rightleftharpoons}} D$, Constales et al. (2012) considered a triple

experiment consisting of three trajectories, defined using the principle of maximum initial concentration:

- the A trajectory: $(A_A(0), B_A(0), C_A(0), D_A(0)) = ((1 - \Delta)/2, (1 + \Delta)/2, 0, 0)$
- the C trajectory: $(A_C(0), B_C(0), C_C(0), D_C(0)) = (0, \Delta, (1 - \Delta)/2, 0)$
- the D trajectory: $(A_D(0), B_D(0), C_D(0), D_D(0)) = (0, \Delta, 0, (1 - \Delta)/2)$

The kinetic equations in normalized form are

$$\frac{dA}{dt} = -k_1^+ A(t) B(t) + k_1^- C(t) \tag{6.143}$$

$$\frac{dB}{dt} = -k_1^+ A(t) B(t) + k_1^- C(t) \tag{6.144}$$

$$\frac{dC}{dt} = k_1^+ A(t) B(t) - \left(k_1^- + k_2^+\right) C(t) + k_2^- D(t) \tag{6.145}$$

$$\frac{dD}{dt} = k_2^+ C(t) - k_2^- D(t) \tag{6.146}$$

Considering the first reaction, $A + B \underset{k_1^-}{\overset{k_1^+}{\rightleftharpoons}} C$, separately, the thermodynamic time variance is

$$\frac{C_A(t)}{A_C(t)B_A(t)} = K_{eq,1} \tag{6.147}$$

Equation (6.147) is also valid for this same reaction as part of the two-step complex reaction in the domain of large time values, near the equilibrium of the complex reaction, as a result of the principle of detailed balance. Intuitively, this same relationship has to be valid in the domain of small time values in which the influence of the second reaction is insignificant. This can be checked using power series solutions to the kinetic equations, by successive derivation and substitution of initial values. Retaining the first two nonzero terms in the power series, Constales et al. (2012) found that

$$C_A(t) = \frac{k_1^+(1-\Delta^2)}{4}t - \frac{k_1^+(1-\Delta^2)\left(k_1^+ + k_1^- + k_2^+\right)}{8}t^2 + \cdots \tag{6.148}$$

$$A_C(t) = \frac{k_1^-(1-\Delta)}{2}t - \frac{k_1^-(1-\Delta)\left(k_1^+\Delta + k_1^- + k_2^+\right)}{8}t^2 + \cdots \tag{6.149}$$

$$B_A(t) = \frac{1+\Delta}{2} - \frac{k_1^+(1-\Delta^2)}{4}t + \cdots \tag{6.150}$$

and thus

$$\frac{C_A(t)}{A_C(t)B_A(t)} = \frac{k_1^+}{k_1^-} - \frac{(k_1^+)^2 k_2^+(1-\Delta)}{12 k_1^-}t^2 + \cdots \tag{6.151}$$

which clearly tends to $\frac{k_1^+}{k_1^-} = K_{eq,1}$ as $t \to 0$, confirming the initial assumption.

The expression for $K_{eq,1}$ can be improved by correcting for the second term in Eq. (6.151). Constales et al. (2012) also developed power series expansions for $D_A(t), A_D(t), C_D(t)$ and $D_C(t)$, which they used, together with the power series of Eqs. (6.148), (6.149), to deduce a power series expansion of the Onsager-like quotient:

$$\frac{D_A(t)C_D(t)A_C(t)}{A_D(t)D_C(t)C_A(t)} = 1 + \frac{k_1^+(1-\Delta)}{6}t + \cdots \tag{6.152}$$

For small t, the quotient in Eq. (6.152) is close to one. For large t, both the numerator and the denominator tend to $D_{eq}(t)C_{eq}(t)A_{eq}(t)$ so the quotient also tends to one. Now Eq. (6.152) can be used to correct the second term in Eq. (6.151), relying on the additional power series expansion

$$A_A(t) = \frac{(1-\Delta)}{2} - \frac{k_1^+(1-\Delta^2)}{4}t + \cdots \tag{6.153}$$

to obtain

$$\frac{3}{2}\frac{D_C(t)}{A_C(t)A_A(t)}\left(\frac{D_A(t)C_D(t)A_C(t)}{A_D(t)D_C(t)C_A(t)} - 1\right)^2 = \frac{\left(k_1^+\right)^2 k_2^+(1-\Delta)}{12k_1^-}t^2 + \cdots \tag{6.154}$$

Combination of Eq. (6.154) with Eq. (6.151) then yields the following expression:

$$K_{eq,1} = \frac{C_A(t)}{A_C(t)B_A(t)} + \frac{3}{2}\frac{D_C(t)}{A_C(t)A_D(t)}\left(\frac{D_A(t)C_D(t)A_C(t)}{A_D(t)D_C(t)C_A(t)} - 1\right)^2 \tag{6.155}$$

The domain of validity of these approximations has yet to be determined. Constales et al. (2012) hypothesized that such results hold for general reaction systems. Fig. 6.18 shows plots of the trajectories $C_A(t), A_C(t)$, and $B_A(t)$ and the right-hand side of Eq. (6.155) with only the first term and with two terms.

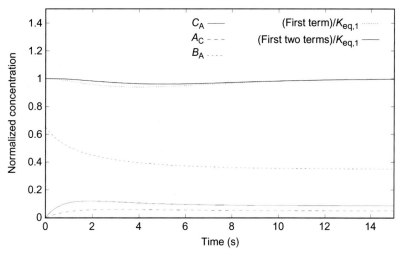

Fig. 6.18

Trajectories C_A, A_C, and B_A versus time and estimate of the ratio $C_A/(A_C B_A K_{eq,1})$, first term in Eq. (6.155) and its first-order correction (first two terms in Eq. 6.155) for the reaction sequence $A + B \rightleftarrows C \rightleftarrows D$; $k_1^+ = 1\,s^{-1}$; $k_1^- = 0.2\,s^{-1}$; $k_2^+ = 0.5\,s^{-1}$; $k_2^- = 0.2\,s^{-1}$ and $\Delta = 0.1$). *Reprinted from Constales, D., Yablonsky, G.S., Marin, G.B., 2012. Thermodynamic time invariants for dual kinetic experiments: nonlinear single reactions and more. Chem. Eng. Sci. 73, 20–29, Copyright (2012), with permission from Elsevier.*

6.7.4 Physicochemical Meaning of New Invariances and Their Novelty

Table 6.7 summarizes the results of all reactions discussed in this section.

Table 6.7 Time invariances for the reactions considered

Reaction	Invariance
$A \underset{k^-}{\overset{k^+}{\rightleftharpoons}} B$	$K_{eq} = \dfrac{k^+}{k^-} = \dfrac{B_A(t)}{A_B(t)}$
$2A \underset{k^-}{\overset{k^+}{\rightleftharpoons}} B$	$K_{eq} = \dfrac{k^+}{k^-} = \dfrac{B_A(t)}{A_A(t)A_B(t)}$
$2A \underset{k^-}{\overset{k^+}{\rightleftharpoons}} 2B$	$K_{eq} = \dfrac{k^+}{k^-} = \dfrac{B_A(t)B_B(t)}{A_A(t)A_B(t)}$
$A \underset{k_1^-}{\overset{k_1^+}{\rightleftharpoons}} B; 2A \underset{k_2^-}{\overset{k_2^+}{\rightleftharpoons}} 2B$	$-\dfrac{k_1^-}{k_2^-} = \dfrac{2\left(K_{eq}^2 A_A(t)A_B(t) - B_A(t)B_B(t)\right)}{K_{eq}A_B(t) - B_A(t)}$
$A + B \underset{k^-}{\overset{k^+}{\rightleftharpoons}} C \, (B - A = \Delta \geq 0)$	$K_{eq} = \dfrac{k^+}{k^-} = \dfrac{C_A(t)}{A_C(t)B_A(t)}$
$A + B \underset{k^-}{\overset{k^+}{\rightleftharpoons}} 2C \, (B - A = \Delta \geq 0)$	$K_{eq} = \dfrac{k^+}{k^-} = \dfrac{C_A(t)C_C(t)}{A_C(t)B_A(t)}$
$A + B \underset{k^-}{\overset{k^+}{\rightleftharpoons}} C + D \left(B - A = \Delta^{left}; D - C = \Delta^{right} \geq 0\right)$	$K_{eq} = \dfrac{k^+}{k^-} = \dfrac{C_A(t)D_C(t)}{A_C(t)B_A(t)}$
$A + Z \underset{k_1^-}{\overset{k_1^+}{\rightleftharpoons}} AZ \underset{k_2^-}{\overset{k_2^+}{\rightleftharpoons}} B + Z$ with quasi-steady state assumed for AZ.	$\dfrac{\left(1 + K_{eq}\right)\left(K_{eq,1}L^- - 1\right)}{\left(K_{eq,2} + L^-\right)\left(1 + K_{eq,1} + K_{eq}\right)}$ $= \dfrac{1}{A_B(t)B_A(t)} \ln \left(\dfrac{1 - \dfrac{B_A(t)}{B_{eq}}}{1 - \dfrac{A_B(t)}{A_{eq}}} \right)$

6.7.4.1 Constraints on dual experiments

There are two important requirements for performing dual experiments. First, the initial conditions have to meet the fundamental requirement that for both experiments the values of the equilibrium concentrations are the same. Therefore, both dual experiments must start with the same amounts of atoms of each type, and thus with the same total mass.

Second, the initial conditions must have extreme values within the given balances. For reversible reactions with one type of reactant and one type of product, for example, $A \rightleftharpoons B$ and $2A \rightleftharpoons B$, one of the dual experiments is performed starting from pure A and the other starting from pure B. For nonlinear reversible reactions with two types of reactants and one or two types of products ($A+B \rightleftarrows C$, $A+B \rightleftarrows 2C$, $A+B \rightleftarrows C+D$), the initial difference between the reactant concentrations must be the same. For example, if we start one dual experiment for the reaction $A+B \rightleftarrows C$ from a mixture of A and B with B in excess, $\Delta = B - A$, with $B > A$, then in the dual experiment starting from C, this same concentration difference Δ has to be maintained. Therefore, the initial conditions of this second experiment must be $A = 0$, $B = \Delta$, so $C = 1 - \Delta$. Other initial concentration differences are given in Table 6.7.

6.7.4.2 Kinetic fingerprints

Based on the theoretical results presented in this section, Constales et al. (2012) have derived kinetic fingerprints for distinguishing between the types of reversible reactions. This could be done because every reaction is characterized by a specific time invariance, which has the appearance of an equilibrium coefficient (see Table 6.7).

Special phase trajectories can also be used for distinguishing between reaction types. For the reaction $A \rightleftarrows B$, a linear dependence is found between the trajectories $B_A(t)$ and $B_B(t)$ (see Fig. 6.13). For the reactions $2A \rightleftarrows B$ and $2A \rightleftarrows 2B$ (see Figs. 6.14 and 6.15), the dependence is no longer linear, but in each case the curves form a family of conic sections, so that the coincidence of data points with one of them is a fingerprint of the corresponding reaction. The two families are highly distinct. For example, inspection of their tangent near equilibrium provides a downward angle of 45 degrees as a fingerprint for the reaction $2A \rightleftarrows 2B$.

Using phase trajectories for distinguishing between reactions is more visual in nature, while the tests of the equilibrium coefficients are more numerical, but they are essentially equivalent.

6.7.5 Concluding Remarks

The construction of thermodynamic time invariances may offer an interesting tool in distinguishing between different reaction types by comparing relationships between observed temporal concentration dependences with the equations summarized in Table 6.7.

The example of temporal behavior of a single-route catalytic reaction shows promising results. Apparently, many results obtained for single-route two-step catalytic reactions can be transferred into results for single-route reactions with a linear mechanism under the assumption of a quasi-steady-state kinetic regime regarding the surface intermediates.

The results obtained for an example of a multistep reaction with a nonlinear step, $A+B \rightleftarrows C \rightleftarrows D$, show that we are still at the beginning of revealing the time invariances for nonlinear mechanisms. What is known now is that the combination of temporal concentration dependences obtained during a pair of dual experiments can be approximated in terms of a power series expansion with the first term being the equilibrium coefficient. However, the domain of validity of these approximations is yet to be determined.

Appendix

The kinetic model for the mechanism of Example 6.2 in Section 6.4.2, $A \underset{k_1^-}{\overset{k_1^+}{\rightleftarrows}} B \overset{k_2^+}{\longrightarrow} C$, can be written as

$$\frac{dc_A}{dt} = -k_1^+ c_A + k_1^- c_B \tag{A6.1}$$

$$\frac{dc_B}{dt} = k^+ c_A - k_1^- c_B - k_2^+ c_B \tag{A6.2}$$

$$c_C = 1 - c_A - c_B \tag{A6.3}$$

Using the Laplace transform we can write

$$\mathcal{L}(c_A) - c_{A0} = -k_1^+ \mathcal{L}(c_A) + k_1^- \mathcal{L}(c_B) \tag{A6.4}$$

$$\mathcal{L}(c_B) - c_{B0} = k_1^+ \mathcal{L}(c_A) - k_1^- \mathcal{L}(c_B) - k_2^+ \mathcal{L}(c_B) \tag{A6.5}$$

$$\mathcal{L}(c_C) - c_{A0} = \frac{1}{s} - \mathcal{L}(c_A) - \mathcal{L}(c_B) \tag{A6.6}$$

Then,

$$\mathcal{L}(c_A) = \frac{(k_1^- + k_2^+ + s)c_{A0} + k_1^- c_{B0}}{D(s)} \tag{A6.7}$$

$$\mathcal{L}(c_B) = \frac{k_1^+ c_{A0} + (k_1^+ + s)c_{B0}}{D(s)} \tag{A6.8}$$

where $D(s)$ is the characteristic polynomial,

$$D(s) = s^2 + (k_1^+ + k_1^- + k_2^+)s + k_1^+ k_2^+ = (s + \lambda_p)(s + \lambda_m) \tag{A6.9}$$

with characteristic roots

$$\lambda_p, \lambda_m = \frac{(k_1^+ + k_1^- + k_2^+) \pm \sqrt{(k_1^+ + k_1^- + k_2^+)^2 - 4k_1^+ k_2^+}}{2} \tag{A6.10}$$

The term $(k_1^+ + k_1^- + k_2^+)^2 - 4k_1^+ k_2^+ = (k_1^+ + k_1^- - k_2^+)^2 + 4k_1^- k_2^+$ is always positive so both λ_p and λ_m are real.

Fig. A6.1 shows the dependence of the characteristic polynomial, $D(s)$, on the Laplace variable s.

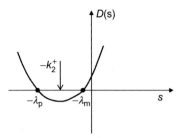

Fig. A6.1
Dependence of characteristic polynomial $D(s)$ on s.

If $s = -k_2^+$, the characteristic polynomial becomes

$$D(s) = \left(-k_2^+\right)^2 + \left(k_1^+ + k_1^- + k_2^+\right)\left(-k_2^+\right) + k_1^+ k_2^+ = -k_1^- k_2^+ < 0 \tag{A6.11}$$

Hence, $\lambda_m < k_2^+ < \lambda_p$. Also, $\lambda_p = k_1^+ + k_1^- + \left(k_2^+ - \lambda_m\right) > k_1^+$ and $\lambda_p \lambda_m = k_1^+ k_2^+$. Hence, $\lambda_m = \dfrac{k_1^+ k_2^+}{\lambda_p}$, but since $\lambda_p > k_2^+$ this implies that $\lambda_m < \dfrac{k_1^+ k_2^+}{k_2^+} = k_1^+$.

Obviously, $\lambda_p > k_2^+ > \lambda_m$ and $\lambda_p > k_1^+ > \lambda_m$. Therefore, we have justified our statements related to Example 6.2 in Section 6.4.2.

Nomenclature

A	coefficient in Eq. (6.31)
A_A, A_B	normalized concentration of A starting from resp. A, B
a_i	activity of component i
$\mathbf{c}$	column vector of concentrations (mol m^{-3})
c_i	concentration of component i (mol m^{-3})
C_V	total heat capacity of mixture at constant volume (J K^{-1})
$C_{V,i}$	molar heat capacity of component i at constant volume (J mol^{-1} K^{-1})
$D_{eff,G}$	effective diffusivity of gas (G m^2 s^{-1})
E_a	activation energy (J mol^{-1})
F_i	ith thermodynamic force
G	Gibbs free energy (J or J m^{-3})
$\Delta_r G$	Gibbs free energy change of reaction (J mol^{-1})
$\Delta_r H$	enthalpy change of reaction (J mol^{-1})
J	thermodynamic flux (mol m^{-2} s^{-1})
K_{eq}	equilibrium coefficient (depends)
$\mathbf{k}$	column vector of kinetic coefficients (first order) (s^{-1})
k	kinetic coefficient (first order) (s^{-1})
L_i	length of zone i (m)

N_c number of components

N_s number of elementary steps

n total number of moles in a system (mol)

n_i number of moles of component i (mol)

p pressure (total) (Pa)

p_i partial pressure of component i (Pa)

q_V volumetric flow rate (m^3 s^{-1})

R_g universal gas constant (J mol^{-1} K^{-1})

r reaction rate for a homogeneous reaction (mol m^{-3} s^{-1})

T temperature (K)

t time (s)

U internal energy (J)

u_i energy of the normal state of component i (J mol^{-1})

V volume (m^3)

$\mathbf{v}$ normal vector

(x, y, z) homogeneous coordinates

x_i conversion of component i

x_i, x_j any variable with respect to component i, j (depends)

Greek Symbols

α E_a^+/E_a^-

α_i absolute value of stoichiometric coefficient of reactant A_i

β_i stoichiometric coefficient of product B_i

γ_i activity coefficient

Δ difference between normalized concentrations

δ_i arbitrary constant

ε_i void fraction of zone i

λ eigenvalue

μ_i chemical potential of component i (J mol^{-1})

τ residence time (s)

τ_i residence time for zone i (s)

Φ thermodynamic potential

Subscripts

0   initial

eq   equilibrium

opt   optimum

s   step

ss   steady state

t   total

Superscripts

+	of forward reaction
−	of reverse reaction
o	standard conditions

References

Astumian, R.D., 2008. Reciprocal relations for nonlinear coupled transport. Phys. Rev. Lett. 101, 046802.

Atkins, P.W., 2010. The Laws of Thermodynamics: A Very Short Introduction. Oxford University Press, Oxford.

Boltzmann, L., 1964. Lectures on Gas Theory. University of California Press, Berkeley, CA.

Boudart, M., 1968. Kinetics of Chemical Processes. Prentice-Hall, Englewood Cliffs, NJ.

Boyd, R.K., 1974. Detailed balance in chemical kinetics as a consequence of microscopic reversibility. J. Chem. Phys. 60, 1214–1222.

Constales, D., Yablonsky, G.S., Galvita, V., Marin, G.B., 2011. Thermodynamic time-invariants: theory of TAP pulse-response experiments. Chem. Eng. Sci. 66, 4683–4689.

Constales, D., Yablonsky, G.S., Marin, G.B., 2012. Thermodynamic time invariants for dual kinetic experiments: nonlinear single reactions and more. Chem. Eng. Sci. 73, 20–29.

Constales, D., Yablonsky, G.S., Marin, G.B., 2015. Predicting kinetic dependences and closing the balance: Wei and Prater revisited. Chem. Eng. Sci. 123, 328–333.

de Groot, S.R., Mazur, P., 1962. Nonequilibrium Thermodynamics. North-Holland, Amsterdam.

Einstein, A., 1916. Strahlungs-Emission und -Absorption nach der Quantentheorie. Verh. Dtsch. Phys. Ges. 18, 318–323.

Feinberg, M., 1972. On chemical kinetics of a certain class. Arch. Ration. Mech. Anal. 46, 1–41.

Fogler, H.S., 2005. Elements of Chemical Reaction Engineering, fourth ed. Prentice Hall, Upper Saddle River, NJ.

Gleaves, J.T., Ebner, J.R., Kuechler, T.C., 1988. Temporal analysis of products (TAP)—a unique catalyst evaluation system with submillisecond time resolution. Catal. Rev. Sci. Eng. 30, 49–116.

Gorban, A.N., 1979. Invariant sets for kinetic equations. React. Kinet. Catal. Lett. 10, 187–190.

Gorban, A.N., 1984. Equilibrium Encircling: Equations of Chemical Kinetics and Their Thermodynamic Analysis. Nauka, Novosibirsk.

Gorban, A., 2013. Thermodynamic tree: the space of admissible paths. SIAM J. Appl. Dyn. Syst. 12, 246–278.

Gorban, A.N., Yablonskii, G.S., Bykov, V.I., 1982. Path to equilibrium. Int. Chem. Eng. 22, 368–375.

Gorban, A.N., Kaganovich, B.M., Filippov, S.I., et al., 2006. Thermodynamic Equilibria and Extrema. Analysis of Attainability Regions and Partial Equilibrium. Springer, New York, NY.

Grmela, M., 2002. Reciprocity relations in thermodynamics. Physica A 309, 304–328.

Kaganovich, B.M., Kuchmenko, E.V., Shamansky, V.A., Shirkalin, I.A., 2005. Thermodynamic modeling of phase transitions in multicomponent systems. Izv. Ross. Akad. Nauk. Energ. 2, 114–121 (in Russian).

Kaganovich, B.M., Keiko, A.V., Shamansky, V.A., Shirkalin, I.A., 2006. Description of non-equilibrium processes in energy problems by equilibrium thermodynamics. Izv. Ross. Akad. Nauk. Energ. 2, 64–75 (in Russian).

Kaganovich, B.M., Keiko, A.V., Shamansky, V.A., 2007. Equilibrium Thermodynamic Modeling of Dissipative Macroscopic Systems. Melentiev Institute of Energy Systems, Irkutsk (in Russian).

Kaganovich, B.M., Keiko, A.V., Shamansky, V.A., 2010. Equilibrium thermodynamic modeling of dissipative macroscopic systems. In: West, D.H., Yablonsky, G.S. (Eds.), Advances in Chemical Engineering. In: Thermodynamics and Kinetics of Complex Systems, vol. 39. Academic Press, New York, NY, pp. 1–74.

Marin, G.B., Yablonsky, G.S., 2011. Kinetics of Chemical Reactions—Decoding Complexity. Wiley-VCH, Weinheim.

Miller, D.G., 1960a. Errata—"Thermodynamics of irreversible processes. The experimental verification of the Onsager reciprocal relations" Chem. Rev. 60, 593.

Miller, D.G., 1960b. Thermodynamics of irreversible processes. The experimental verification of the Onsager reciprocal relations. Chem. Rev. 60, 15–37.

Onsager, L., 1931a. Reciprocal relations in irreversible processes. I. Phys. Rev. 37, 405–426.

Onsager, L., 1931b. Reciprocal relations in irreversible processes. II. Phys. Rev. 38, 2265–2279.

Ozer, M., Provaznik, I., 2005. A comparative tool for the validity of rate kinetics in ion channels by Onsager reciprocity theorem. J. Theor. Biol. 233, 237–243.

Schmidt, L.D., 1998. The Engineering of Chemical Reactions. Oxford University Press, Oxford.

Shinnar, R., 1988. Thermodynamic analysis in chemical process and reactor design. Chem. Eng. Sci. 43, 2303–2318.

Shinnar, R., Feng, C.A., 1985. Structure of complex catalytic reactions: thermodynamic constraints on kinetic modeling and catalyst evaluation. Ind. Eng. Chem. Res. 24, 153–170.

Smith, J.M., Van Ness, H.C., Abbott, M.M., 2005. Introduction to Chemical Engineering Thermodynamics, seventh ed. McGraw-Hill, New York, NY.

Stöckel, H., 1983. Linear and nonlinear generalizations of Onsager's reciprocity relations. Treatment of an example of chemical reaction kinetics. Fortschr. Phys. 31, 165–184.

Tolman, R.C., 1938. The Principles of Statistical Mechanics. Oxford University Press, Oxford.

van Kampen, N.G., 1973. Nonlinear irreversible processes. Physica 67, 1–22.

Vuddagiri, S.R., Hall, K.R., Eubank, Ph.T., 2000. Dynamic modeling of reaction pathways on the Gibbs energy surface. Ind. Eng. Chem. Fundam. 39, 508–517.

Wegscheider, R., 1901. Über simultane Gleichgewichte und die Beziehungen zwischen Thermodynamik und Reactionskinetik homogener Systeme. Monatsh. Chem. 22, 849–906.

Wei, J., Prater, C.D., 1962. The structure and analysis of complex reaction systems. In: Eley, D.D. (Ed.), Advances in Catalysis. Academic Press, New York, NY, pp. 203–392.

Yablonskii, G.S., Bykov, V.I., Gorban, A.N., et al., 1991. Kinetic models of catalytic reactions. Comprehensive Chemical Kinetics, vol. 32 Elsevier, Amsterdam.

Yablonsky, G., Ray, A., 2008. Equilibrium and optimum: how to kill two birds with one stone. IJCRE 6, A36.

Yablonsky, G.S., Constales, D., Marin, G.B., 2011a. Equilibrium relationships for non-equilibrium chemical dependences. Chem. Eng. Sci. 66, 111–114.

Yablonsky, G.S., Gorban, A.N., Constales, D., Galvita, V.V., Marin, G.B., 2011b. Reciprocal relations between kinetic curves. Europhy. Lett. 93, 20004–20007.

Zel'dovich, Ya.B., 1938. Proof of a unique solution to the mass action law. Zh. Tekh. Fiz. 11, 685–687 (in Russian).

## CHAPTER 7

Stability of Chemical Reaction Systems

7.1 Stability—General concept

7.1.1 Introduction

The concept of stability is extremely important for a wide variety of dynamical systems, ranging from purely mathematical to natural and social systems. Physical and engineering systems, such as chemical reaction systems, are usually designed to operate at an equilibrium state (closed systems) or steady state (open systems), which must be stable for successful operation. This stability depends on whether a small perturbation is damped, leading to a stable state, or amplified, which results in an unstable state. Typically, the simple mechanical analogy of the ball on top of a hill or at the bottom of a valley is used to clarify the stability concept (Fig. 7.1). The system in Fig. 7.1A is considered to be unstable because after only a very small perturbation, the ball lying on top of the hill will continue to move away from its original position. If the ball is initially at the bottom of a valley and the hills surrounding the valley are infinitely high (Fig. 7.1B), upon any perturbation the ball will oscillate around its original position eventually coming to a halt at the bottom of the valley. This corresponds to "global" stability. In the case of a finite height of the hills (Fig. 7.1C and D), the ball will return to its original position upon a small perturbation, but if the perturbation reaches a certain threshold value, the ball will slip over the hill to the next valley. This is "local" stability.

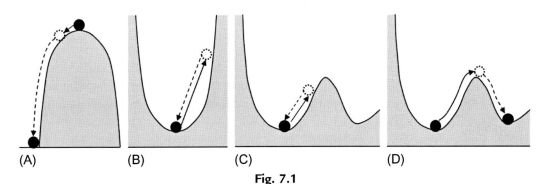

(A) (B) (C) (D)

Fig. 7.1

Instability and stability. The system is (A) unstable; (B) globally stable; (C) locally stable with small perturbation; (D) locally stable with perturbation beyond the threshold value.

Advanced Data Analysis and Modelling in Chemical Engineering. http://dx.doi.org/10.1016/B978-0-444-59485-3.00007-2

7.1.2 Non-Steady-State Models

The dynamic behavior of chemical reaction systems is described in terms of non-steady-state models, which in their simplest form are sets of ordinary differential equations of the type

$$\frac{d\mathbf{c}}{dt} = f(\mathbf{c}, \mathbf{k}) \tag{7.1}$$

in which $\mathbf{c}$ is a vector of concentrations and $\mathbf{k}$ is a vector of kinetic parameters. The space of vectors $\mathbf{c}$ is the phase space of Eq. (7.1), the space in which all possible states of the reaction system are represented with every possible state corresponding to one unique point. The points are specified by the state variables $c_1, c_2, \ldots, c_{N_c}$, where N_c is the number of components present in the chemical reaction mixture. In the case of a nonisothermal system, this set of equations has to be supplemented with an equation of the type

$$\frac{dT}{dt} = f(\mathbf{c}, \mathbf{k}, \mathbf{h}) \tag{7.2}$$

where $\mathbf{h}$ are heat parameters of the model, that is, heat capacities, heat-transfer coefficients, and heats of reaction.

Eq. (7.1) has only one solution, $\mathbf{c}(t, \mathbf{k}, \mathbf{c_0})$, for *any* nonnegative initial condition $\mathbf{c}(0, \mathbf{k}, \mathbf{c_0}) = \mathbf{c_0}$, which is natural for chemical kinetics. The changing state of the system with time traces a path $\mathbf{c}(t, \mathbf{k}, \mathbf{c_0})$ at fixed $\mathbf{k}$ and $\mathbf{c_0}$ and $t \in [0, \infty]$ through the phase space and is called a phase trajectory. A phase trajectory represents the set of chemical compositions of the reaction mixture that can be reached from *a particular* initial condition.

From a physicochemical point of view, we are only interested in positive trajectories, which in mathematics are called "semitrajectories." However, in some cases, values of $\mathbf{c}(t, \mathbf{k}, \mathbf{c_0})$ on negative trajectories with $t \in [-\infty, 0]$ and whole trajectories with $t \in [-\infty, \infty]$ are informative regarding the behavior in the physicochemically relevant (positive) concentration domain.

As Eq. (7.1) for each initial condition has only one, unique solution, every point in the phase space is passed by one and only one of the phase trajectories, which neither cross one another nor merge.

A family of phase trajectories starting from different initial states is called a phase diagram or phase portrait. A phase portrait graphically shows how the system moves from the initial states and reveals important aspects of the dynamics of the system. Certain phase portraits display one or more attractors, which are long-term stable sets of states toward which a system tends to evolve (is attracted) dynamically, from a wide variety of initial conditions, the basin of attraction. The simplest type of attractor is a so-called rest point. This is a point for which the derivatives of all the state variables are zero:

$$\frac{d\mathbf{c}}{dt} = f(\mathbf{c}, \mathbf{k}) = 0 \tag{7.3}$$

In mathematics this is also called a fixed point, an equilibrium point, a stationary point, or a singular point. If the system is placed in such a point, it will remain there if not disturbed:

$$\mathbf{c}(t, \mathbf{k}, \mathbf{c_0}) \equiv \mathbf{c_0} \tag{7.4}$$

For a closed chemical system, that is, a system without exchange of matter with the environment, the rest point is termed the equilibrium point, the equilibrium state, or just equilibrium, with concentration vector $\mathbf{c_{eq}}$. For an open chemical system, in which there is exchange of matter, the rest point is termed the steady state, with corresponding concentration vector $\mathbf{c_{ss}}$. Since a closed system is a particular case of an open system, equilibrium is a particular case of steady state.

A special mathematical problem is that of determining the number of rest points. Rest points can be classified as internal with all variables nonzero or as boundary with at least one variable equal to zero. In a closed chemical system, the internal rest point is unique. Zel'dovich (1938) provided the first version of the proof of this uniqueness. Gorban (1980) analyzed the boundary equilibrium points in a closed chemical system. For an open chemical system, isothermal or nonisothermal, multiple internal steady states may exist.

Besides the rest points, there are trajectories that reflect the movement of a chemical system to the rest point or around it. Some of these trajectories, called limit cycles, are closed and represent a mathematical image of oscillations, to or from which all trajectories nearby will converge or diverge. In the case of convergence, the limit cycle is a periodic attractor. The term "limit cycle" refers to the cyclical behavior. In a nonlinear system, limit cycles describe the amplitudes and periods of self-oscillations, which have been observed experimentally in many chemical systems.

In the mathematical theory of dynamical systems, the more general concept of ω-limit sets has been developed. These ω-limit sets help in gaining an overall understanding of how a dynamical system behaves, particularly in the long term, by providing an asymptotic description of the dynamics. An ω-limit set combines a set of rest points with a set of specific phase trajectories, particularly trajectories around the rest points. In the simplest case, the ω-limit set $\omega(\mathbf{k}, \mathbf{c_0})$ consists of just a rest point.

The obvious advantage of a qualitative analysis of differential equations is that it can be used for sketching phase portraits without actually solving the equations. It is very useful in the analysis of many nonlinear models if no analytical solution can be obtained and the capabilities of computational methods are limited.

An important problem in the qualitative analysis of nonlinear systems is elucidating the structure of the ω-limit sets. Unfortunately, no general method exists yet for solving this problem. Until the 1960s, only two cases had been studied extensively:

- systems with many linear variables and/or a few nonlinear variables, but near the rest point
- nonlinear systems with only two variables

In 1963, Lorenz (Lorenz, 1963), driven by the wish to understand the foundations of weather forecasting, presented the following seemingly simple set of ordinary differential equations

$$\begin{cases} \dfrac{dx}{dt} = \sigma(y - x) \\[2mm] \dfrac{dy}{dt} = \rho x - xz - y \\[2mm] \dfrac{dz}{dt} = xy - \beta z \end{cases} \qquad (7.5)$$

where x, y, and z are variables and σ, ρ, and β are parameters.

This model with only three variables, whose only nonlinearities are xy and xz, exhibited dynamic behavior of unexpected complexity (Fig. 7.2). It was especially surprising that this deterministic model was able to generate chaotic oscillations. The corresponding limit set was called the "Lorenz attractor" and limit sets of similar type are called "strange attractors." Trajectories within a strange attractor appear to hop around randomly but, in fact, are organized by a very complex type of stable order, which keeps the system within certain ranges.

Although the Lorenz model is not a model of chemical kinetics, there is some similarity; in both model types the right-hand side is of the polynomial type with first- and second-order terms. In this chapter, we will present results of the analysis of a nonlinear model—also with three variables—the "catalytic oscillator" model.

7.1.3 Local Stability—Rigorous Definition

Let $c(t, \mathbf{k}, \mathbf{c_0})$ be a solution of Eqs. (7.3) and (7.4). This solution is called *Lyapunov stable* if for any infinitesimal perturbation $\varepsilon > 0$, values of $\delta > 0$ exist such that the inequality

$$|\mathbf{c_0} - \tilde{\mathbf{c}}_0| < \delta \qquad (7.6)$$

results in

$$|\mathbf{c}(t, \mathbf{k}, \mathbf{c_0}) - \mathbf{c}(t, \mathbf{k}, \tilde{\mathbf{c}}_0)| < \varepsilon \qquad (7.7)$$

Here, $\mathbf{c_0}$ and $\tilde{\mathbf{c}}_0$ are sets of initial concentrations ($t = 0$) in, respectively, the unperturbed and the perturbed case. The solution $\mathbf{c}(t, \mathbf{k}, \mathbf{c_0})$ is called asymptotically stable if it is Lyapunov stable and a value of δ exists such that the inequality of Eq. (7.7) results in

$$\lim_{t \to \infty} |\mathbf{c}(t, \mathbf{k}, \mathbf{c_0}) - \mathbf{c}(t, \mathbf{k}, \tilde{\mathbf{c}}_0)| \to 0 \qquad (7.8)$$

so all systems starting out near $\mathbf{c_0}$ converge to $\mathbf{c_0}$.

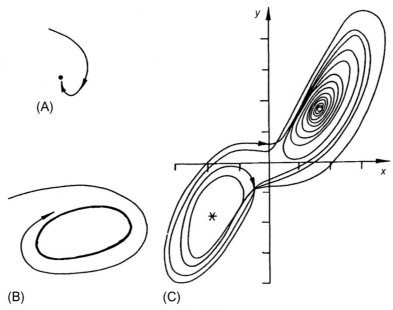

Fig. 7.2

Examples of ω-limit sets; (A) rest point, (B) limit cycle, (C) Lorenz attractor (projection on the (x, y) plane; $\sigma = 10$, $\rho = 30$, $\beta = 8/3$). *Reprinted from Yablonskii, G.S., Bykov, V.I., Gorban, A.N., Elokhin, V.I., 1991. Kinetic models of catalytic reactions. In: Compton, R.G. (Ed.), Comprehensive Chemical Kinetics, vol. 32. Elsevier, Amsterdam, Copyright (1991), with permission from Elsevier.*

This definition is related to a phase trajectory and as a rest point is a particular type of phase trajectory, this definition also applies to rest points. A rest point is Lyapunov stable if for any $\varepsilon_c > 0$ a value of $\delta > 0$ exists such that after a deviation from this point within δ, the system remains close to it, within the value of ε_c, for a long period of time. A rest point is asymptotically stable if it is Lyapunov stable and values of $\delta > 0$ exist such that after a deviation from this point within δ, the system approaches the rest point at $t \to \infty$.

7.1.4 Analysis of Local Stability

7.1.4.1 Procedure

For the analysis of local Lyapunov stability of rest points, a traditional and very reliable ritual exists, which in fact is an extension of the approach demonstrated in Section 3.6.2.
This ritual is a combination of a number of steps:

1. Linearization of the system
   a. Introducing new variables $\boldsymbol{\xi} = \mathbf{c} - \tilde{\mathbf{c}}$ that are equal to the deviation of the current concentration from the rest point
   b. Expressing Eq. (7.1) as a function of ξ

c. Expanding the right-hand side of Eq. (7.1) into a series over the powers of ξ and discarding the nonlinear terms

The linearized system obtained can be represented as

$$\frac{d\boldsymbol{\xi}}{dt} = \mathbf{J}\boldsymbol{\xi} \tag{7.9}$$

where $\mathbf{J}$ is the matrix of partial derivatives, the so-called Jacobian:

$$\mathbf{J} = \frac{\partial \mathbf{f}(\mathbf{c}, \mathbf{k})}{\partial \mathbf{c}}\bigg|_{\mathbf{c}=\tilde{\mathbf{c}}} \tag{7.10}$$

2. Derivation of the characteristic equation

$$\det(\mathbf{J} - \lambda \mathbf{I}) = 0 \tag{7.11}$$

where $\mathbf{I}$ is the unit matrix and λ are the characteristic roots.

3. Analysis of the roots of the characteristic equation

The stability of the rest points of Eq. (7.1) depends on the roots of the characteristic equation, Eq. (7.11). The rest point is asymptotically stable if the real parts of all of the roots of Eq. (7.11) are negative, and unstable if the real part of at least one of the roots is positive. If some roots are purely imaginary and the others have a negative real part, the rest point of Eq. (7.1) is Lyapunov stable but not asymptotically stable.

7.1.4.2 Analysis of local stability in a system with two variables

We will illustrate the procedure described above with an example of a model with two variables:

$$\begin{cases} \dfrac{dc_1}{dt} = f_1(c_1, c_2) \\[2mm] \dfrac{dc_2}{dt} = f_2(c_1, c_2) \end{cases} \tag{7.12}$$

This set of equations can be linearized by introducing the variables $\xi_1 = c_1 - \tilde{c}_1$ and $\xi_2 = c_2 - \tilde{c}_2$, resulting in

$$\begin{cases} \dfrac{d\xi_1}{dt} = a_{11}\xi_1 + a_{12}\xi_2 \\[2mm] \dfrac{d\xi_2}{dt} = a_{21}\xi_1 + a_{22}\xi_2 \end{cases} \tag{7.13}$$

where the matrix of the coefficients is

$$\mathbf{J} = \begin{bmatrix} a_{11} & a_{12} \\ a_{21} & a_{22} \end{bmatrix} \tag{7.14}$$

with

$$a_{11} = \frac{\partial f_1(c_1, c_2)}{\partial c_1}\bigg|_{\tilde{c}} \quad ; \quad a_{12} = \frac{\partial f_1(c_1, c_2)}{\partial c_2}\bigg|_{\tilde{c}}$$

$$a_{21} = \frac{\partial f_2(c_1, c_2)}{\partial c_1}\bigg|_{\tilde{c}} \quad ; \quad a_{22} = \frac{\partial f_2(c_1, c_2)}{\partial c_2}\bigg|_{\tilde{c}} \tag{7.15}$$

The solution of Eq. (7.13) is of the form

$$\begin{cases} \xi_1 = A e^{\lambda t} \\ \xi_2 = B e^{\lambda t} \end{cases} \tag{7.16}$$

Substitution of Eq. (7.16) into Eq. (7.13) and division by $e^{\lambda t}$ yields

$$\lambda A = a_{11}A + a_{12}B \tag{7.17}$$

$$\lambda B = a_{21}A + a_{22}B \tag{7.18}$$

Rearranging Eq. (7.18) we can express coefficient B in terms of A:

$$B = A\frac{\lambda - a_{11}}{a_{12}} \tag{7.19}$$

and after substitution of Eq. (7.19) into Eq. (7.17) and rearrangement, we obtain

$$\lambda^2 - (a_{11} + a_{22})\lambda + (a_{11}a_{22} - a_{12}a_{21}) = 0 \tag{7.20}$$

This is the characteristic equation which in the form of Eq. (7.11) can be written as

$$\begin{bmatrix} a_{11} - \lambda & a_{12} \\ a_{21} & a_{22} - \lambda \end{bmatrix} = 0 \tag{7.21}$$

Solving Eq. (7.20) yields

$$\lambda_{1,2} = \frac{a_{11} + a_{22}}{2} \pm \sqrt{\left(\frac{a_{11} + a_{22}}{2}\right)^2 + a_{12}a_{21} - a_{11}a_{22}} \tag{7.22}$$

It is important to stress that the characteristic roots can be estimated based on the matrix of coefficients of the linearized system, Eq. (7.14). The sum of the characteristic roots is equal to the sum of the diagonal elements of this matrix, the trace,

$$\lambda_1 + \lambda_2 = a_{11} + a_{22} = \mathrm{Tr}(\mathbf{J}) \tag{7.23}$$

and their product is equal to its determinant:

$$\lambda_1 \lambda_2 = a_{11}a_{22} - a_{12}a_{21} = \det(\mathbf{J}) \tag{7.24}$$

which is the free term in the characteristic equation, Eq. (7.20). These summation and multiplication rules always hold for linear systems.

According to the theory of linear differential equations, the complete solution of Eq. (7.13) can be found to be

$$\begin{cases} \xi_1 = b_{11}e^{\lambda_1 t} + b_{12}e^{\lambda_2 t} \\ \xi_2 = b_{21}e^{\lambda_1 t} + b_{22}e^{\lambda_2 t} \end{cases} \tag{7.25}$$

with coefficients b_{ij} determined by the initial conditions.

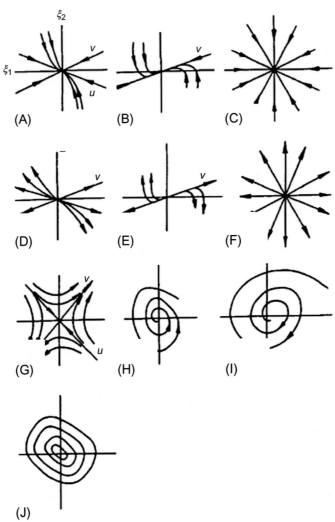

Fig. 7.3

Types of rest points on the plane. (A), (B), (C) stable nodes; (D), (E), (F) unstable nodes; (G) saddle point; (H) stable focus; (I) unstable focus; (J) center; line u: $\dfrac{\xi_1}{\xi_2} = \dfrac{b_{11}}{b_{21}}$, line v: $\dfrac{\xi_1}{\xi_2} = \dfrac{b_{12}}{b_{22}}$.

Adapted from Yablonskii, G.S., Bykov, V.I., Gorban, A.N., Elokhin, V.I., 1991. Kinetic models of catalytic reactions. In: Compton, R.G. (Ed.), Comprehensive Chemical Kinetics, vol. 32. Elsevier, Amsterdam, Copyright (1991), with permission from Elsevier.

The characteristic roots determine not only the local stability of the rest point but also the nature of the dynamics near this point. Fig. 7.3 (Yablonskii et al., 1991) shows different types of rest points.

Let us now analyze the linearized equation, Eq. (7.13). For this purpose we will consider two cases: (i) the roots of the characteristic equation are real and (ii) these roots are imaginary.

Real roots

In the case of real roots, there are a number of possible situations:

1. If $\lambda_1, \lambda_2 < 0$ and $\lambda_1 \neq \lambda_2$, ξ_1 and ξ_2 both are sums of exponents decreasing with time. Therefore, the rest point is a stable node. The nature of the trajectories near a stable node is illustrated in Fig. 7.3A for $\lambda_1 < \lambda_2$. The straight lines u and v are specified, respectively, by the equations

$$\frac{\xi_1}{\xi_2} = \frac{b_{11}}{b_{21}} \tag{7.26}$$

and

$$\frac{\xi_1}{\xi_2} = \frac{b_{12}}{b_{22}} \tag{7.27}$$

2. In the case that $\lambda_1 = \lambda_2 = \lambda < 0$, there are two possibilities. The first is that the matrix of coefficients can be reduced to the form

$$\mathbf{J} = \begin{bmatrix} a_{11} & a_{12} \\ a_{21} & a_{22} \end{bmatrix} = \begin{bmatrix} \lambda & 1 \\ 0 & \lambda \end{bmatrix} \tag{7.28}$$

by linear transformation of variables. In that case

$$\begin{cases} \xi_1 = b_{11}e^{\lambda t} + b_{12}te^{\lambda t} \\ \xi_2 = b_{21}e^{\lambda t} + b_{22}te^{\lambda t} \end{cases} \tag{7.29}$$

and the nature of the trajectories is as shown in Fig. 7.3B, where v is again specified by Eq. (7.27).

The second possibility is that the matrix of coefficients of the system represented by Eq. (7.13) is a diagonal one:

$$\mathbf{J} = \begin{bmatrix} a_{11} & a_{12} \\ a_{21} & a_{22} \end{bmatrix} = \begin{bmatrix} \lambda & 0 \\ 0 & \lambda \end{bmatrix} \tag{7.30}$$

Then the solution is of the form of Eq. (7.16) and the trajectories behave as shown in Fig. 7.3C. In both cases, the rest point is a stable node.

3. If $\lambda_1, \lambda_2 > 0$ and $\lambda_1 > \lambda_2$, the phase trajectories extend far from the rest point. This is an unstable node (Fig. 7.3D).

4. If $\lambda_1 = \lambda_2 = \lambda > 0$, the direction of motion is reversed compared with case 2. The phase trajectories extend far from the rest points, which are unstable nodes (Fig. 7.3E and F).

5. If the roots are of different signs, for example, $\lambda_1 > 0$ and $\lambda_2 < 0$, ξ_1 and ξ_2 both are sums of exponents with different signs. The rest point is unstable, since with time the term with the positive exponent becomes predominant. Such a rest point is shown in Fig. 7.3G and is called a saddle point. A saddle point can be reached, but via two trajectories only. Therefore, rigorously speaking it should be treated as semistable.

Imaginary roots

For analyzing the case where there are imaginary roots, we must transform Eq. (7.13) into a second-order equation. Differentiating the first equation and eliminating ξ_2 yields

$$\frac{d^2 \xi_1}{dt^2} + 2\varphi \frac{d\xi_1}{dt} + \omega_0^2 \xi_1 = 0, \tag{7.31}$$

where $2\varphi = a_{11} + a_{22}$ and $\omega_0^2 = a_{11}a_{22} - a_{12}a_{21}$. Eq. (7.22) can now be written as follows:

$$\lambda_{1,2} = -\varphi \pm \sqrt{\varphi^2 - \omega_0^2} \tag{7.32}$$

As these roots have imaginary parts, $\omega_0^2 - \varphi^2 = \omega^2 > 0$ and

$$\lambda_{1,2} = -\varphi \pm i\omega \tag{7.33}$$

It can be shown readily that the solution of Eq. (7.31) is of the form

$$\xi_1 = e^{-\varphi t}(b_1 \cos \omega t + b_2 \sin \omega t) \tag{7.34}$$

Now there are three possible situations:

1. If $\varphi > 0$, the solution of Eq. (7.34) is an image of damped oscillations. The phase trajectories are converging spirals and the rest point is called a stable focus (Fig. 7.3H).
2. If $\varphi < 0$, the phase trajectories are diverging spirals and the rest point is called an unstable focus (Fig. 7.3I).
3. If $\varphi = 0$, the solution is related to undamped oscillations with a frequency ω. The phase trajectories are concentric circles (Fig. 7.3J) and the rest point is called a center. This is a very sensitive rest point; even in the case of very small variations of the parameters, the phase portrait changes. In contrast, for nonsensitive rest points, the phase portrait is not affected by small parameter variations.

7.1.5 Global Dynamics

Now let us define the global stability of rest points. The rest point c_0 is called a global asymptotically stable rest point within the phase space D if it is Lyapunov stable and for any condition $d_0 \in D$ the solution $c(t, k, d_0)$ approaches c_0 at $t \rightarrow \infty$. An analysis of the problem of global stability is much more difficult than that of local stability. In most cases, global stability can be proved by using a properly selected function called the Lyapunov function.

If a mathematical model has a Lyapunov function, this means that the rest point is stable. Let us consider the function $V(c_1, c_2, \ldots)$ having first-order partial derivatives $\partial V/\partial c_i$, which can be treated as derivatives of the solution of Eq. (7.1):

$$\frac{dV(\mathbf{c}(t))}{dt} = \sum_i \frac{\partial V}{\partial c_i} \frac{dc_i}{dt} = \sum_i \frac{\partial V}{\partial c_i} f_i(\mathbf{c}, \mathbf{k}) \tag{7.35}$$

Intuitively, the physical meaning of the Lyapunov function $V(\mathbf{c})$ in dissipative systems is that of an energy, for example the Gibbs free energy, negative entropy, or mechanical energy.

In practice, various different stabilities are applied, but we will discuss only one version of a method based on the Lyapunov function. Let $dV(\mathbf{c})/dt \leq 0$ and only at the rest point under consideration $dV(\mathbf{c})/dt = 0$. Then, the minimum, $V(\mathbf{c}) = V(\mathbf{c})_{\min}$, occurs at the rest point $\bar{\mathbf{c}}$ and for some $\varepsilon > V(\mathbf{c})_{\min}$ the set specified by the inequality $V(\mathbf{c_0}) < \varepsilon$, with $\mathbf{c_0}$ a set of initial conditions, is finite. Thus, for any set of initial conditions, the solution of Eq. (7.1) is

$$\lim_{t \to \infty} \mathbf{c}(t, \mathbf{k}, \mathbf{c_0}) = \bar{\mathbf{c}} \tag{7.36}$$

In a closed physicochemical system, the Lyapunov function is the Gibbs free energy. The rest point of such a system, that is, equilibrium or steady state, is unique and stable.

7.2 Thermodynamic Lyapunov Functions

7.2.1 Introduction

Special attention has to be paid to thermodynamic Lyapunov functions (Gorban, 1984). Basic variables describing a chemical state are amounts of components i, which can be expressed as amount of substance, n_i, or concentration, c_i. If the chemical system is homogeneous, the conditions are fixed and the equation of state is known, all system characteristics can be expressed using the vector of amounts of substance, $\mathbf{n}$, and some constant parameters, for example

- isothermal isochoric conditions: $\mathbf{n}$, T, V
- isothermal isobaric conditions: $\mathbf{n}$, T, p
- thermally isolated systems at isochoric conditions: $\mathbf{n}$, U, V
- thermally isolated systems at isobaric conditions: $\mathbf{n}$, p, H

The enthalpy H is given by $H = U + pV$, with U the internal energy, p the total pressure, and V the total volume.

As discussed previously, every elementary step s is characterized by its rate:

$$r_s(\mathbf{c}, T) = r_s^+(\mathbf{c}, T) - r_s^-(\mathbf{c}, T) \tag{7.37}$$

where $r_s^+(\mathbf{c}, T)$ and $r_s^-(\mathbf{c}, T)$ are the rates of, respectively, the forward and the reverse reaction:

$$r_s^+(\mathbf{c}, T) = k_s^+(T) \prod_i c_i^{\alpha_{si}} \tag{7.38}$$

$$r_s^+(\mathbf{c}, T) = k_s^-(T) \prod_i c_i^{\beta_{si}} \tag{7.39}$$

with α_{si} and β_{si} the partial reaction orders. Eqs. (7.38) and (7.39) are traditional mass-action-law equations. The rate coefficients are given by the Arrhenius equation:

$$k_s^+(T) = k_{s0}^+ T^{n_s^+} \exp\left(-\frac{E_{a,s}^+}{R_g T}\right) \tag{7.40}$$

$$k_s^-(T) = k_{s0}^- T^{n_s^-} \exp\left(-\frac{E_{a,s}^-}{R_g T}\right) \tag{7.41}$$

where $E_{a,s}^+$ and $E_{a,s}^-$ are the activation energies of the forward and the reverse reaction, k_{s0}^+ and k_{s0}^- are the preexponential factors, and $T^{n_s^+}$ and $T^{n_s^-}$ are factors describing the temperature dependences of the preexponential factors.

Every component i is characterized by its chemical potential μ_i, which is an intensive characteristic, that is, it is independent on the volume of the system. The function $\mu_i(\mathbf{c}, T)$ is defined within the domain $T > 0$, $c_i > 0$, $c_j \geq 0$, $j \neq i$ and has derivatives in the same domain. At points where $c_i = 0$, this function typically has a logarithmic peculiarity, for $c_i \to 0$, $\mu_i \to -\infty$ (similar to $\ln c_i$). For any positive T and c_i, the following condition holds:

$$r_s(\mathbf{c}, T) \sum_i \gamma_{si} \mu_i(\mathbf{c}, T) \leq 0 \tag{7.42}$$

where $\gamma_{si} = \beta_{si} - \alpha_{si}$ is the stoichiometric coefficient of component i in step s. The rate of step s only equals zero at such positive values of T and c_i for which

$$\sum_i \gamma_{si} \mu_i(\mathbf{c}, T) = 0 \tag{7.43}$$

Equations of chemical kinetics can be written as

$$\frac{d\mathbf{n}}{dt} = V \sum_s \boldsymbol{\gamma}_s r_s(\mathbf{c}, T) \tag{7.44}$$

The fact that chemical potentials meet requirements (7.43) and (7.44) has many consequences.

The first is a version of the principle of detailed balance. Assuming that at some condition $\mathbf{c} = \mathbf{c}_{eq}$, $T = T_{eq}$, the following equality is fulfilled:

$$r_1(\mathbf{c}_{eq}, T_{eq}) = r_2(\mathbf{c}_{eq}, T_{eq}) = \cdots = r_\kappa(\mathbf{c}_{eq}, T_{eq}) = 0.$$

If the stoichiometric vector $\boldsymbol{\gamma}_{\kappa+1}$ is a linear combination of $\gamma_1, \ldots, \gamma_\kappa$, then $r_s(\mathbf{c}_{eq}, T_{eq}) = 0$. The point $(\mathbf{c}_{eq}, T_{eq})$ at which the rates of all steps equal zero is a point of the detailed balance.

If for a given reaction mechanism, the vectors $\boldsymbol{\gamma}_s$ are linearly independent, the rest point of Eq. (7.43) is a point of the detailed balance. Adding a new step with a stoichiometric vector that is a linear combination of the stoichiometric vectors of the original mechanism does not change the point of the detailed balance. This is a very important property of a closed chemical system.

If all of the chemical potentials are multiplied with the same positive function $f(\mathbf{c}, T)$, the conditions of Eqs. (7.42) and (7.43) will still be valid. In this case, a pseudopotential μ_i' arises instead of the potential.

Based on Eqs. (7.42) and (7.43), a generalized kinetic law can be constructed:

$$r_s(\mathbf{c}, T) = r_s^o(\mathbf{c}, T) \left[\exp\left(\sum_i \alpha_i \tilde{\mu}_i \right) - \exp\left(\sum_i \beta_i \tilde{\mu}_i \right) \right] \tag{7.45}$$

where $r^o(\mathbf{c}, T)$ is an arbitrary positive kinetic function and $\tilde{\mu}_i = \dfrac{\mu_i}{R_g T}$ is the dimensionless potential or pseudopotential. Eq. (7.45) is commonly known as the Marcelin-de Donder equation and the characteristic $a_i = \exp(\tilde{\mu}_i)$ is termed the activity. For ideal systems, this activity is equal to the concentration. In that case, the Marcelin-de Donder expression is transformed to the mass-action law.

7.2.2 Dissipativity of Thermodynamic Lyapunov Functions

In an isolated system (U, V = constant), the entropy $S(\mathbf{n}, U, V)$ increases in the course of any chemical step:

$$r_s(\mathbf{c}, T) \sum_i \gamma_{si} \frac{\partial S}{\partial n_i} \geq 0 \tag{7.46}$$

For the regimes defined in Section 7.2.1, a thermodynamic Lyapunov function, L_x, can be constructed, for example

$$\left(\frac{\partial L_x(\mathbf{n})}{\partial n_i} \right)_{constants} = \frac{\mu_i(\mathbf{c}, T)}{R_g T} = \tilde{\mu}_i(\mathbf{c}, T) \tag{7.47}$$

We will use the basic thermodynamic equation

$$dU + pdV - TdS - \sum_i \mu_i dn_i = 0 \tag{7.48}$$

and rewrite it as

$$\sum_i \tilde{\mu}_i dn_i = \frac{1}{R_g T} dU + \frac{p}{R_g T} dV - \frac{1}{R_g} dS \tag{7.49}$$

The corresponding thermodynamic Lyapunov function $L_x(n, U, T)$ at constant V and T can be found from

$$dL_{V,T} = \left(\frac{1}{R_g T} dU + \frac{p}{R_g T} dV - \frac{1}{R_g} dS \right)_{V,T} \tag{7.50}$$

The function

$$L_{V,T} = \frac{U - TS}{R_g T} = \frac{A}{R_g T} \tag{7.51}$$

meets the condition of Eq. (7.50). Here $A = U - TS$ is the Helmholtz free energy.

For studying systems at constant pressure, Eq. (7.49) has to be rewritten using the enthalpy $H = U + pV$. Then

$$dH = dU + pdV + Vdp \tag{7.52}$$

and substitution of Eq. (7.52) in Eq. (7.49) yields

$$\sum_i \tilde{\mu}_i dn_i = \frac{1}{R_g T} dH + \frac{V}{R_g T} dp - \frac{1}{R_g} dS \tag{7.53}$$

For an isobaric thermally isolated system $L_{p,H}(\mathbf{n}, p, H)$

$$dL_{p,H} = \left(\frac{1}{R_g T} dH + \frac{V}{R_g T} dp - \frac{1}{R_g} dS \right)_{p,H} \tag{7.54}$$

The function meeting this condition is

$$L_{p,H} = -\frac{S}{R_g} \tag{7.55}$$

This is the same state function as $L_{U,V}$.

Eq. (7.54) for constant p and T leads to

$$L_{p,T} = \frac{H - TS}{R_g T} = \frac{G}{R_g T} \tag{7.56}$$

Here $G = H - TS$ is the Gibbs free energy function.

For any classical conditions, the derivative of the thermodynamic Lyapunov function will be presented as follows:

$$\frac{dL_x}{dt} = -V\sum_s \left(r_s^+ - r_s^-\right) \ln\frac{r_s^+}{r_s^-} \qquad (7.57)$$

It has been proposed to apply this characteristic to the analysis of complex chemical reactions (Bykov et al., 1977; Dimitrov et al., 1977). Using Eq. (7.57), it is possible to calculate the contribution of a step s to the whole rate of change of the thermodynamic Lyapunov function, for instance the Gibbs free energy rate of change. This characteristic together with the calculated contribution of step s to the complete rates of change of the components have been used for the simplification of complex reaction mechanisms (Bykov et al., 1977), in particular the mechanism for the homogeneous oxidation of hydrogen (Dimitrov et al., 1977; Dimitrov, 1982). Gorban (1984) has analyzed the thermodynamic Lyapunov functions in detail.

In addition to considering Lyapunov functions, it is also useful to investigate so-called ω-invariant sets. A set S in the phase space is called ω-invariant if for any solution of Eq. (7.3) $\mathbf{c}(t_0)$ lies within S, that is, $\mathbf{c}(t_0) \in S$. It then follows that $\mathbf{c}(t_0) \in S$ for any $t > t_0$. This ω-invariant set is a kind of 'bag'; once entered, the solution will not leave this bag.

7.2.3 Chemical Oscillations

Isothermal chemical oscillations discovered by Belousov and Zhabotinsky became a sensation in chemistry in the 1950–1970s. At that time, oscillations were well known in physics, in mechanical pendula, and in electronic circuits, but not in chemistry. In the field of chemical engineering too a big interest was expressed in new ideas of non-steady-state technology.

How can oscillations, in particular chemical oscillations, be explained within the mathematical dynamic theory? Unfortunately, there is still no rigorous theory for distinguishing multidimensional models of self-sustained oscillations. A typical strategy is first finding the models and parametric domains in which these oscillations *do not exist*. For instance, according to the so-called Poincaré-Bendixson criterion (which is only valid for systems with two variables), if the sum

$$a_{11} + a_{22} = \frac{\partial f_1(c_1, c_2)}{\partial c_1} + \frac{\partial f_2(c_1, c_2)}{\partial c_2} \qquad (7.58)$$

for Eq. (7.12) does not change sign in a certain region of the phase plane, this region does not contain closed phase trajectories, which are the phase trajectories related to oscillations. Furthermore, the closed phase trajectories around center-type rest points (Fig. 7.3J) do not represent self-sustained oscillations, since these trajectories are highly sensitive to the system parameters and initial conditions.

A typical mathematical model for self-sustained oscillations is the *stable limit cycle*, which can be defined as an isolated closed phase trajectory with its inner and outer sides approximated by spiral-shaped trajectories. One can say that a stable limit cycle is a mathematical image of a self-sustained oscillation. Just as there is no general mathematical theory of oscillations, a general theory of limit cycles does not exist, although many efficient approaches or, rather, recipes have been proposed for finding them. The most popular is determining the cases in which the rest point is *unique and unstable*. For physicochemical reasons, a phase trajectory reflecting a change of chemical composition cannot go to infinity because of the law of mass conservation. At the same time, this phase trajectory cannot reach the rest point because it is unstable. This is a typical scenario by which self-sustained oscillations (stable limit cycles) arise.

The generation of self-sustained oscillations is a particular case of bifurcation. The term "bifurcation" is often used in connection with the mathematical study of dynamical systems. It denotes a sudden qualitative change in the behavior of a system upon the smooth variation of a parameter, the so-called bifurcation parameter, and is applied to the point of the fundamental reconstruction of the phase portrait where the bifurcation parameter attains its critical value. The simplest examples of bifurcation are the appearance of a new rest point in the phase space, the loss of the rest-point stability, and the appearance of a new limit cycle. Bifurcation relates to physicochemical phenomena such as "ignition" and "extinction," that is, a jump-like transition from one steady state to another one, the appearance of oscillations, or a chaotic regime, and so on.

It is important to stress that an ideal model of such nonlinear (or critical) phenomena must reflect the main presented facts and, at the same time, must not be too complex. Such a model can be selected from the large collection of models that can be found in literature.

7.3 Multiplicity of Steady States in Nonisothermal Systems

7.3.1 Introduction

Experimental evidence of the existence of multiple steady states was found in the 1930s by Semenov and Zel'dovich and by Frank-Kamenetskii (1969), who described these results in detail in his classical monograph. Multiple steady states were first observed in combustion processes. The simplest model for describing multiple steady states was developed for a nonisothermal continuous stirred-tank reactor (CSTR), taking into account chemical, thermal, and transport processes occurring simultaneously. The basic explanation for this phenomenon is that the rate of heat generation (Q_{hg}) increases exponentially with increasing temperature, whereas the rate of heat removal (Q_{hr}) shows a linear dependence (see Fig. 7.4). Therefore, both rates may be equal at more than one temperature, resulting in multiple steady states.

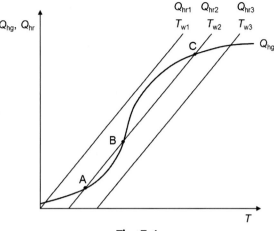

Fig. 7.4
Dependence of the rate of heat generation Q_{hg} and the rate of heat removal Q_{hr} on the reaction temperature in a CSTR at three different wall temperatures with $T_{w1} < T_{w2} < T_{w3}$. Certain Q_{hr} lines can intersect the Q_{hg} curve in as much as three different points (A, B and C).

The Q_{hr1} line represents the situation in which only a small amount of heat is generated by the reaction so that the temperature of the reaction mixture, and thus the steady-state conversion, is low. In contrast, if the rate of heat generation by the reaction is so large that the Q_{rg} curve crosses the Q_{hr3} line, the reaction temperature is so high that the steady-state conversion will be complete. In the intermediate situation, represented by the Q_{hr2} line, there are three possible steady states, points A, B, and C. However, point B is an unstable steady state; if the temperature only drops slightly below the steady-state temperature, the rate of heat generation by the reaction is smaller than the rate of heat removal, resulting in a further temperature decrease until the steady state represented by point A is reached. Similarly, in the case of a small temperature rise, the temperature will continue to increase until point C is reached.

7.3.2 Mathematical Model of the Nonisothermal CSTR

A model of a nonisothermal CSTR in which an irreversible first-order reaction, $A \rightarrow B$, takes place can be formulated as follows (Frank-Kamenetskii, 1969; Bykov and Tsybenova, 2011):

$$V_r \frac{dc_A}{dt} = -V_r k(T)c_A + q_V(c_{A0} - c_A) \tag{7.59}$$

$$c_p \rho V_r \frac{dT}{dt} = (-\Delta_r H)V_r k(T)c_A + c_p \rho q_V(T_0 - T) + hA_r(T_w - T) \tag{7.60}$$

where c_A is the reactant concentration, T and T_w are the temperatures of the reaction mixture and the reactor wall, $k(T) = k_0 \exp(-E_a/R_g T)$ is the reaction rate coefficient, V_r is the reaction volume, q_V is the volumetric flow rate, ρ and c_p are the density and specific heat capacity at constant pressure of the reaction mixture, $(-\Delta_r H)$ is the heat of reaction, A_r is the reactor surface area available for heat exchange, and h is the heat-transfer coefficient.

This set of two equations contains many kinetic, geometric, and thermal parameters. The phase variables of this system are the reactant concentration and the temperature. We can introduce the following dimensionless parameters (Frank-Kamenetskii, 1969):

$$T^* = \frac{1}{hA_r + c_p \rho q_V}\left(hA_r T_0 + c_p \rho q_V T_w\right); \quad \alpha^* = \frac{hA_r + c_p \rho q_V}{A_r} \tag{7.61}$$

$$\mathrm{Da} = \frac{V_r}{q_V} k(T_0) \tag{7.62}$$

$$\mathrm{Se} = \frac{(-\Delta_r H)\rho}{\alpha^*(A_r/V_r)R_g(T^*)^2}\frac{E_a}{} k_0 \exp\left(-\frac{E_a}{R_g T}\right) \tag{7.63}$$

$$\beta = \frac{(-\Delta_r H)c_{A0}}{c_p \rho T_0}; \quad \gamma = \frac{E_a}{R_g T_0} \tag{7.64}$$

$$x = \frac{c_{A0} - c_A}{c_{A0}}; \quad y = \frac{E_a}{R_g T^2}(T - T^*) \tag{7.65}$$

$$t^* = k_0 t \exp\left(-\frac{E_a}{R_g T^*}\right) \tag{7.66}$$

where x is the conversion, Da is the (first) Damköhler number, and Se is the Semenov number.

We assume that the inlet conditions are the same as the initial conditions:

$$c_A(0) = c_{A0}; \quad c_p \rho q_V(T_0 - T(0)) = hA_r(T(0) - T_w) \tag{7.67}$$

Using the dimensionless parameters, for an exothermic first-order reaction, this model can be written as

$$\frac{dx}{dt^*} = (1-x)\exp\left(\frac{y}{1+\beta y}\right) - \frac{x}{\mathrm{Da}} = f_1(x, y) \tag{7.68}$$

$$\gamma\frac{dy}{dt^*} = (1-x)\exp\left(\frac{y}{1+\beta y}\right) - \frac{y}{\mathrm{Se}} = f_2(x, y) \tag{7.69}$$

The steady-state values can be found by putting dx/dt^* and dy/dt^* equal to zero:

$$(1-x)\exp\left(\frac{y}{1+\beta y}\right) - \frac{x}{\mathrm{Da}} = 0 \tag{7.70}$$

$$(1-x)\exp\left(\frac{y}{1+\beta y}\right) - \frac{y}{\mathrm{Se}} = 0 \tag{7.71}$$

from which it follows that

$$x = \frac{\text{Da}}{\text{Se}} y = \nu y \tag{7.72}$$

Substituting Eq. (7.72) into Eq. (7.71) yields the following nonlinear equation in y:

$$(1 - \nu y) \exp\left(\frac{y}{1 + \beta y}\right) = \frac{y}{\text{Se}} \tag{7.73}$$

Equality (7.73) is an equation of steady states in which the left-hand side represents heat removal,

$$Q_{\text{hr}}(y) = (1 - \nu y) \exp\left(\frac{y}{1 + \beta y}\right) = (1 - \nu y) e(y) \tag{7.74}$$

and the right-hand side represents heat generation.

$$Q_{\text{hg}}(y) = \frac{y}{\text{Se}} \tag{7.75}$$

Solutions of Eq. (7.73) can be studied using a Semenov diagram, which reflects the dependences of the rate of heat generation and the rate of heat removal on the temperature. This diagram is similar to the dependence shown in Fig. 7.4, but with dimensionless coordinates. The points of intersection of $Q_{\text{hr}}(y)$ and $Q_{\text{hg}}(y)$ in Eq. (7.73) determine the temperature of the steady state; knowing this temperature, the corresponding steady-state conversion can be calculated using Eq. (7.72).

The condition

$$\frac{1}{\text{Se}} = \frac{dQ_{\text{hg}}(y)}{dy} > \frac{dQ_{\text{hr}}(y)}{dy} \tag{7.76}$$

is necessary and sufficient for uniqueness of the steady state. This condition means that the rate of heat removal must be higher than that of heat generation.

7.3.3 Types of Stability of the Steady State

The type of stability of a steady state can be determined from the roots of the characteristic polynomial:

$$\lambda^2 - (a_{11} + a_{22})\lambda + a_{11}a_{22} - a_{12}a_{21} = \lambda^2 - \sigma\lambda + \Delta = 0 \tag{7.77}$$

where the elements of the Jacobian matrix **J** are given by

$$a_{11} = \frac{\partial f_1}{\partial x} = -e(y) - \frac{1}{\text{Da}} \tag{7.78}$$

$$a_{12} = \frac{\partial f_1}{\partial y} = \frac{(1-x)e(y)}{(1+\beta y)^2} \qquad (7.79)$$

$$a_{21} = \frac{\partial f_2}{\partial x} = -\frac{e(y)}{\gamma} \qquad (7.80)$$

$$a_{22} = \frac{\partial f_2}{\partial y} = \frac{(1-x)e(y)}{(1+\beta y)^2} - \frac{1}{\gamma Se} \qquad (7.81)$$

with f_1 and f_2 given by, respectively, Eqs. (7.68) and (7.69).

The roots of the characteristic polynomial, Eq. (7.77), are expressed through its coefficients:

$$\lambda_{1,2} = \frac{\sigma}{2} \pm \sqrt{\frac{\sigma^2}{4} - \Delta} \qquad (7.82)$$

The parametric dependences of the steady states are solutions of the equation

$$F(y) = Q_{hr}(y) - Q_{hg}(y) = 0 \qquad (7.83)$$

which can be written as

$$F(y, \text{Da}, \text{Se}, \beta) = 0 \qquad (7.84)$$

Then the expressions for the parameters as a function of y can be found:

$$\beta = \frac{1}{\ln[y/(\text{Se} - \text{Day})]} - \frac{1}{y} \qquad (7.85)$$

$$\text{Se} = \left(\text{Da} + \frac{1}{e(y)}\right) y \qquad (7.86)$$

$$\text{Da} = \frac{\text{Se}}{y} - \frac{1}{e(y)} \qquad (7.87)$$

and

$$\nu = \frac{\text{Da}}{\text{Se}} = \frac{1}{y} - \frac{1}{e(y)\text{Se}} \qquad (7.88)$$

Subsequently, the dimensionless temperature y is determined as a function of the parameter ν.

The boundary between the domain with one steady state and the domain with three steady states is called the *line of multiplicity*, L_Δ ($\Delta = 0$) For finding this dependence in the plane (Da, Se), we need to solve the set of equations

$$f_1(x, y, \text{Da}, \text{Se}) = 0 \qquad (7.89)$$

$$f_2(x, y, \text{Da}, \text{Se}) = 0 \qquad (7.90)$$

$$\Delta(y, \text{Da}, \text{Se}) = 0 \qquad (7.91)$$

where Δ is determined using elements of the Jacobian matrix.

Combining Eq. (7.84) and Eqs. (7.89)–(7.91), we obtain the equation for the multiplicity line in the plane (Da,Se),

$$
L_\Delta(\text{Da, Se}) : \begin{cases} \text{Da}(y) = \dfrac{y - (1+\beta y)^2}{(1+\beta y)^2 e(y)} \\[3mm] \text{Se}(y, \text{Da}) = \left(\text{Da} + \dfrac{1}{e(y)}\right) y \end{cases}
\tag{7.92}
$$

Lines of multiplicity for other parameter planes can be found in a similar way. For the plane (Se, ν),

$$
L_\Delta(\text{Se}, \nu) : \begin{cases} \text{Se}(y) = \dfrac{y^2}{(1+\beta y)^2 e(y)} \\[3mm] \nu(y, \text{Se}) = \dfrac{1}{y} - \dfrac{1}{e(y)\text{Se}} \end{cases}
\tag{7.93}
$$

Lines of neutrality, $L_\sigma(\sigma = 0)$, determine the type of stability of a steady state. In order to find L_σ, the following equations need to be solved:

$$
F(y, \text{Da}, \text{Se}) = 0
\tag{7.94}
$$

$$
\sigma(y, \text{Da}, \text{Se}) = 0
\tag{7.95}
$$

Simultaneously solving Eqs. (7.94), (7.95), and (7.85)–(7.88) will explicitly yield different lines L_σ.

$$
L_\sigma(\text{Se}, \nu) : \begin{cases} \text{Se}(y) = \dfrac{\text{Da}\left(y - (1+\beta y)^2\right)}{\gamma(1+\beta y)^2 (\text{Da}\,e(y) + 1)} \\[3mm] \nu(y, \text{Se}) = \dfrac{1}{y} - \dfrac{1}{e(y)\text{Se}} \end{cases}
\tag{7.96}
$$

$$
L_\sigma(\text{Da}, \text{Se}) : \begin{cases} \text{Da}(y) = \dfrac{a(y) \pm \sqrt{a^2(y) - 1}}{e(y)}; \quad a(y) = \dfrac{1}{2\gamma(1+\beta y)^2} - \dfrac{1}{2\gamma y} - 1 \\[3mm] \text{Se}(y, \text{Da}) = \left(\text{Da} + \dfrac{1}{e(y)}\right) y \end{cases}
\tag{7.97}
$$

$$
L_\sigma(\gamma, \text{Da}) : \begin{cases} \gamma(y, \text{Da}) = \dfrac{\text{Da}\,e(y)\left(y - (1+\beta y)^2\right)}{y(1+\beta y)^2 (\text{Da}\,e(y) + 1)^2} \\[3mm] \text{Da}(y) = \dfrac{\text{Se}}{y} - \dfrac{1}{e(y)} \end{cases}
\tag{7.98}
$$

$$L_\sigma(\gamma, \mathrm{Se}): \begin{cases} \gamma(y, \mathrm{Se}) = \dfrac{(e(y)\mathrm{Se} - y)\left(y - (1 + \beta y)^2\right)}{(e(y)\mathrm{Se}(1 + \beta y))^2} \\[4mm] \mathrm{Se}(y, \mathrm{Da}) = \left(\mathrm{Da} + \dfrac{1}{e(y)}\right)y \end{cases} \tag{7.99}$$

$$L_\sigma(\gamma, \beta): \begin{cases} \gamma(y, \beta) = \dfrac{\mathrm{Da}\, e(y)\left(y - (1 + \beta y)^2\right)}{y(1 + \beta y)^2 (\mathrm{Da}\, e(y) + 1)^2} \\[4mm] \beta(y) = \dfrac{1}{\ln\left[y/(\mathrm{Se} - \mathrm{Da}y)\right]} - \dfrac{1}{y} \end{cases} \tag{7.100}$$

$$L_\sigma(\gamma, \nu): \begin{cases} \gamma(y, \nu) = \dfrac{\nu e(y)\left(y - (1 + \beta y)^2\right)}{y(1 + \beta y)^2 (\mathrm{Da}\, e(y) + 1)^2} \\[4mm] \nu(y, \mathrm{Se}) = \dfrac{1}{y} - \dfrac{1}{\mathrm{Se}\, e(y)} \end{cases} \tag{7.101}$$

The line of multiplicity and the line of neutrality together form a parametric portrait. In this portrait, parametric domains with a different number of steady states that differ by their stability as well can be distinguished. In the current model, the complete parametric portrait consists of six domains. Furthermore, six types of phase portraits can be distinguished, corresponding to six parametric portrait domains. The first domain (I) is characterized by a unique stable steady state. This behavior is observed at low and high temperatures. The second domain (II) relates to a unique unstable steady state. This is a domain of oscillations. The domains III, IV, and V all have three steady states. Domain III has one unstable steady state at low temperature and two stable steady states at high temperature. Domains IV and V both have one stable and one unstable steady state. Finally, domain VI has three steady states, one of which is stable and two are unstable. In this domain, oscillations are also observed.

7.4 Multiplicity of Steady States in Isothermal Heterogeneous Catalytic Systems

7.4.1 Historical Background

Multiplicity of steady states not only occurs in nonisothermal systems. In the 1950s and 1960s, this phenomenon was found in many isothermal heterogeneous catalytic oxidation systems such as the platinum-catalyzed oxidation of hydrogen and carbon monoxide.

Currently, a vast amount of information on the occurrence of multiple steady states in heterogeneous catalysis has been accumulated. Different values of the reaction rate can correspond to the same composition of the reaction mixture or to the same temperature.

Typically, multiplicity of steady states is accompanied by hysteresis, a term derived from the Greek word meaning "lagging behind." A characteristic mark of hysteresis is that the output—here the reaction rate—forms a loop, which can be oriented clockwise or counterclockwise. Hysteresis causes the reaction rate to jump up or fall down, signifying, respectively, ignition and extinction of the reaction. Fig. 7.5 shows some examples of kinetic dependences involving hysteresis.

Liljenroth (1918) was the first to describe a nonlinear phenomenon in heterogeneous catalysis. He studied the platinum-catalyzed oxidation of ammonia and analyzed the stability of the steady state of this process. Apparently, Davies (1934) was the first to systematically observe nonlinear phenomena and to point out the chemical nature of the rate hysteresis. In the 1950s, Boreskov and colleagues (Boreskov et al., 1953; Kharkovskaya et al., 1959) studied the oxidation of hydrogen over metal catalysts and found that in a certain range of reaction parameters, very different values of the steady-state reaction rate corresponded with the same gas composition. These data were obtained at isothermal conditions. For more historical information on multiplicity of steady states, see Yablonskii et al. (1991) and Marin and Yablonsky (2011).

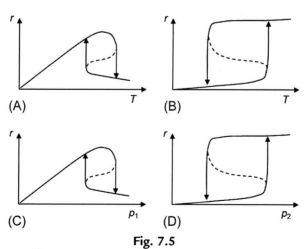

Fig. 7.5

Examples of kinetic dependences with a multiplicity of steady states. Solid lines correspond to stable branches of the reaction rate, dashed lines to unstable branches; (A), (B) r versus T; (C) r versus p_1 at constant p_2; (D) r versus p_2 at constant p_1. (A) and (C) are examples of clockwise hysteresis, (B) and (D) of counterclockwise hysteresis.

7.4.2 The Simplest Catalytic Trigger

The simplest mechanism for a catalytic reaction permitting three steady states is the following adsorption mechanism (also called Langmuir-Hinshelwood mechanism) involving two types of intermediates:

$$
\begin{aligned}
&(1)\ A_2 + 2Z\ \rightleftarrows\ 2AZ\\
&(2)\ B + Z\ \ \ \rightleftarrows\ BZ\\
&(3)\ AZ + BZ\ \rightarrow\ AB + 2Z
\end{aligned}
\tag{7.102}
$$

The well-known adsorption mechanism for the oxidation of carbon monoxide over platinum at constant partial pressures of oxygen and carbon monoxide,

$$
\begin{aligned}
&(1)\ O_2 + 2Pt\ \ \rightleftarrows\ 2PtO\\
&(2)\ CO + Pt\ \ \ \rightleftarrows\ PtCO\\
&(3)\ PtO + PtCO\ \rightarrow\ CO_2 + 2Pt
\end{aligned}
\tag{7.103}
$$

is an example of this type of mechanism.

Let us start our analysis by assuming all steps in the mechanism represented by Eq. (7.102) are irreversible (Bykov et al., 1981; Bykov and Yablonskii, 1987), (1) $A_2 + Z \rightarrow 2AZ$; (2) $B + Z \rightarrow BZ$; (3) $AZ + BZ \rightarrow AB + 2Z$. Then, the steady-state model is given by

$$
\frac{d\theta_{AZ}}{dt} = 2k_1^+ p_{A_2}(1 - \theta_{AZ} - \theta_{BZ})^2 - k_3^+ \theta_{AZ}\theta_{BZ} = 0
\tag{7.104}
$$

$$
\frac{d\theta_{BZ}}{dt} = k_2^+ p_B(1 - \theta_{AZ} - \theta_{BZ}) - k_3^+ \theta_{AZ}\theta_{BZ} = 0
\tag{7.105}
$$

where θ_{AZ} and θ_{BZ} are the normalized surface concentrations of A and B and $\theta_Z = 1 - \theta_{AZ} - \theta_{BZ}$ is the normalized surface concentration of free active sites Z. p_{A_2} and p_B are the partial pressures of A_2 and B and k_1^+, k_2^+, and k_3^+ are the rate coefficients of (forward) reactions 1, 2, and 3.

In our model, two boundary steady states exist: (1) $\theta_{AZ} = 1$ and $\theta_{BZ} = \theta_Z = 0$, which corresponds to complete surface coverage by component A_2 (eg, oxygen) and (2) $\theta_{AZ} = \theta_Z = 0$ and $\theta_{BZ} = 1$, which corresponds to complete surface coverage by component B (eg, CO). In both cases, the steady-state reaction rate is zero:

$$
r = 2k_1^+ p_{A_2}(1 - \theta_{AZ} - \theta_{BZ})^2 = k_2^+ p_B(1 - \theta_{AZ} - \theta_{BZ}) = k_3^+ \theta_{AZ}\theta_{BZ} = 0
\tag{7.106}
$$

In addition to these boundary steady states, two internal steady states exist with nonzero values of the reaction rate. These are found as follows. Subtraction of Eq. (7.105) from Eq. (7.104) yields

$$
2k_1^+ p_{A_2}(1 - \theta_{AZ} - \theta_{BZ})^2 - k_2^+ p_B(1 - \theta_{AZ} - \theta_{BZ}) = 0
\tag{7.107}
$$

from which it follows that

$$1 - \theta_{AZ} - \theta_{BZ} = \theta_Z = \frac{k_2^+ p_B}{2k_1^+ p_{A_2}} \tag{7.108}$$

Substitution of Eq. (7.108) in Eq. (7.106) for the steady-state reaction rate yields

$$r = \frac{\left(k_2^+ p_B\right)^2}{2k_1^+ p_{A_2}} = k_3^+ \theta_{AZ} \theta_{BZ} \tag{7.109}$$

and thus

$$\theta_{AZ} \theta_{BZ} = \frac{\left(k_2^+ p_B\right)^2}{2k_1^+ k_3^+ p_{A_2}} \tag{7.110}$$

Then from Eq. (7.108) it follows that

$$\theta_{BZ} = 1 - \theta_{AZ} - \frac{k_2^+ p_B}{2k_1^+ p_{A_2}} \tag{7.111}$$

so that Eq. (7.110) can be transformed into

$$\theta_{AZ} \left(1 - \theta_{AZ} - \frac{k_2^+ p_B}{2k_1^+ p_{A_2}}\right) = \frac{\left(k_2^+ p_B\right)^2}{2k_1^+ k_3^+ p_{A_2}} \tag{7.112}$$

This quadratic equation can be written as

$$\theta_{AZ}^2 - \sigma \theta_{AZ} + \Delta = 0 \tag{7.113}$$

with $\sigma = 1 - \dfrac{k_2^+ p_B}{2k_1^+ p_{A_2}}$ and $\Delta = \dfrac{\left(k_2^+ p_B\right)^2}{2k_1^+ k_3^+ p_{A_2}}$.

Then,

$$\theta_{AZ} = \frac{\sigma}{2} \pm \sqrt{\frac{\sigma^2}{4} - \Delta} \tag{7.114}$$

and with Eq. (7.111) we find the same result for θ_{BZ}.

Eq. (7.114) has two solutions in the domain of values having a physical meaning $(0 \le \theta_{AZ} \le 1)$ under the condition that

$$\left(1 - \frac{k_2^+ p_B}{2k_1^+ p_{A_2}}\right)^2 \ge \frac{2\left(k_2^+ p_B\right)^2}{k_1^+ k_3^+ p_{A_2}} \tag{7.115}$$

Thus, this model has four steady states, two boundary ones and two internal ones. The two boundary steady states (I and IV) are symmetrical and so are the two internal steady states (II and III). Mathematically, this property is the result of the symmetry of Eqs. (7.104)

Table 7.1 Steady-state characteristics for the reaction mechanism represented by Eq. (7.102)

Steady State	Normalized Concentrations	Reaction Rate r	Comments
I	$\theta_{AZ,I} = 1$ $\theta_{BZ,I} = 0$ $\theta_{Z,I} = 0$	0	Unstable steady state; catalyst surface completely covered by component A_2 (intermediate AZ)
II	$\theta_{AZ,II} = \dfrac{\sigma + \sqrt{\sigma^2 - 4\Delta}}{2}$ $\theta_{BZ,II} = \dfrac{\sigma - \sqrt{\sigma^2 - 4\Delta}}{2}$ $\theta_{Z,II} = 1 - \sigma$	$\dfrac{\left(k_2^+ p_B\right)^2}{2k_1^+ p_{A_2}}$	Stable steady state; catalyst surface contains intermediates AZ and BZ and free active sites Z
III	$\theta_{AZ,III} = \dfrac{\sigma - \sqrt{\sigma^2 - 4\Delta}}{2}$ $\theta_{BZ,III} = \dfrac{\sigma + \sqrt{\sigma^2 - 4\Delta}}{2}$ $\theta_{Z,III} = 1 - \sigma$	$\dfrac{\left(k_2^+ p_B\right)^2}{2k_1^+ p_{A_2}}$	Unstable steady state; catalyst surface contains intermediates AZ and BZ and free active sites Z
IV	$\theta_{AZ,I} = 0$ $\theta_{BZ,I} = 1$ $\theta_{Z,I} = 0$	0	Stable steady state; catalyst surface completely covered by component B (intermediate BZ)

and (7.105) with respect to the variables θ_{AZ} and θ_{BZ}. Table 7.1 shows the characteristics of the four steady states. It has been shown that steady states II and III are stable whereas steady states I and IV are unstable. The steady-state reaction rates for steady states II and III follow from the first equality of Eq. (7.109). A detailed analysis of the stability of the steady states for this model can be found in the monographs by Yablonskii et al. (1991) and Marin and Yablonsky (2011).

Every stable steady state is characterized by a basin of attraction, which is a set of initial conditions—or compositions—starting from which the mixture reaches this steady state. Clearly, unstable steady states do not have such a basin.

7.4.3 Influence of Reaction Reversibility on Hysteresis

Reversibility of one or more reaction steps of the adsorption mechanism represented by Eq. (7.102) may significantly influence the reaction kinetics, particularly the steady-state kinetic behavior. As a result of reversibility, the steady states corresponding to complete coverage of the catalyst surface with surface intermediate AZ or BZ—the boundary steady states—disappear. At certain conditions, only three internal steady states, two stable and one unstable, exist. Fig. 7.6 shows a typical dependence of the steady-state reaction rate on the partial pressure of component B for the adsorption mechanism with the first and second steps being reversible and the third step irreversible. At low temperature, the reaction rate is characterized by a multiplicity of steady-state rates; three steady-state reaction rates are observed, two stable and one unstable (see Fig. 7.6A, curve (1), and Fig. 7.6B, curves (1)

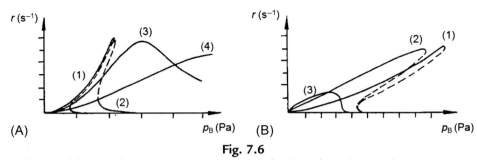

Fig. 7.6

Dependence of the steady-state reaction rate on p_B for the adsorption mechanism represented by Eq. (7.101) at constant values of p_{A_2} and $k_1^+ = k_2^+$ for (A) constant k_1^- and k_2^- increasing from (1) to (4) and (B) constant k_2^- and k_1^- increasing from (1) to (3). *Reprinted from Yablonskii, G.S., Bykov, V.I., Gorban, A.N., Elokhin, V.I., 1991. Kinetic models of catalytic reactions. In: Compton, R.G. (Ed.), Comprehensive Chemical Kinetics, vol. 32. Elsevier, Amsterdam, Copyright (1991), with permission from Elsevier.*

and (2)). Component A dominates on the upper branch and B dominates on the lower branch. In this case, the transition from one branch to another one occurs in a dramatic way: via extinction (from the upper branch to the lower one) or via ignition (from the lower branch to the upper one). Finally, the multiplicity of steady states is exhibited as a counterclockwise hysteresis for the kinetic dependence $r(p_B)$ (Fig. 7.6A, curve (1) and Fig. 7.6B, curves (1) and (2)) and clockwise for $r(p_{A_2})$ (not shown). A temperature rise increases the desorption coefficient and kills the multiplicity of steady states. The kinetic dependence is gradually transformed into a dependence with a maximum rate (Fig. 7.6A, curve (3) and Fig. 7.6B, curve (3)). Finally, the temperature rise causes the maximum of the reaction rate to vanish completely (Fig. 7.6A, curve (4)).

Fig. 7.7A shows the steady-state reaction rate in the three-dimensional space. This rate is characterized by a special point called a cusp point, a term taken from catastrophe theory meaning a singular point of the geometric curve. Fig. 7.7B shows two curves that are loci of turning points. The curve on the left is the locus of extinction points and can be defined as the locus of turning points from the upper to the lower branch. The curve on the right is the locus of ignition points and can be defined as the locus of turning points from the lower to the upper branch.

7.4.4 Transient Characteristics and "Critical Slowing Down"

The transient characteristics, $\theta_{AZ}(t)$ and $\theta_{BZ}(t)$, of the adsorption model (Eq. (7.102)) show different relaxation times, which can differ by orders of magnitude. The trajectories are often characterized by fast initial motion and slow motion in the vicinity of a slow trajectory. This slow trajectory is a simple case of the slow manifolds known in the literature (Gorban and Karlin, 2003) and is located in the region between the two null clines $d\theta_{AZ}/dt = 0$ and $d\theta_{BZ}/dt = 0$. Both null clines are second-order curves with an axis of symmetry $\theta_{AZ} = \theta_{BZ}$

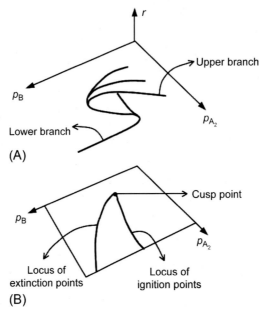

Fig. 7.7
Surface of the steady-state reaction rate at constant temperature for the adsorption
mechanism represented by Eq. (7.101); (A) reaction rate in the three-dimensional space;
(B) loci of turning points. *Adapted from Yablonskii, G.S., Bykov, V.I., Gorban, A.N., Elokhin, V.I., 1991.
Kinetic models of catalytic reactions. In: Compton, R.G. (Ed.), Comprehensive Chemical Kinetics, vol. 32. Elsevier,
Amsterdam, Copyright (1991), with permission from Elsevier.*

on which an unstable steady state is located. In the case that a phase portrait has multiple
steady states, a trajectory may rapidly approach an unstable steady state and then slowly relax
toward one of the stable steady states (Fig. 7.8).

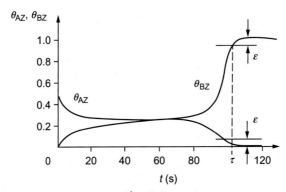

Fig. 7.8
Example of critical slowing down for the adsorption mechanism represented by Eq. (7.102);
relaxation from the initial state $\theta_{AZ}(0) = 0.5; \theta_{BZ}(0) = 0$; ε is the vicinity of the final state and τ is the
time required to reach this vicinity. *Reprinted from Yablonskii, G.S., Bykov, V.I., Gorban, A.N., Elokhin, V.I.,
1991. Kinetic models of catalytic reactions. In: Compton, R.G. (Ed.), Comprehensive Chemical Kinetics,
vol. 32. Elsevier, Amsterdam, Copyright (1991), with permission from Elsevier.*

If such critical slowing down occurs, the observed relaxation time is unexpectedly long and considerably longer than the relaxation time corresponding to the lowest rate coefficient. The main nontrivial explanation of this phenomenon is the following. In the region of critical effects, one of the parameters achieves its bifurcation value while one of the characteristic roots changes from a negative to a positive value. Experimentally, the phenomenon of very slow relaxation toward a stable steady state has been observed in many heterogeneous catalytic oxidation reactions (Yablonskii et al., 1991; Marin and Yablonsky, 2011). Gorban and colleagues (Gorban and Cheresiz, 1981; Gorban and Karlin, 2003) have developed a comprehensive mathematical theory of slow relaxation using concepts from the theory of dynamical systems, such as and ω-limit sets.

7.4.5 General Model of the Adsorption Mechanism

In the adsorption mechanism, two gaseous reactants are adsorbed through two independent reactions. Then the two surface intermediates interact and a gaseous product is released. The general form of this mechanism can be represented as

$$
\begin{aligned}
&(1) \quad A_m + mZ && \rightleftarrows \quad mAZ \\
&(2) \quad B_n + nZ && \rightleftarrows \quad nBZ \\
&(3) \quad pAZ + qBZ && \rightleftarrows \quad A_pB_q + (p+q)Z
\end{aligned}
\tag{7.116}
$$

The kinetic model corresponding to this mechanism is given by the following set of nonlinear differential equations:

$$
\frac{d\theta_{AZ}}{dt} = m\bar{k}_1^+ \theta_Z^m - mk_1^- \theta_{AZ}^m - pk_3^+ \theta_{AZ}^p \theta_{BZ}^q + q\bar{k}_3^- \theta_Z^{p+q}
\tag{7.117}
$$

$$
\frac{d\theta_{BZ}}{dt} = n\bar{k}_2^+ \theta_Z^n - nk_2^- \theta_{BZ}^n - qk_3^+ \theta_{AZ}^p \theta_{BZ}^q + q\bar{k}_3^- \theta_Z^{p+q}
\tag{7.118}
$$

with apparent parameters $\bar{k}_1^+ = k_1^+ p_{A_m}, \bar{k}_2^+ = k_2^+ p_{B_n}$ and $\bar{k}_3^- = k_3^- p_{A_pB_q}$ and constant partial pressures of A_m, B_n and A_pB_q.

The nonlinearity of the model is caused by both the adsorption steps and the interaction between surface intermediates. The corresponding set of algebraic equations $\dfrac{d\theta_{AZ}}{dt} = 0$ and $\dfrac{d\theta_{BZ}}{dt} = 0$ determines the number of steady states and the corresponding values of the normalized surface concentrations. A necessary condition for the nonuniqueness of steady states is the presence of a step in which different surface intermediates react. This result was obtained in a physically meaningful domain of normalized surface concentrations ($\theta_Z, \theta_{AZ}, \theta_{BZ} \geq 0$; $\theta_Z + \theta_{AZ} + \theta_{BZ} = 1$).

The sufficient condition for the uniqueness of an internal steady state is the relationship

$$
m = n \geq p, q.
\tag{7.119}
$$

If the condition of Eq. (7.119) is not satisfied, multiplicity of steady states is possible.

For the adsorption mechanism, the presence of an interaction step between different intermediates is only one necessary condition for the occurrence of multiple steady states; another requirement is that the reaction orders of the two adsorption steps must be different. Consequently, competition between surface intermediates is the main factor leading to multiplicity of steady states.

The reversibility of the adsorption steps affects the number of steady states. Analysis shows that boundary steady states are absent if both adsorption steps are reversible. If one of the adsorption steps is irreversible, there is one boundary steady state with the irreversibly adsorbed intermediate occupying all active catalyst sites. If both adsorption steps are irreversible, two boundary steady states exist, one with complete coverage of the catalyst surface by AZ and one with complete coverage by BZ. If the reaction orders of these irreversible adsorption steps are equal, there is a line of internal steady states connecting the two boundary steady states. In this case, the steady state is very sensitive to the initial conditions and experimental data may be irreproducible.

From a physicochemical point of view, elementary reactions with an overall kinetic order of more than three can be disregarded because such reactions are impossible. Table 7.2 shows the results of a comprehensive analysis by Yablonskii et al. (1991) of realistic adsorption mechanisms, assuming that the interaction step between the surface intermediates is irreversible ($k_3^- = 0$) and that the stoichiometric coefficients equal either one or two and the

Table 7.2 Characteristics of "realistic" adsorption mechanisms: necessary conditions for steady states (ss) and their number

Partial Reaction Orders	Reversibility of Adsorption Steps			
	I $k_1^- = 0; k_2^- = 0$	IIa $k_1^- = 0; k_2^- \neq 0$	IIb $k_1^- \neq 0; k_2^- = 0$	III $k_1^- \neq 0; k_2^- \neq 0$
(1) $m=1; n=1;$ $p=1; q=1$	Two ss; Line of ss for $\bar{k}_1^+ = \bar{k}_2^+$	One bss[a] and one iss[b] at $\bar{k}_1^+ > \bar{k}_2^+$	One bss and one iss at $\bar{k}_2^+ > \bar{k}_1^+$	No bss and one iss
(2) $m=2; n=2;$ $p=1; q=1$	Same as I.1	Same as IIa.1	Same as IIb.2	Same as III.1
(3) $m=2; n=2;$ $p=2; q=1$	Same as I.1 but with $\bar{k}_1^+ = 2\bar{k}_2^+$	Same as IIa.1 but with $\bar{k}_1^+ > 2\bar{k}_2^+$	Same as IIb.2 but with $2\bar{k}_2^+ > \bar{k}_1^+$	Same as III.1
(4) $m=1; n=1;$ $p=2; q=1$	Same as I.3	One bss and two iss at $\bar{k}_1^+ > 2\bar{k}_2^+$	One bss and one iss	No bss and one or three iss
(5) $m=2; n=1;$ $p=1; q=1$	Two bss and two iss at $2\bar{k}_1^+ > \bar{k}_2^+$	One bss and two iss at $2\bar{k}_1^+ > \bar{k}_2^+$	One bss and one or three iss at $2\bar{k}_1^+ > \bar{k}_2^+$	Same as III.4
(6) $m=2; n=1;$ $p=2; q=1$	I.5 but with $\bar{k}_1^+ > \bar{k}_2^+$	IIa.5 but with $\bar{k}_1^+ > \bar{k}_2^+$	IIb.5 but with $\bar{k}_1^+ > \bar{k}_2^+$	Same as III.4
(7) $m=2; n=1;$ $p=1; q=2$	I.5 but with $4\bar{k}_1^+ > \bar{k}_2^+$	IIa.5 but with $4\bar{k}_1^+ > \bar{k}_2^+$	IIb.5 but with $4\bar{k}_1^+ > \bar{k}_2^+$	Same as III.4

[a]bss = boundary steady state.
[b]iss = internal steady state.

Table 7.3 Adsorption mechanisms that can explain the multiplicity of steady states

A	(1) $A + Z \rightleftarrows AZ$ (2) $B + Z \rightleftarrows BZ$ (3) $2AZ + BZ \rightarrow A_2B + 3Z$
B	(1) $A_2 + 2Z \rightleftarrows 2AZ$ (2) $B + Z \rightleftarrows BZ$ (3) $AZ + BZ \rightarrow AB + 2Z$
C	(1) $A_2 + 2Z \rightleftarrows 2AZ$ (2) $B + Z \rightleftarrows BZ$ (3) $2AZ + BZ \rightarrow A_2B + 3Z$
D	(1) $A_2 + 2Z \rightleftarrows 2AZ$ (2) $B + Z \rightleftarrows BZ$ (3) $AZ + 2BZ \rightarrow AB_2 + 3Z$

sum $p + q \leq 3$. These results are based on the algebraic analysis of the number of positive roots in the interval [0,1] using Descartes's rule of signs and Sturm's theorem.

The main result is that real multiplicity of steady states—the existence of two internal stable steady states and one internal unstable steady state—can be explained using one of the four mechanisms shown in Table 7.3. If the experimentally observed steady-state reaction rate is characterized by two different branches, and multiplicity of steady states is observed, one of these four mechanisms can be used to interpret the data. We prefer to use mechanism B because its steps are characterized by overall reaction orders that are not larger than two. In fact, this mechanism is identical to the mechanism represented by Eq. (7.102), an example of which is the adsorption mechanism for the oxidation of carbon monoxide (Eq. (7.103)). We will return to this mechanism in Chapter 11, in which the problem of critical simplification is discussed.

7.5 Chemical Oscillations in Isothermal Systems

7.5.1 Historical Background

The discovery of isothermal chemical oscillations was one of the big scientific sensations of the 20th century. Belousov and Zhabotinsky found self-sustained oscillations in the cerium-ion catalyzed oxidation of citric acid by bromate, now known as the Belousov-Zhabotinsky reaction. This reaction became one of the starting points for Prigogine and his coworkers in Brussels for studying complicated dynamic behavior of chemical mixtures that are far from equilibrium (Prigogine and Lefever, 1968; Glansdorff and Prigogine, 1971; Nicolis and Prigogine, 1977). They used a model with a specific type of mechanism involving autocatalytic reactions, later named the Brusselator, for the quantitative interpretation of chemical oscillations in isothermal systems. An autocatalytic reaction is a reaction in which at least one of the reactants is also a product, for example, $A + B \rightarrow 2\,A$. Prigogine received the 1977 Nobel Prize in Chemistry for this work. In the early 1970s, the research by Prigogine's group inspired

Field et al. (1972) at the University of Oregon to perform a systematic and detailed thermodynamic and kinetic analysis of the Belousov-Zhabotinsky reaction. They suggested a detailed reaction mechanism that could explain the oscillations. Field and Noyes (1974) later simplified this mechanism and named it the Oregonator, after Prigogine's Brusselator.

The group of Wicke (Beusch et al., 1972a; Beusch et al., 1972b) was the first to observe kinetic self-sustained oscillations in solid-catalyzed reactions, when studying the oxidation of carbon monoxide over platinum-based catalysts. At certain conditions, oscillations were observed in the production rate of carbon dioxide. Based on these experimental data, Beusch et al. (1972b) assumed that the only factor responsible for the rate oscillations was the complex surface-reaction mechanism and that neither external nor internal mass- or heat-transfer limitations played a role. Slin'ko and Slin'ko (1978, 1982) performed comprehensive experimental and theoretical investigations of kinetic self-sustained oscillations, in particular in the oxidation of hydrogen over nickel and platinum catalysts (see also Slin'ko and Jaeger, 1994).

Self-sustained rate oscillations have been observed in many heterogeneous catalytic reactions. The largest amount of information on kinetic self-sustained oscillations was obtained for the oxidation of carbon monoxide over noble metals, particularly by the group of Ertl, who in 2007 received a Nobel Prize in Chemistry for his work regarding this phenomenon. Other examples are the oxidation of cyclohexane (Berman and Elinek, 1979), propylene (Sheintuch and Luss, 1981), and ethylene (Vayenas et al., 1981); the cooxidation of carbon monoxide and 1-butene (Mukesh et al., 1982, 1983, 1984); and reactions of nitric oxide with ammonia (Nowobilski and Takoudis, 1985), carbon monoxide (Bolten et al., 1985), and carbon monoxide and oxygen (Regalbuto and Wolf, 1986).

7.5.2 Model of Catalytic Chemical Oscillations

Now the question is how to construct the simplest model of a chemical oscillator, in particular, a catalytic oscillator. It is quite easy to include an autocatalytic reaction in the adsorption mechanism, for example $A+B \rightarrow 2A$. The presence of an autocatalytic reaction is a typical feature of the known Brusselator and Oregonator models that have been studied since the 1970s. Autocatalytic processes can be compared with biological processes, in which species are able to give birth to similar species. Autocatalytic models resemble the famous Lotka-Volterra equations (Berryman, 1992; Valentinuzzi and Kohen, 2013), also known as the predator-prey or parasite-host equations.

Let us again consider a model with two variables, Eq. (7.12):

$$\begin{cases} \dfrac{dc_1}{dt}=f_1(c_1,c_2) \\ \dfrac{dc_2}{dt}=f_2(c_1,c_2) \end{cases} \qquad (7.12)$$

In the presence of autocatalytic steps, the derivatives $\partial f_1(c_1, c_2)/\partial c_1$ and $\partial f_2(c_1, c_2)/\partial c_2$ can be positive in some parametric domain, whereas in the absence of autocatalytic steps these derivatives are always negative. This results in more complicated behavior of autocatalytic processes compared to traditional chemical processes. Thus, the mass-action-law model with two variables and without autocatalytic steps obeys the Poincaré-Bendixson criterion:

$$\frac{\partial f_1(c_1, c_2)}{\partial c_1} + \frac{\partial f_2(c_1, c_2)}{\partial c_2} \leq 0 \tag{7.120}$$

Based on this criterion, it can be concluded that two-dimensional models that do not include autocatalytic reactions are not able to reproduce the experimentally observed isothermal kinetic oscillations.

Seemingly, the description of catalytic oscillations at least requires a two-variable model with autocatalytic reactions or a three-variable model without autocatalytic reactions. Eigenberger (1978a,b) made a step toward the construction of a three-variable model by adding a so-called buffer step (4 in the following list) to an adsorption mechanism:

$$
\begin{array}{lll}
(1) & 2A + 2Z & \rightleftarrows & 2AZ \\
(2) & B + 2Z & \rightleftarrows & BZ_2 \\
(3) & 2AZ + BZ_2 & \rightarrow & 2C + 4Z \\
(4) & B + Z & \rightleftarrows & (BZ)^*
\end{array}
\tag{7.121}
$$

for the overall reaction $2A + B \rightarrow 2C$. In the buffer step, $(BZ)^*$ is produced, a nonreactive form that does not participate in the main catalytic cycle.

This is a realistic model because gaseous reactants can be adsorbed in a number of forms, some of which are nonreactive. An example is the oxidation of carbon monoxide over platinum, which in the above mechanism translates to $A = CO$, $B = O_2$, and $C = CO_2$. Although oxygen typically adsorbs irreversibly on platinum, Eigenberger assumed step (4) to be reversible. Based on this assumption, which was not substantiated, Eigenberger transformed the initial model to a model similar to that for a mechanism with autocatalytic reactions, which was studied by Prigogine and colleagues (Prigogine and Lefever, 1968; Glansdorff and Prigogine, 1971; Nicolis and Prigogine, 1977) and found oscillations using computer calculations.

Bykov et al. (1978) have presented a detailed analysis of the following adsorption mechanism with buffer step (also see Yablonskii et al., 1991; Marin and Yablonsky, 2011):

$$
\begin{array}{lll}
(1) & A_2 + 2Z & \rightleftarrows & 2AZ \\
(2) & B + Z & \rightleftarrows & BZ \\
(3) & AZ + BZ & \rightarrow & AB + 2Z \\
(4) & B + Z & \rightleftarrows & (BZ)^*
\end{array}
\tag{7.122}
$$

The corresponding model for this mechanism is

$$\frac{d\theta_{AZ}}{dt} = 2\bar{k}_1^+ \theta_Z^2 - 2k_1^- \theta_{AZ}^2 - k_3^+ \theta_{AZ}\theta_{BZ} = f_1\left(\theta_{AZ}, \theta_{BZ}, \theta_{(BZ)^*}\right) \tag{7.123}$$

$$\frac{d\theta_{BZ}}{dt} = \bar{k}_2^+ \theta_Z - k_2^- \theta_{BZ} - k_3^+ \theta_{AZ}\theta_{BZ} = f_2\left(\theta_{AZ}, \theta_{BZ}, \theta_{(BZ)*}\right) \qquad (7.124)$$

$$\frac{d\theta_{(BZ)*}}{dt} = \bar{k}_4^+ \theta_Z - k_4^- \theta_{(BZ)*} = f_3\left(\theta_{AZ}, \theta_{BZ}, \theta_{(BZ)*}\right) \qquad (7.125)$$

with $\bar{k}_1^+ = k_1^+ p_{A_2}, \bar{k}_2^+ = k_2^+ p_B$ and $\bar{k}_4^+ = k_4^+ p_B$.

The steady states for this model can be determined from the set of algebraic equations

$$f_1\left(\theta_{AZ}, \theta_{BZ}, \theta_{(BZ)*}\right) = f_2\left(\theta_{AZ}, \theta_{BZ}, \theta_{(BZ)*}\right) = f_3\left(\theta_{AZ}, \theta_{BZ}, \theta_{(BZ)*}\right) = 0 \qquad (7.126)$$

By expressing $f_1 = f_2 = 0$ in implicit form, we obtain θ_{AZ}, θ_{BZ}, and θ_Z as functions of $\theta_{(BZ)*}$.
From Eq. (7.126), it follows that

$$\theta_Z\left(\theta_{(BZ)*}\right) = \frac{k_4^-}{\bar{k}_4^+}\theta_{(BZ)*} = \alpha\theta_{(BZ)*} \qquad (7.127)$$

The steady states correspond to the points of intersection of this straight line with the curve $\theta_Z = \theta_Z\left(\theta_{(BZ)*}\right)$ in the plane $(\theta_Z, \theta_{(BZ)*})$. This curve is plotted in accordance with the solutions of $f_1\left(\theta_{AZ}, \theta_{BZ}, \theta_{(BZ)*}\right) = f_2\left(\theta_{AZ}, \theta_{BZ}, \theta_{(BZ)*}\right) = 0$ with respect to θ_{AZ} and θ_{BZ}, which are the steady states of the model representing the three-step adsorption mechanism (steps (1)–(3) of Eq. (7.122)). For this three-step submechanism, a parametric domain can be distinguished in which three different steady states exist. In this domain, the curve $\theta_Z = \theta_Z\left(\theta_{(BZ)*}\right)$ has a typical S-shaped form, see Fig. 7.9.

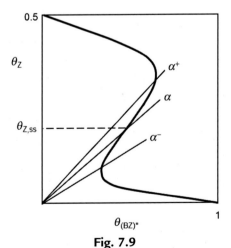

Fig. 7.9

Normalized steady-state concentration of free active sites Z as a function of the normalized steady-state concentration of the "buffer" intermediate $(BZ)*$, curve $\theta_Z = \theta_Z\left(\theta_{(BZ)*}\right)$. The straight lines represent Eq. (7.127); $\alpha = k_4^-/\bar{k}_4^+$ with $\alpha^+ > \alpha > \alpha^-$. Intersections of the dependences represents the (unstable) steady states $\theta_{Z,ss}$. *Reprinted from Yablonskii, G.S., Bykov, V.I., Gorban, A.N., Elokhin, V.I., 1991. Kinetic models of catalytic reactions. In: Compton, R.G. (Ed.), Comprehensive Chemical Kinetics, vol. 32. Elsevier, Amsterdam, Copyright (1991), with permission from Elsevier.*

The reduced set of equations for the time derivatives of θ_{AZ} and θ_{BZ}, Eqs. (7.123) and (7.124), with not three but two variables, in which $\theta_{(BZ)*}$ is regarded as a parameter, is a catalytic trigger. For this system, a unique steady state is stable. If there are three steady states, the two outer steady states are stable and the middle steady state is unstable.

The complete set of equations with three variables, Eqs. (7.123)–(7.125), including the buffer step, may have a unique unstable solution. In view of the Poincaré-Bendixson theorem, this is a necessary and sufficient condition for the occurrence of oscillations. For this set of equations, the solution is considered to be in the so-called reaction simplex S:

$$S\left(\theta_{AZ}, \theta_{BZ}, \theta_{(BZ)*}\right) : \theta_{AZ} \geq 0,\, \theta_{BZ} \geq 0,\, \theta_{(BZ)*} \geq 0,\, \theta_{AZ} + \theta_{BZ} + \theta_{(BZ)*} \leq 1 \qquad (7.128)$$

Let $\left(\theta_{AZ,ss}, \theta_{BZ,ss}, \theta_{(BZ)*,ss}\right) = (\theta_{ss})$ be the steady-state solution of the set of equations, Eqs. (7.123)–(7.125). For this system, the characteristic equation is

$$\lambda^3 + \sigma\lambda^2 + \delta\lambda + \Delta = 0 \qquad (7.129)$$

where $\sigma = -\operatorname{Tr}(\mathbf{J})$, $\delta = a_{11} + a_{22} + a_{33} = \operatorname{Tr}(\operatorname{adj}\mathbf{J})$, and $\Delta = -\det \mathbf{J}$. $\mathbf{J} = [a_{ij}]$ $(i, j = 1, 2, 3)$ is the matrix of the corresponding linearized set of equations at the steady state; a_{11}, a_{22}, and a_{33} are the principal minors of matrix $\mathbf{J}$; and $\operatorname{adj}\mathbf{J}$ is the adjoint of $\mathbf{J}$. The values of the nonpositive roots λ_1 and λ_2 of matrix $\mathbf{J}$ are determined by the relationship between σ, δ, and Δ, with $\sigma < 0$. It can be shown (Yablonskii et al., 1991) that a necessary and sufficient condition for the instability of the steady state is that δ is negative. For $\delta = 0$, the parameters are at their critical values, that is, the value for which the real parts of the characteristic roots λ_1 and λ_2 change sign.

When the parameters for reaction steps (1)–(3) are properly chosen, a parametric domain for step (4) can be found that results in a unique unstable steady state, resulting in oscillation (see Fig. 7.10). Comparison with Fig. 7.9 shows that the oscillations are observed in the domain close to the hysteresis in the curve $\theta_Z(\theta_{(BZ)*})$. The shape of the limit cycles in Fig. 7.10 very much depends on the values of $\bar{k}_4^+$ and k_4^-. The lower these values, the closer the shape of

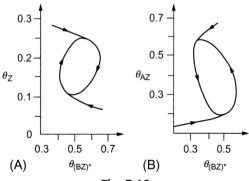

Fig. 7.10

Examples of limit cycles on (A) $(\theta_{(BZ)*}, \theta_Z)$ and (B) $(\theta_{(BZ)*}, \theta_{AZ})$ phase-space projections. *Reprinted from Yablonskii, G.S., Bykov, V.I., Gorban, A.N., Elokhin, V.I., 1991. Kinetic models of catalytic reactions. In: Compton, R.G. (Ed.), Comprehensive Chemical Kinetics, vol. 32. Elsevier, Amsterdam, Copyright (1991), with permission from Elsevier.*

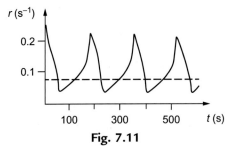

Fig. 7.11

Self-sustained oscillations of the reaction rate; the dashed line marks the value of the reaction rate at the unstable steady state. *Reprinted from Yablonskii, G.S., Bykov, V.I., Gorban, A.N., Elokhin, V.I., 1991. Kinetic models of catalytic reactions. In: Compton, R.G. (Ed.), Comprehensive Chemical Kinetics, vol. 32. Elsevier, Amsterdam, Copyright (1991), with permission from Elsevier.*

the limit cycle to the hysteresis in the curve $\theta_Z(\theta_{(BZ)*})$. The corresponding self-sustained oscillations of the reaction rate are shown in Fig. 7.11.

On increasing $\bar{k}_4^+$ to the limit value $(\bar{k}_4^+)^*$, the frequency of the oscillations increases, while the amplitude remains nearly constant. At $\bar{k}_4^+ \geq (\bar{k}_4^+)^*$, the oscillations vanish jump-like and the system enters the stable steady state. Thus, the four-step mechanism with a buffer step, Eq. (7.122), described by Eqs. (7.123)–(7.125) can be termed the simplest catalytic oscillator based on mass-action-law assumptions.

A computational and qualitative analysis of a number of modifications of the four-step mechanism and the corresponding model shows that three factors are of importance with respect to the occurrence of self-sustained oscillations:

(1) The presence of a nonlinear step such as $AZ+BZ \rightarrow AB+2Z$, that is, the interaction between different surface intermediates. This step is responsible for nonlinear phenomena such as multiplicity of steady states, rate oscillations, and discontinuities in kinetic dependences (see Yablonskii et al., 1991; Marin and Yablonsky, 2011).
(2) The presence of a buffer step, which is responsible for slow motion.
(3) Reversibility of at least one step, eg, $A+Z \rightleftarrows AZ$.

One of the basic types of self-sustained oscillations in chemical kinetics is relaxation oscillation. This term usually means the occurrence of self-sustained oscillations whose period can be divided into several parts corresponding to slow and fast changes of the process characteristics. The simplest model for describing relaxation oscillation was proposed by Gol'dshtein et al. (1986). This model is based on the combination of a two-step mechanism and a buffer step:

$$
\begin{aligned}
&(1) \ A+Z && \rightleftarrows \ AZ \\
&(2) \ AZ+2Z && \rightarrow \ A+3Z \\
&(3) \ Z && \rightleftarrows \ (Z)^*
\end{aligned}
\qquad (7.130)
$$

where the second step is autocatalytic and $(Z)^*$ is a nonreactive surface substance formed in the buffer step.

The model corresponding to this mechanism is

$$\frac{d\theta_Z}{dt} = -\bar{k}_1^+ \theta_Z + (k_1^- + k_2^+ \theta_Z^2)(1 - \theta_{(Z)*} - \theta_Z) - k_3^+ \theta_Z + k_3^- \theta_{(Z)*} = f_1(\theta_Z, \theta_{(Z)*}) \qquad (7.131)$$

$$\frac{d\theta_{(Z)*}}{dt} = k_3^+ \theta_Z - k_3^- \theta_{(Z)*} = f_2(\theta_Z, \theta_{(Z)*}) \qquad (7.132)$$

The buffer step (3) is assumed to be slow, so $k_3^+, k_3^- \ll k_1^+, k_1^-, k_2^+$. The ratio k_3^+/k_1^+ is a small parameter in this model. The fast variable is θ_Z, while $\theta_{(Z)*}$ is the slow one.

All three factors mentioned above are present in this simple autocatalytic model.

7.6 General Procedure for Parametric Analysis

We are going to describe a general procedure for the parametric analysis of a typical chemical dynamical system based on the examples presented in this chapter. This procedure has also been laid out in detail by Bykov and Tsybenova (2011) and Bykov et al. (2015). The studied process occurs at external conditions that are characterized by a number of parameters. These parameters relate to the system of ordinary differential equations

$$\frac{dx_i}{dt} = f_i(x_1, \cdots x_i, \cdots x_n, P_1, \cdots P_j, \cdots P_m) \qquad (7.133)$$

where x_i are phase variables (concentrations and temperature) and P_j are parameters, such as reactor volume and inlet and initial concentrations and temperature.

The first step of the parametric analysis of Eq. (7.133) is the determination of its steady states, which are solutions of the algebraic set of equations

$$f_i(x_1, \cdots x_i, \cdots x_n, P_1, \cdots P_j, \cdots P_m) = 0 \qquad (7.134)$$

In some situations, the solution of Eq. (7.134) can be obtained explicitly,

$$x_i = \varphi_i(P_1, \cdots P_j, \cdots P_m) \qquad (7.135)$$

but generally this is impossible and the parametric dependences have to be found by using some kind of computational procedure.

There are cases in which the set of steady-state equations can be transformed to one equation:

$$f(x, P_1 \cdots P_j, \cdots P_m) = 0 \qquad (7.136)$$

in which x is one of the phase variables. This nonlinear equation may have several steady states and, in this domain, a hysteresis of the steady-state dependences on the parameters may be observed.

The second step in this procedure is a study of the stability of the steady state(s). To this end, the Jacobian matrix **J** has to be formulated with elements

$$a_{ij} = \frac{\partial f_i}{\partial x_j}(x_{ss}) \qquad (7.137)$$

where x_{ss} is the steady-state variable. The stability of a steady state is determined by the properties of the eigen roots $\lambda_i (i = 1,...,n)$ of the Jacobian matrix. If all λ_i have nonzero real parts, the steady state is "robust." Its stability depends on the sign of $\text{Re}(\lambda_i)$.

The most important part of this analysis is determining the parametric dependences $\varphi_i(P_1 \cdots P_j, \cdots P_m)$. In some cases, based on Eq. (7.135) the following explicit dependence can be written:

$$P_j = \eta_i(x); \quad j = 1, \cdots, m \qquad (7.138)$$

where x is the same variable as in Eq. (7.136).

As a result, a function

$$x = \varphi(P_j) \qquad (7.139)$$

can be found, which is the reversal of the unknown parametric dependence.

Varying one of the parameters, for example, P_1, one can find its bifurcational value at which the number of steady states and their stability change. Varying a second parameter, for example, P_2, bifurcation curves as functions of P_1 and P_2 can be plotted. For the two-dimensional case, there are two main bifurcational dependences (also see Section 7.3.3):

- the line of multiplicity of steady states, L_Δ;
- the line of neutrality, L_σ.

The stability of a steady state is determined by the eigen roots of the second-order characteristic equation:

$$\lambda^2 - \sigma\lambda + \Delta = 0 \qquad (7.140)$$

where $\sigma = a_{11} + a_{22}$ and $\Delta = a_{11}a_{22} - a_{12}a_{21}$.

Let us now choose two parameters, P_1 and P_2, and determine L_Δ and L_σ in the plane of these parameters. The boundaries of the domain of multiplicity are defined as solutions of the set of equations

$$\begin{cases} f(x, P_1, P_2) = 0 \\ \Delta(x, P_1, P_2) = 0 \end{cases} \qquad (7.141)$$

In many cases, taking into account Eq. (7.138), Eq. (7.141) can be formulated as

$$\begin{cases} P_2 = \xi_2(x) \\ P_1 = \xi_1(x, \xi_2(x)) \end{cases} \tag{7.142}$$

It is possible to determine the parametric portrait of the system by analyzing the mutual rearrangement of the lines L_Δ and L_σ on the plane (P_1, P_2). The number of steady states and their stability can be estimated using this parametric portrait.

The phase portraits of the model represented by Eq. (7.133) provide us with useful information on the dynamic behavior of this model. These phase portraits are calculated by numerical integration for a given set of parameters at different initial conditions. Every parametric domain, which is determined by the mutual rearrangement of the lines L_Δ and L_σ, relates to a special type of phase portrait. For a model with two variables, it is possible to present all types of phase portraits. However, for many models with three variables, the number of phase portraits is enormous and the corresponding classification of phase trajectories is quite difficult.

Typically, temporal dependences $x_i = x_i(t, P_1, \cdots P_j, \cdots P_m)$ are computed for the set of ordinary differential equations, Eq. (7.133) with a given set of parameters. Technical difficulties are usually related to the "stiffness" of the system.

Let us analyze the set of nonlinear equations

$$F(\mathbf{x}, P) = 0 \tag{7.143}$$

where F is a vector-function, $\mathbf{x}$ is a vector of variables, and P is a parameter. Eq. (7.143) determines the implicit dependence

$$\mathbf{x} = \mathbf{x}(P) \tag{7.144}$$

and the main goal of a parametric analysis is to determine this dependence.

Nonlinear sets of equations are usually solved using iterative methods based on parameter continuation. For a particular parameter value P_0, a mapping is repeatedly applied to an initial guess $\mathbf{x}_0$. If the method converges, and is consistent, then in the limit the iteration approaches a solution of $F(\mathbf{x}, P_0) = 0$. A general scheme of the method of parameter continuation is the following: substitution of Eq. (7.143) into Eq. (7.144) and taking the derivative yields

$$\mathbf{J} \frac{d\mathbf{x}}{dP} + \frac{\partial F}{\partial P} = 0 \tag{7.145}$$

where $\mathbf{J}$ is the Jacobian matrix,

$$\mathbf{J} = \frac{\partial F(\mathbf{x}, P)}{\partial \mathbf{x}} \tag{7.146}$$

Using Eq. (7.144) taken as a set of linear equations with respect to dx/dP, one can write equations of change along the parameter:

$$\frac{d\mathbf{x}}{dP} = -\mathbf{J}^{-1}\frac{\partial F}{\partial P} \tag{7.147}$$

The parametric dependence $\mathbf{x}(P)$ is a solution of the set of ordinary differential equations, Eq. (7.147) at a given set of initial conditions $\mathbf{x}(P_0) = \mathbf{x}_0$.

At the bifurcation points of the system, Eq. (7.142), the Jacobian matrix becomes singular. Hence, for numerical integration of Eq. (7.147), a parameterization procedure along the curve $\mathbf{x}(P)$ can be used in the corresponding space of dimension $\dim(\mathbf{x}+1)$. This integration requires the use of special methods, in particular, methods based on the calculation of the Jacobian matrix. In this case, we have to find the partial derivatives $\dfrac{\partial}{\partial \mathbf{x}}\left(\mathbf{J}^{-1}\dfrac{\partial F}{\partial \mathbf{x}}\right)$, which creates additional problems. The set of equations, Eq. (7.147), is written on the basis of the solution, Eq. (7.145), for example using the Gauss method regarding $\dfrac{d\mathbf{x}}{dP}$. When solving Eq. (7.147), the accuracy of the initial conditions is very important. Generally, the computer realization of the parameter-continuation method is quite complicated. The degree of this complexity depends on the number of variables in Eq. (7.143). Because these numerical calculations are typically carried out with fixed, limited precision, additional considerations relating to floating-point accuracy arise, and the difficulties are exacerbated precisely at the most interesting points of the calculations.

Let us analyze the particular case in which the set of equations, Eq. (7.145), can be formulated as just one equation,

$$f(x, P) = 0 \tag{7.148}$$

where f is not a vector function but a scalar function and x is the only variable. Differentiation of the identity $f(x(P), P) = 0$ leads to

$$\frac{dx}{dP} = -\frac{\partial f}{\partial P}\frac{\partial x}{\partial f} \tag{7.149}$$

or

$$\frac{dP}{dx} = -\frac{\partial f}{\partial x}\frac{\partial P}{\partial f} \tag{7.150}$$

The parametric dependence $P(x)$ is the reversal of $x(P)$. They can be found by numerical integration of Eq. (7.149) or (7.150).

At the turning points, $\dfrac{\partial f}{\partial x} = 0$ or $\dfrac{\partial f}{\partial P} = 0$, so the condition

$$\left| \frac{\partial f}{\partial P} \frac{\partial x}{\partial f} \right| < 1 \tag{7.151}$$

has to be fulfilled to make a choice to move along either P or x.

For numerical integration, Eq. (7.149) or (7.150) is chosen depending on the condition of Eq. (7.151). The advantages of this modified method of parameter continuation for the scalar equation, Eq. (7.148) are obvious. In this case, it is not necessary to calculate the matrix, which is the inverse of the Jacobian matrix.

The transformation of the original set of equations to just one equation is not always possible. Let us analyze another case, in which the original set is transformed into a set of two equations:

$$\begin{cases} f_1(x, y, P) = 0 \\ f_2(x, y, P) = 0 \end{cases} \tag{7.152}$$

where x and y are variables and f_1 and f_2 are nonlinear scalar functions. Eq. (7.152) implicitly determines the following dependences:

$$\begin{cases} x = x(P) \\ y = y(P) \end{cases} \tag{7.153}$$

One of the goals of the parametric analysis of Eq. (7.152) is to determine the functions represented by Eq. (7.153). Substitution of Eq. (7.153) into Eq. (7.152) and differentiation of the resulting identities yields

$$\begin{cases} \dfrac{\partial f_1}{\partial x} \dfrac{dx}{dP} + \dfrac{\partial f_1}{\partial y} \dfrac{dy}{dP} + \dfrac{\partial f_1}{\partial P} = 0 \\[2mm] \dfrac{\partial f_2}{\partial x} \dfrac{dx}{dP} + \dfrac{\partial f_2}{\partial y} \dfrac{dy}{dP} + \dfrac{\partial f_2}{\partial P} = 0 \end{cases} \tag{7.154}$$

Defining

$$\Delta = \begin{bmatrix} \dfrac{\partial f_1}{\partial x} & \dfrac{\partial f_1}{\partial y} \\[2mm] \dfrac{\partial f_2}{\partial x} & \dfrac{\partial f_2}{\partial y} \end{bmatrix}, \Delta_1 = \begin{bmatrix} -\dfrac{\partial f_1}{\partial P} & \dfrac{\partial f_1}{\partial y} \\[2mm] -\dfrac{\partial f_2}{\partial P} & \dfrac{\partial f_2}{\partial y} \end{bmatrix}, \Delta_2 = \begin{bmatrix} \dfrac{\partial f_1}{\partial x} & -\dfrac{\partial f_1}{\partial P} \\[2mm] \dfrac{\partial f_2}{\partial x} & \dfrac{\partial f_2}{\partial P} \end{bmatrix},$$

Eq. (7.155) can be written as

$$\begin{cases} \dfrac{dx}{dP} = \dfrac{\Delta_1}{\Delta} \\[3mm] \dfrac{dy}{dP} = \dfrac{\Delta_2}{\Delta} \end{cases} \tag{7.155}$$

The initial conditions are $x(P_0) = x_0$ and $y(P_0) = y_0$. At the turning points of the curve described by Eq. (7.152), one of the determinants Δ, Δ_1, and Δ_2 equals zero. Hence, in addition to Eq. (7.155), two sets of differential equations can be analyzed:

$$\begin{cases} \dfrac{dP}{dx} = \dfrac{\Delta}{\Delta_1} \\ \dfrac{dy}{dx} = \dfrac{\Delta_2}{\Delta_1} \end{cases} \tag{7.156}$$

and

$$\begin{cases} \dfrac{dx}{dy} = \dfrac{\Delta_1}{\Delta_2} \\ \dfrac{dP}{dy} = \dfrac{\Delta}{\Delta_2} \end{cases} \tag{7.157}$$

Therefore, movement along the line of the parametric dependence Eq. (7.153) can be realized by the integration of one of three sets of equations: Eq. (7.154), Eq. (7.156), or Eq. (7.157). Three different situations can be distinguished. If $|\Delta_1| < |\Delta|$ and $|\Delta_2| < |\Delta|$, P is chosen as a parameter and Eq. (7.155) is integrated; if $|\Delta| < |\Delta_1|$ and $|\Delta_2| < |\Delta_1|$, the parameter is x and Eq. (7.156) is integrated; finally, if $|\Delta| < |\Delta_2|$ and $|\Delta_1| < |\Delta_2|$, the parameter is y and Eq. (7.157) is integrated. We always choose the path of the largest denominator. These criteria can be applied when all variables and all parameters have commensurate variations.

The method described is one possible modification of the parameter-continuation method.

Nomenclature

A_r	reactor surface available for heat exchange (m^2)
$\mathbf{c}$	vector of concentrations (mol m^{-3})
c_i	concentration of component i (mol m^{-3})
c_p	specific heat capacity at constant pressure (J mol^{-1} K^{-1})
D	phase space (mol m^{-3})
Da	(first) Damköhler number (–)
d	element of phase space (mol m^{-3})
E_a	activation energy (J mol^{-1})
H	enthalpy (J)
$\Delta_r H$	enthalpy change of reaction (J mol^{-1})
k	reaction rate coefficient ((m^3 mol^{-1})^n s^{-1})
k_0	preexponential factor (same as k)
n	amount of substance (mol)
P_i	parameter i (varies)
p	pressure (Pa)

p	partial pressure of component i (Pa)
Q_{hg}	rate of heat generation (J s^{-1})
Q_{hr}	rate of heat removal (J s^{-1})
q_V	volumetric flow rate (m^3 s^{-1})
R_g	universal gas constant (J mol^{-1} K^{-1})
r	reaction rate (s^{-1})
Se	Semenov number (–)
T	temperature (K)
t	time (s)
t^*	new time scale $= (\varepsilon/\Gamma_t)t$ (mol^{-1} m$_{cat}^2$ s)
U	internal energy (J)
V	volume (m^3)
$V(x)$	Lyapunov function (varies)
V_r	volume of reaction mixture (m^3)
x	conversion (–)
x,y,z	variables (varies)

Greek symbols

β	dimensionless number (–)
δ	value of concentration upon perturbation (mol m^{-3})
ε	concentration perturbation (mol m^{-3})
ε_c	boundary on deviation in concentration (mol m^{-3})
θ_j	normalized concentration of surface intermediate j (–)
λ	characteristic root (s^{-1})
μ_i	chemical potential of component i (J mol^{-1})
ν	ratio Da/Se (–)
ξ	deviation of the concentration $\tilde{c}$ from the rest point (mol m^{-3})
ρ	fluid density (mol m^{-3})
τ_{ss}	relaxation time (s)

Subscripts

0	initial
eq	equilibrium
i	component i
j	surface intermediate j
min	minimum
s	step
ss	steady state

Superscripts

+   forward reaction

−   reverse reaction

α_{si} partial order of forward reaction of step s in reactant i

β_{si} partial order of reverse reaction of step s in product i

$\gamma\beta_{si}$ $\beta_{si} - \alpha_{si}$

References

Berman, A.D., Elinek, A.V., 1979. On dynamics of homogeneous-heterogeneous reactions. Dokl. Akad. Nauk. SSSR 248, 643–647.

Berryman, A.A., 1992. The orgins and evolution of predator-prey theory. Ecology 73, 1530–1535.

Beusch, H., Fieguth, P., Wicke, E., 1972a. Thermisch und kinetisch verursachte Instabilitäten im Reaktionsverhalten einzelner Katalysatorkörner. Chem. Ing. Techn. 44, 445–451.

Beusch, H., Fieguth, P., Wicke, E., 1972b. Unstable behavior of chemical reactions of single catalyst particles. Adv. Chem. Ser. 109, 615–621.

Bolten, H., Hahn, Th., LeRoux, J., Lintz, H.-G., 1985. Bistability of the reaction rate in the oxidation of carbon monoxide by nitrogen monoxide on polycrystalline platinum. Surf. Sci. 160, L529–L532.

Boreskov, G.K., Slin'ko, M.G., Filippova, A.G., 1953. Catalytic activity of nickel, palladium, and platinum in hydrogen oxidation. Dokl. Akad. Nauk SSSR 92, 353–355.

Bykov, V.I., Serafimov, L.A., Tsybenova, S.B., 2015. Emergency starting regimes of a continuous stirred tank reactor. Theor. Found Chem. Eng. 49, 361–369.

Bykov, V.I., Tsybenova, S.B., 2011. Non-linear Models of Chemical Kinetics. Krasand, Moscow.

Bykov, V.I., Yablonskii, G.S., 1987. Steady-state multiplicity in heterogeneous catalytic reactions. Int. Chem. Eng. 21, 142–154.

Bykov, V.I., Yablonskii, G.S., Akramov, T.A., Slin'ko, M.G., 1977. The rate of the free energy decrease in the course of the complex chemical reaction. Dokl. Akad. Nauk. USSR 234, 621–634.

Bykov, V.I., Yablonskii, G.S., Elokhin, V.I., 1981. Steady state multiplicity of the kinetic model of CO oxidation reaction. Surf. Sci. Lett. 107, L334–L338.

Bykov, V.I., Yablonskii, G.S., Kim, V.F., 1978. One simple model for kinetic autooscillations in the catalytic reaction of CO oxidation. Dokl. Akad. Nauk SSSR 242, 637–639.

Davies, W., 1934. The rate of heating of wires by surface combustion. Philos. Mag. 17, 233–251.

Dimitrov, V.I., 1982. Simple Kinetics. Nauka, Novosibirsk.

Dimitrov, V.I., Bykov, V.I., Yablonskii, G.S., 1977. On Characteristics of Complex Chemical Reactions. Nauka, Moscow. pp. 565–570 (in Russian).

Eigenberger, G., 1978a. Kinetic instabilities in heterogeneously catalyzed reactions—I: Rate multiplicity with Langmuir-type kinetics. Chem. Eng. Sci. 33, 1255–1261.

Eigenberger, G., 1978b. Kinetic instabilities in heterogeneously catalyzed reactions—II: Oscillatory instabilities with Langmuir-type kinetics. Chem. Eng. Sci. 33, 1263–1268.

Field, R.J., Körös, E., Noyes, R.M., 1972. Oscillations in chemical systems. 2. Thorough analysis of temporal oscillation in bromate-cerium-malonic acid system. J. Am. Soc. 94, 8649–8664.

Field, R.J., Noyes, R.M., 1974. Oscillations in chemical systems. IV. Limit cycle behavior in a model of a real chemical reaction. J. Chem. Phys. 60, 1877–1884.

Frank-Kamenetskii, D.A., 1969. Diffusion and Heat Transfer in Chemical Kinetics. Plenum Press, New York.

Glansdorff, P., Prigogine, I., 1971. Thermodynamic Theory of Structure, Stability, and Fluctuations. Wiley, New York.

Gol'dshtein, V.M., Sobolev, V.A., Yablonskii, G.S., 1986. Relaxation self-oscillations in chemical kinetics: A model, conditions for realization. Chem. Eng. Sci. 41, 2761–2766.

Gorban, A.N., 1980. On the problem of boundary equilibrium points. React. Kinet. Catal. Lett. 15, 315–319.

Gorban, A.N., 1984. Equilibrium Encircling: Equations of Chemical Kinetics and their Thermodynamic Analysis. Nauka, Novosibirsk.

Gorban, A.N., Cheresiz, V.M., 1981. Slow relaxations of dynamical systems and bifurcations of ω-limit sets. Dokl. Akad. Nauk SSSR 261, 1050–1054.

Gorban, A.N., Karlin, I.V., 2003. Method of invariant manifold for chemical kinetics. Chem. Eng. Sci. 58, 4751–4768.

Kharkovskaya, E.N., Boreskov, G.K., Slin'ko, M.G., 1959. Dokl. Akad. Nauk SSSR 127, 145–148.

Liljenroth, F.G., 1918. Starting and stability phenomena of ammonia-oxidation and similar reactions. Chem. Met. Eng. 19, 287–293.

Lorenz, E.N., 1963. Deterministic nonperiodic flow. J. Atmos. Sci. 20, 130–141.

Marin, G.B., Yablonsky, G.S., 2011. Kinetics of Chemical Reactions—Decoding Complexity. Wiley-VCH, Weinheim, Germany.

Mukesh, D., Cutlip, M.C., Goodman, M., Kenney, C.N., Morton, W., 1982. The stability and oscillations of carbon monoxide oxidation over platinum supported catalyst. Effect of butene. Chem. Eng. Sci. 37, 1807–1810.

Mukesh, D., Kenney, C.N., Morton, W., 1983. Concentration oscillations of carbon monoxide, oxygen and 1-butene over a platinum supported catalyst. Chem. Eng. Sci. 38, 69–77.

Mukesh, D., Morton, W., Kenney, C.N., Cutlip, M.B., 1984. Island models and the catalytic oxidation of carbon monoxide and carbon monoxide-olefin mixtures. Surf. Sci. 138, 237–257.

Nicolis, G., Prigogine, I., 1977. Self-Organization in Nonequilibrium Systems. Wiley-Interscience, New York.

Nowobilski, P., Takoudis, C.G., 1985. Reaction between nitric oxide and ammonia on polycrystalline Pt: A mathematical model of rate oscillations. Chem. Eng. Sci. 40, 1751–1757.

Prigogine, I., Lefever, R., 1968. Symmetry breaking instabilities in dissipative systems. II. J. Chem. Phys. 48, 1695–1700.

Regalbuto, J., Wolf, E.E., 1986. FTIR studies of self-sustained oscillations during the CO-NO-O_2 reaction on Pt/SiO_2 catalysts. Chem. Eng. Commun. 41, 315–326.

Sheintuch, M., Luss, D., 1981. Reaction rate oscillations during propylene oxidation on platinum. J. Catal. 68, 245–248.

Slin'ko, M.G., Slin'ko, M.M., 1978. Self-oscillations of heterogeneous catalytic reaction rates. Catal. Rev. Sci. Eng. 17, 119–153.

Slin'ko, M.M., Jaeger, N.I., 1994. Oscillatory Heterogeneous Catalytic Systems. Elsevier, Amsterdam.

Slin'ko, M.M., Slin'ko, M.G., 1982. Rate oscillations in heterogeneous catalyzed reactions. Kinet. Katal. 23, 1421–1428.

Valentinuzzi, M.E., Kohen, A.J., 2013. The mathemization of biology and medicine. IEEE Pulse, 50–56.

Vayenas, C.G., Georgakis, C., Michaels, J., Tormo, J., 1981. The role of PtOx in the isothermal rate oscillations of ethylene oxidation on platinum. J. Catal. 67, 348–361.

Yablonskii, G.S., Bykov, V.I., Gorban, A.N., Elokhin, V.I., 1991. Kinetic models of catalytic reactions. In: Compton, R.G. (Ed.), Comprehensive Chemical Kinetics, vol. 32. Elsevier, Amsterdam.

Zel'dovich, Ya.B., 1938. Proof of a unique solution to the mass action law. Zh. Tekh. Fiz. 11, 685–687 (in Russian).

## CHAPTER 8

Optimization of Multizone Configurations

In this chapter, the problem of optimization of multizone configurations will be dealt with. When talking about a zone, we consider a thin active zone in which perfect mixing is assumed. Typically, thin catalyst zones are used in temporal analysis of products (TAP) studies, but other types of experiments involving thin active zones are also conceivable.

Thin active zones are separated by inactive transport zones. Here we consider transport to be purely diffusional, and the inactive zone to be purely diffusive. We also assume that the diffusion coefficient does not depend on the gas composition.

8.1 Reactor Model

For gas-solid systems studied under high-vacuum conditions, we assume Knudsen flow, and the diffusion coefficient depends only on the temperature and the molar mass of the gas.

In our model, we assume that every active zone is kinetically identical under the assumption of first-order kinetics. Every active zone has the same adsorption rate coefficient k_a and the same thickness ΔL_z, so

$$k_a \Delta L_z = \kappa = \text{constant} \tag{8.1}$$

We also assume that for every inactive diffusion zone, the effective diffusion coefficient D_{eff} is the same. We apply the general model of the TAP reactor as a system of partial differential equations (see Chapter 5). In the Laplace domain, each zone is represented by a zone transfer matrix (Constales et al., 2006). The thin zone of activity κ corresponds to the matrix:

$$\mathbf{A} = \begin{bmatrix} 1 & 0 \\ \kappa & 1 \end{bmatrix} \tag{8.2}$$

and the inactive diffusion zone of length L_i corresponds to the matrix:

$$\mathbf{B} = \begin{bmatrix} \cosh\left(L_i\sqrt{\dfrac{\varepsilon_b s}{D_{\text{eff}}}}\right) & \dfrac{\sinh\left(L_i\sqrt{\dfrac{\varepsilon_b s}{D_{\text{eff}}}}\right)}{\sqrt{\varepsilon_b s D_{\text{eff}}}} \\ \sqrt{\varepsilon_b s D_{\text{eff}}}\sinh\left(L_i\sqrt{\dfrac{\varepsilon_b s}{D_{\text{eff}}}}\right) & \cosh\left(L_i\sqrt{\dfrac{\varepsilon_b s}{D_{\text{eff}}}}\right) \end{bmatrix} \tag{8.3}$$

where ε_b is the bed voidage, that is, the fraction of the bed volume that is occupied by the voids.

Advanced Data Analysis and Modelling in Chemical Engineering. http://dx.doi.org/10.1016/B978-0-444-59485-3.00008-4

The goal of our analysis is to calculate and then optimize the conversion of a chemical transformation. For this, only the zeroth moment of the outlet flux is required, which can be obtained by setting $s = 0$ in the Laplace transform. Consequently, we only need Eq. (8.3) with $s = 0$:

$$\mathbf{B}_{s=0} = \begin{bmatrix} 1 & \dfrac{L_i}{D_{\text{eff}}} \\ 0 & 1 \end{bmatrix} \tag{8.4}$$

To combine the active zone with the subsequent inactive diffusion zone, we multiply matrix **A** with $\mathbf{B}_{s=0}$:

$$\mathbf{AB}_{s=0} = \begin{bmatrix} 1 & \dfrac{L_i}{D_{\text{eff}}} \\ \kappa & 1 + \dfrac{\kappa L_i}{D_{\text{eff}}} \end{bmatrix} \tag{8.5}$$

For representing the entire reactor, we have to multiply all matrices from left to right for $i = 1, \ldots, N_z$, with N_z the total number of active zones. Then the expression for the zeroth moment M_0,

$$\frac{1}{M_0} = \frac{1}{1 - X} \tag{8.6}$$

with X the reactant conversion, is determined by the (2,2) element of the resulting matrix. For $N_z = 1$,

$$\frac{1}{M_0} = 1 + \frac{\kappa}{D_{\text{eff}}} L_1 \tag{8.7}$$

For $N_z = 2$,

$$\frac{1}{M_0} = 1 + \frac{\kappa}{D_{\text{eff}}} (L_1 + 2L_2) + \left(\frac{\kappa}{D_{\text{eff}}}\right)^2 L_1 L_2 \tag{8.8}$$

For $N_z = 3$,

$$\frac{1}{M_0} = 1 + \frac{\kappa}{D_{\text{eff}}} (L_1 + 2L_2 + 3L_3) + \left(\frac{\kappa}{D_{\text{eff}}}\right)^2 (L_1 L_2 + 2L_1 L_3 + 2L_2 L_3) + \left(\frac{\kappa}{D_{\text{eff}}}\right)^3 L_1 L_2 L_3 \tag{8.9}$$

In general form, for N_z zones,

$$\frac{1}{M_0} = 1 + \sum_{i=1}^{N_z} \left(\frac{\kappa}{D_{\text{eff}}}\right)^i \sum_{1 \le j_1 < \cdots < j_i \le N_z} L_{j_1} L_{j_2}, \ldots, L_{j_i} \; j_1 \prod_{k=2}^{i} (j_k - j_{k-1}) \tag{8.10}$$

where the term "1" represents $\left(\dfrac{\kappa}{D_{\text{eff}}}\right)^0$.

In these equations, L_i is the distance between active zone i and active zone $i+1$, or the distance between active zone i and the reactor exit, if i is the last zone or if there is only a single active zone, that is, the active zone with zone number N_z.

8.2 Maximizing the Conversion

Now, what are the conditions for a maximum conversion? First, note that maximizing the conversion is equivalent to maximizing $1/M_0$ (see Eq. 8.6), since this is a strictly increasing function over the interval $0 \le X < 1$. In general, we must therefore maximize Eq. (8.10) for all choices $L_1 \ge 0, L_2 \ge 0, \ldots, L_{N_z} \ge 0$ such that $L_1 + L_2 + \cdots + L_{N_z} = L$, is the total length of the reactor. This analysis can be performed in two extreme cases: (1) κ is very small (but positive) and (2) κ is very large (but finite).

(1) In the case of very small κ, all terms involving κ^2, κ^3, and so on may be neglected, so the expression for $1/M_0$ simplifies to

$$\frac{1}{M_0} = 1 + \frac{\kappa}{D_{\text{eff}}}(L_1 + 2L_2 + \cdots + N_z L_{N_z}) \tag{8.11}$$

Since $L_1 + L_2 + \cdots + L_{N_z} = L$, the largest value of $1/M_0$ is achieved by setting $L_1 = L_2 = \cdots = L_{N_z-1} = 0$ and $L_{N_z} = L$. Thus, we combine all active zones into one active zone and place this at the very entrance of the reactor.

(2) For very large κ, all terms except the one involving κ^{N_z} may be neglected, and the expression for $1/M_0$ simplifies to

$$\frac{1}{M_0} = \left(\frac{\kappa}{D_{\text{eff}}}\right)^{N_z} L_1 L_2, \ldots, L_{N_z} \tag{8.12}$$

and the largest value of $1/M_0$ is achieved by setting $L_1 = L_2 = \cdots = L_{N_z} = \dfrac{L}{N_z}$. This is a property of arithmetic and geometric means of $L_1, \ldots, L_{N_z}$. These means are equal only if $L_1 = L_2 = \cdots = L_{N_z}$.

In all these equations, the thickness of the active catalyst zone, ΔL_z, is considered to be negligible compared with the length of the reactor: this value is implemented into the value of the adsorption rate coefficient, $\kappa = k_a \Delta L_z$.

Both optimal configurations are shown in Fig. 8.1.

For the physicochemical interpretation of Eq. (8.10), representing a cascade of thin active catalyst zones, it makes sense to compare this equation with the equation for a cascade of imaginary continuous stirred-tank reactors (CSTRs). Each imaginary CSTR in this cascade is assumed to have the same conversion as a unitary reactor consisting of one active zone and a

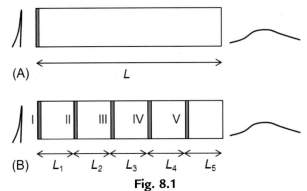

Fig. 8.1

Optimum configuration of active catalyst zones for (A) a low-activity catalyst (small κ) and (B) a high-activity catalyst (large κ); $L_1 = L_2 = L_3 = L_4 = L_5 = L/5$.

corresponding inactive diffusion zone. The zeroth moment of the first (or one) imaginary CSTR then is given by Eq. (8.7).

For $N_z = 2$,

$$\frac{1}{M_0} = 1 + \frac{\kappa}{D_{\text{eff}}}(L_1 + L_2) + \left(\frac{\kappa}{D_{\text{eff}}}\right)^2 L_1 L_2 \tag{8.13}$$

For $N_z = 3$,

$$\frac{1}{M_0} = 1 + \frac{\kappa}{D_{\text{eff}}}(L_1 + L_2 + L_3) + \left(\frac{\kappa}{D_{\text{eff}}}\right)^2 (L_1 L_2 + L_1 L_3 + 2L_2 L_3) + \left(\frac{\kappa}{D_{\text{eff}}}\right)^3 L_1 L_2 L_3 \tag{8.14}$$

and for the entire cascade

$$\frac{1}{M_0} = \prod_{i=1}^{N_z}\left(1 + \frac{\kappa}{D_{\text{eff}}}L_i\right) = \sum_{i=1}^{N_z}\left(\frac{\kappa}{D_{\text{eff}}}\right)^i \sum_{1 \le j_1 < \cdots < j_i \le N_z} L_{j_1} L_{j_2}, \ldots, L_{j_i} \tag{8.15}$$

which is the same as Eq. (8.10) but without the factor $j_1 \prod_{k=2}^{i}(j_k - j_{k-1})$. In fact, this factor is the main distinguishing feature of thin-zone configurations, which, in contrast with the cascade of CSTRs, include diffusion zones. This factor can therefore be attributed to the backmixing phenomenon, which is specific for diffusion zones and absent in a cascade of CSTRs.

Based on Eq. (8.15), it is easy to show that the optimal configuration of a cascade of imaginary CSTRs is one with equal distances between these CSTRs, independent of whether κ is small or large. This is very different from the case analyzed previously, in which a unitary reactor consists of an active zone and a diffusion zone.

8.3 Optimal Positions of Thin Active Zones

In the case of *one active thin zone*, its optimal position is at the very entrance of the reactor, and the conversion will be governed by Eq. (8.7), with $L_1 = L$ as the total length of the reactor.

If there are *two active thin zones*, we have to maximize Eq. (8.8) to determine the optimal position of the zones subject to the constraints $L_1 \geq 0$, $L_2 \geq 0$, and $L_1 + L_2 = L$. If we disregard conditions $L_1 \geq 0$ and $L_2 \geq 0$, the maximum conversion is obtained for

$$L_1 = \frac{1}{2}\left(L - \frac{D_{\text{eff}}}{\kappa}\right) \quad \text{or} \quad L_1 = \frac{L}{2}\left(1 - \frac{1}{\text{Da}}\right) \tag{8.16}$$

and

$$L_2 = \frac{1}{2}\left(L + \frac{D_{\text{eff}}}{\kappa}\right) \quad \text{or} \quad L_2 = \frac{L}{2}\left(1 + \frac{1}{\text{Da}}\right) \tag{8.17}$$

where $\text{Da} = \dfrac{\kappa L}{D_{\text{eff}}}$, the second Damköhler number (actually Da_{II}), which is the ratio of the reaction rate to the diffusion rate. Obviously, both L_1 and L_2 tend to reach $L/2$ for sufficiently large Da. The physical meaning of Eqs. (8.16), (8.17) is that Da must be larger than one. Otherwise, the maximum conversion is reached for $L_1 = 0$ and $L_2 = L$, that is, all of the catalyst is placed at the very entrance of the reactor.

For the situation that there are *three active thin zones*, we need to maximize Eq. (8.9) with constraints $L_1 \geq 0$, $L_2 \geq 0$, $L_3 \geq 0$ and $L_1 + L_2 + L_3 = L$. If the catalyst activity is low so κ is small, all zones have to be attached and placed at the very entrance of the reactor ($L_1 = 0$, $L_2 = 0$, $L_3 = L$). By setting $L_1 = 0$ in Eq. (8.9), we obtain conditions for the detachment of the second zone and from these we can determine the optimal position of the second zone:

$$L_2 = \frac{1}{2}\left(L - \frac{D_{\text{eff}}}{2\kappa}\right) \quad \text{or} \quad L_2 = \frac{L}{2}\left(1 - \frac{1}{2\text{Da}}\right) \tag{8.18}$$

and

$$L_3 = \frac{1}{2}\left(L + \frac{D_{\text{eff}}}{2\kappa}\right) \quad \text{or} \quad L_3 = \frac{L}{2}\left(1 + \frac{1}{2\text{Da}}\right) \tag{8.19}$$

So, the detachment of the second zone occurs when $\text{Da} = \dfrac{1}{2}$. Finally, maximizing Eq. (8.9) without considerations about the inequalities $L_i \geq 0$ for the zone lengths, we obtain the solution

$$L_1 = \frac{L}{6}\left(1 - \frac{7}{\text{Da}} + a\right) \tag{8.20}$$

$$L_2 = \frac{L}{3}\left(2 + \frac{4}{\text{Da}} - a\right) \tag{8.21}$$

$$L_3 = L - L_1 - L_2 = \frac{L}{6}\left(1 - \frac{1}{\text{Da}} + a\right) \tag{8.22}$$

with $a = \sqrt{1 + \dfrac{10}{\text{Da}} + \dfrac{13}{\text{Da}^2}}$.

8.4 *Numerical Experiments in Computing Optimal Active Zone Configurations*

For convenience, let us assume that $L = 1$ and $D_{eff} = 1$, so $Da = \dfrac{\kappa L}{D_{eff}} = \kappa$. Fig. 8.2 shows numerically computed optimal configurations for numbers of active zones ranging from $N_z = 2$ to $N_z = 7$. The horizontal axis represents the Damköhler number Da ($= \kappa$) and the vertical axis represents the ratio L_i/L for $i = 1, \ldots, N_z$, the dimensionless distance between the successive active zones or between the last zone and the reactor outlet.

For $N_z = 2$ and $N_z = 3$, Fig. 8.2A and B, we recover the results from the previous rigorous analysis: for small values of Da, all L_i equal zero, except L_{N_z} which equals one. This means that for small values of Da, the optimal configuration of zones is one in which all zones are as close as possible to the reactor inlet.

For $N_z = 2$, L_1 stops being zero and detaches when Da becomes equal to one. For Da > 1, the optimal configuration, therefore, consists of the first active zone always at the reactor inlet and the second zone at some distance L_1 from the reactor inlet (with $L_1 > 0$). As the value of Da tends to ∞, L_1 tends from 0 to $L/2$ and L_2 tends from L to $L/2$, which is the limit value of the optimal configuration. At all values of Da, $L_1 \leq L_2$.

For $N_z = 3$, the theoretical analysis is also confirmed. In this case, the configuration with all zones as close as possible to the reactor inlet is optimal for Da $< 1/2$. At Da $= 1/2$, L_2 detaches and up to Da $= 3/2$ the optimal configuration has two active zones as close as possible to the reactor inlet and the third at a distance L_2 from the inlet. When Da $= 3/2$, L_1 detaches and for all values Da $> 3/2$ the optimal configuration consists of three separate active zones: the first zone at the reactor inlet, the second at a distance L_1, and the third at a distance L_2 from the first zone. As Da $\to \infty$, the three distances tend to $L/3$ and in the limit the optimal zone configuration again is one with equal distances between the zones. Again, at all values of Da, $L_1 \leq L_2 \leq L_3$.

In the case of four active zones, $N_z = 4$, L_3, L_2, and L_1 successively detach as Da increases (see Fig. 8.2C). Remarkably, L_1 detaches exactly at the point where L_2 and L_3 become equal. Furthermore, the equality of L_2 and L_3 is preserved for all values of Da larger than two. This means that at Da ≥ 2, the optimal configuration consists of four separate active zones: the first zone is always at the inlet of the reactor, the distance L_1 between the first and second zone is smallest, the distances L_2 and L_3 between, respectively, the second and third zone and the third and fourth zone are equal and larger than L_1, and the distance L_4 between the fourth zone and the reactor outlet is largest: $L_1 < L_2 = L_3 < L_4$. Again, when Da $\to \infty$, all L_i tend to the same value, in this case $L/4$.

So far, every new case with a different number of zones was characterized by the manifestation of a new phenomenon. This also happens for $N = 5$. Fig. 8.2D shows the successive detachment of L_4, L_3, L_2, and L_1 with increasing value of Da. Again, detachment coincides

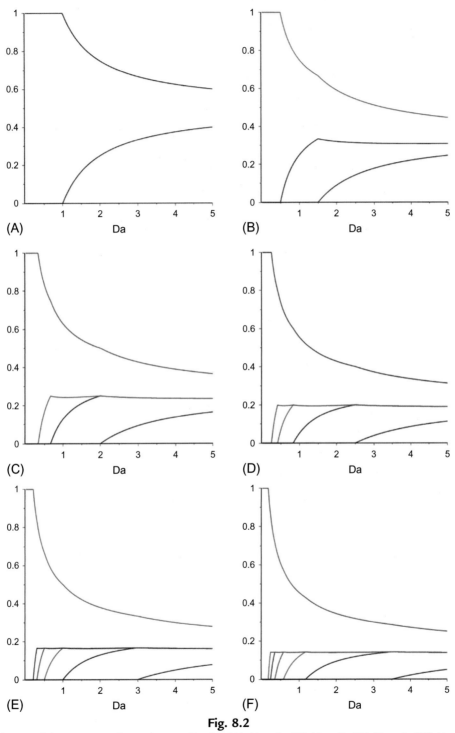

Fig. 8.2
Dependence of the zone configuration on Da for (A) $N_z = 2$, (B) $N_z = 3$, (C) $N_z = 4$, (D) $N_z = 5$, (E) $N_z = 6$, and (F) $N_z = 7$. In all figures the topmost line represents L_{N_z}, and of the other lines the rightmost (bottom line) represents L_1, the one to its left L_2, etc.

with merging: the detachment of L_2 coincides with the merging of L_3 and L_4 and then the detachment of L_1 coincides with the merging of L_2 with the merged L_3 and L_4 ("triple merging").

For $N_z = 6$ and $N_z = 7$, Figs. 8.2E and F show similar properties of successive detachments coinciding with mergers. From this observation, we can conjecture that the general structure of optimal zone configurations for more than one active zone ($N_z \geq 2$) is the following:

- At the inlet of the reactor there is at least one active zone, which can be a combination of a number of zones.
- The remaining part of the reactor must have one of the following configurations:
  - a single active zone some distance from the reactor inlet
  - two separate active zones inside the reactor with the distance between two successive zones increasing along the reactor
  - three or more, say q, active zones inside the reactor. In this case, there is a group of q zones with equal distances between them.

With increasing Da, the zones become more equally distributed over the reactor.

8.5 Equidistant Configurations of the Active Zones

First, in the design of configurations in which the distances between the active zones are all equal, the first zone always has to be placed at the very entrance of the reactor. This may be technically infeasible because of problems with the injection device or requirements such as isothermicity and uniformity of flow. In such cases, a free space before the first active zone is required. However, in our analysis, we do not include the length of this free space in the value of L, which therefore still represents the usable reactor length.

8.5.1 General Equation

The question may arise as to why it is necessary to put the first active zone at the very entrance of the reactor. The answer is that in accordance with Eq. (8.10) we would like to maximize $L_1, L_2, \ldots, L_{N_z}$ while at the same time fitting all zones into the total reactor length L, that is, $\sum L_i = L_1 + L_2 + \cdots + L_{N_z} \leq L$. Now, if this sum were smaller than L, all L_i could be stretched a little and the value of $1/M_0$ from Eq. (8.10) would increase. Therefore, the maximum value of $1/M_0$ is achieved if the sum of the distances strictly equals L (so is not smaller than L) and thus the first zone must be placed at the very entrance of the reactor.

Let us now derive a general equation for configurations in which the distances between the thin active zones are equal. For simplicity we again set $L=1$ and $D_{\text{eff}}=1$.

Each active zone can be represented by a matrix:

$$\mathbf{A} = \begin{bmatrix} 1 & 0 \\ Da & 1 \end{bmatrix} \tag{8.23}$$

and each inactive zone by a matrix:

$$\mathbf{B} = \begin{bmatrix} 1 & \dfrac{1}{N_z} \\ 0 & 1 \end{bmatrix} \tag{8.24}$$

The square root of matrix **B** is

$$\sqrt{\mathbf{B}} = \begin{bmatrix} 1 & \dfrac{1}{2N_z} \\ 0 & 1 \end{bmatrix} \tag{8.25}$$

with inverse

$$\left(\sqrt{\mathbf{B}}\right)^{-1} = \begin{bmatrix} 1 & -\dfrac{1}{2N_z} \\ 0 & 1 \end{bmatrix} \tag{8.26}$$

The value of $1/M_0$ is given by the (2,2) element of the matrix $(\mathbf{AB})^{N_z}$.

For $N_z = 1$, a sequence of transformations leads to

$$\mathbf{AB} = \mathbf{A}\sqrt{\mathbf{B}}\sqrt{\mathbf{B}} = \left(\sqrt{\mathbf{B}}\right)^{-1}\left(\sqrt{\mathbf{B}}\,\mathbf{A}\sqrt{\mathbf{B}}\right)\sqrt{\mathbf{B}} \tag{8.27}$$

which is an obvious identity.

Similarly, for $N_z = 2$:

$$\begin{aligned}
(\mathbf{AB})^2 = \mathbf{AB} \cdot \mathbf{AB} &= \left(\sqrt{\mathbf{B}}\right)^{-1}\left(\sqrt{\mathbf{B}}\,\mathbf{A}\sqrt{\mathbf{B}}\right)\sqrt{\mathbf{B}}\left(\sqrt{\mathbf{B}}\right)^{-1}\left(\sqrt{\mathbf{B}}\,\mathbf{A}\sqrt{\mathbf{B}}\right)\sqrt{\mathbf{B}} \\
&= \left(\sqrt{\mathbf{B}}\right)^{-1}\left(\sqrt{\mathbf{B}}\,\mathbf{A}\sqrt{\mathbf{B}}\right)^2\sqrt{\mathbf{B}}
\end{aligned} \tag{8.28}$$

and for $N_z = 3$:

$$\begin{aligned}
(\mathbf{AB})^3 = (\mathbf{AB})^2 \cdot \mathbf{AB} &= \left(\sqrt{\mathbf{B}}\right)^{-1}\left(\sqrt{\mathbf{B}}\,\mathbf{A}\sqrt{\mathbf{B}}\right)^2\sqrt{\mathbf{B}}\left(\sqrt{\mathbf{B}}\right)^{-1}\left(\sqrt{\mathbf{B}}\,\mathbf{A}\sqrt{\mathbf{B}}\right)\sqrt{\mathbf{B}} \\
&= \left(\sqrt{\mathbf{B}}\right)^{-1}\left(\sqrt{\mathbf{B}}\,\mathbf{A}\sqrt{\mathbf{B}}\right)^3\sqrt{\mathbf{B}}
\end{aligned} \tag{8.29}$$

The generalized equation for N_z thin active zones is

$$(\mathbf{AB})^{N_z} = \left(\sqrt{\mathbf{B}}\right)^{-1}\left(\sqrt{\mathbf{B}}\,\mathbf{A}\sqrt{\mathbf{B}}\right)^{N_z}\sqrt{\mathbf{B}} \tag{8.30}$$

Using Eqs. (8.23), (8.25), we obtain

$$Q = \sqrt{B}\, A\sqrt{B} = \begin{pmatrix} 1 + \dfrac{Da}{2N_z} \dfrac{1}{2N_z}\left(2 + \dfrac{Da}{2N_z}\right) \\ Da \qquad 1 + \dfrac{Da}{2N_z} \end{pmatrix} \tag{8.31}$$

In this matrix, the (1,1) element is equal to the (2,2) element and larger than 1. The determinant of this matrix is given by

$$\Delta = \left(1 + \dfrac{Da}{2N_z}\right)^2 - \dfrac{Da}{2N_z}\left(2 + \dfrac{Da}{2N_z}\right) = 1 \tag{8.32}$$

Based on the elements of matrix Q (Eq. 8.31), we can define parameters α and β as follows:

$$\alpha = \operatorname{arcosh}(Q_{11}) = \operatorname{arcosh}\left(1 + \dfrac{Da}{2N_z}\right) \tag{8.33}$$

$$\beta = \dfrac{Da}{\sinh \alpha} \tag{8.34}$$

In terms of these parameters, matrix Q becomes

$$Q = \begin{bmatrix} \cosh \alpha & \dfrac{\sinh \alpha}{\beta} \\ \beta \sinh \alpha & \cosh \alpha \end{bmatrix} \tag{8.35}$$

and its determinant is given by

$$\Delta = \cosh^2\alpha - \sinh^2\alpha \tag{8.36}$$

In order to calculate $(AB)^{N_z}$ for different values of N_z, matrix Q must be raised to a power N_z. For $N_z = 2$, this leads to

$$Q^2 = \begin{bmatrix} \cosh \alpha & \dfrac{\sinh \alpha}{\beta} \\ \beta \sinh \alpha & \cosh \alpha \end{bmatrix} \begin{bmatrix} \cosh \alpha & \dfrac{\sinh \alpha}{\beta} \\ \beta \sinh \alpha & \cosh \alpha \end{bmatrix} \tag{8.37}$$

and in accordance with the rules of matrix multiplication:

$$Q^2 = \begin{bmatrix} \cosh^2\alpha + \sinh^2\alpha & \dfrac{\cosh \alpha \sinh \alpha}{\beta} + \dfrac{\sinh \alpha \cosh \alpha}{\beta} \\ \beta \sinh \alpha \cosh \alpha + \beta \sinh \alpha \cosh \alpha & \sinh^2\alpha + \cosh^2\alpha \end{bmatrix}$$
$$= \begin{bmatrix} \cosh 2\alpha & \dfrac{\sinh 2\alpha}{\beta} \\ \beta \sinh 2\alpha & \cosh 2\alpha \end{bmatrix} \tag{8.38}$$

For $N_z = 3$:

$$\mathbf{Q}^3 = \begin{bmatrix} \cosh 2\alpha & \dfrac{\sinh 2\alpha}{\beta} \\ \beta \sinh 2\alpha & \cosh 2\alpha \end{bmatrix} \begin{bmatrix} \cosh \alpha & \dfrac{\sinh \alpha}{\beta} \\ \beta \sinh \alpha & \cosh \alpha \end{bmatrix} = \begin{bmatrix} \cosh 3\alpha & \dfrac{\sinh 3\alpha}{\beta} \\ \beta \sinh 3\alpha & \cosh 3\alpha \end{bmatrix} \tag{8.39}$$

In generalized form:

$$\mathbf{Q}^{N_z} = \begin{bmatrix} \cosh N_z\alpha & \dfrac{\sinh N_z\alpha}{\beta} \\ \beta \sinh N_z\alpha & \cosh N_z\alpha \end{bmatrix} \tag{8.40}$$

This general result can be established rigorously by induction.

As shown previously, Eqs. (8.30), (8.31),

$$(\mathbf{AB})^{N_z} = \left(\sqrt{\mathbf{B}}\right)^{-1} \mathbf{Q}^{N_z} \sqrt{\mathbf{B}} \tag{8.41}$$

Substituting the equations for $\left(\sqrt{\mathbf{B}}\right)^{-1}, \mathbf{Q}^{N_z}$, and $\sqrt{\mathbf{B}}$, we find the matrix:

$$(\mathbf{AB})^{N_z} = \begin{bmatrix} \cosh N_z\alpha - \dfrac{\beta}{2N_z}\sinh N_z\alpha & \dfrac{\sinh N_z\alpha}{\beta}\left(1 - \dfrac{\beta^2}{4N_z^2}\right) \\ \beta \sinh N_z\alpha & \cosh N_z\alpha + \dfrac{\beta}{2N_z}\sinh N_z\alpha \end{bmatrix} \tag{8.42}$$

For calculating $1/M_0$, we only need the (2,2) element. With $\beta = \dfrac{\mathrm{Da}}{\sinh \alpha}$ (Eq. 8.34),

$$\frac{1}{M_0} = \cosh N_z\alpha + \frac{\mathrm{Da}}{2N_z}\frac{\sinh N_z\alpha}{\sinh \alpha} \tag{8.43}$$

with $\alpha = \operatorname{arcosh}\left(1 + \dfrac{\mathrm{Da}}{2N_z}\right)$ (Eq. 8.33).

In the mathematical theory of orthogonal polynomials, the expression

$$T_n(x) = \cosh\left(n \operatorname{arcosh} x\right) \tag{8.44}$$

is known as a Chebyshev polynomial of the first kind of degree n, and the expression

$$U_{n-1}(x) = \frac{\sinh\left(n \operatorname{arcosh} x\right)}{\sinh\left(\operatorname{arcosh} x\right)} \tag{8.45}$$

is known as a Chebyshev polynomial of the second kind of degree $(n-1)$. Consequently, the result in Eq. (8.43) can be rewritten as

$$\frac{1}{M_0} = T_{N_z}\left(1 + \frac{\mathrm{Da}}{2N_z}\right) + \frac{\mathrm{Da}}{2N_z}U_{N_z-1}\left(1 + \frac{\mathrm{Da}}{2N_z}\right) \tag{8.46}$$

Thus, the conversion is a polynomial function of degree N_z of the parameter $\dfrac{\mathrm{Da}}{2N_z}$.

8.5.2 Influence of Da on Conversion Characteristics With Increasing Number of Thin Zones

We can use Eq. (8.46) to analyze the influence of Da on the conversion characteristics for a given number of equally spaced thin zones, N_z.

For $N_z = 1$,

$$\frac{1}{M_0} = 1 + \mathrm{Da} \qquad (8.47)$$

Thus, with Eq. (8.6),

$$\frac{1}{M_0} - 1 = \frac{X}{1-X} = \mathrm{Da} \qquad (8.48)$$

This is a direct confirmation of the validity of our result.

For $N_z = 2$,

$$\frac{1}{M_0} = 1 + \frac{3}{2}\mathrm{Da} + \frac{1}{4}\mathrm{Da}^2 \qquad (8.49)$$

and

$$\frac{1}{M_0} - 1 = \frac{X}{1-X} = \frac{3}{2}\mathrm{Da}\left(1 + \frac{1}{6}\mathrm{Da}\right) \qquad (8.50)$$

which is always larger than Da and even larger than 1.5 Da. It can be less than 2 Da if Da < 2. Otherwise it is larger than 2 Da.

For $N_z = 3$,

$$\frac{1}{M_0} = 1 + 2\mathrm{Da} + \frac{5}{9}\mathrm{Da}^2 + \frac{1}{27}\mathrm{Da}^3 \qquad (8.51)$$

so

$$\frac{1}{M_0} - 1 = \frac{X}{1-X} = 2\mathrm{Da}\left(1 + \frac{1}{6}\mathrm{Da}\right)\left(1 + \frac{1}{9}\mathrm{Da}\right) \qquad (8.52)$$

which is always larger than 2 Da but only exceeds 3 Da for $\mathrm{Da} > \frac{3}{2}\sqrt{37} - \frac{15}{2} \approx 1.62$.

In generalized form, the expression for $1/M_0$ is a finite series expansion:

$$\begin{aligned}
\frac{1}{M_0} &= \sum_{i=0}^{N_z} \frac{1}{(2i!)} \frac{(N_z + i)!}{(N_z - i)!} \frac{\mathrm{Da}^i}{N_z^i} \\
&= 1 + \frac{1}{2!} \frac{(N_z + 1)!}{(N_z - 1)!} \frac{\mathrm{Da}}{N_z} + \frac{1}{4!} \frac{(N_z + 2)!}{(N_z - 2)!} \frac{\mathrm{Da}^2}{N_z^2} + \frac{1}{6!} \frac{(N_z + 3)!}{(N_z - 3)!} \frac{\mathrm{Da}^3}{N_z^3} + \cdots
\end{aligned} \qquad (8.53)$$

For $N_z = 2$, Eq. (8.53) becomes

$$\frac{1}{M_0} = 1 + \frac{1}{2!} \frac{3!}{1!} \frac{Da}{2} + \frac{1}{4!} \frac{4!}{0!} \frac{Da^2}{4} = 1 + \frac{3}{2} Da + \frac{1}{4} Da^2 \qquad (8.54)$$

the same result as obtained earlier (Eq. 8.49).

Based on this analysis, a reasonable conclusion is that increasing the number of thin active zones for a certain value of Da results in a higher value of the apparent kinetic parameter $\frac{X}{1-X}$. This value always exceeds $\frac{N_z + 1}{2} Da$. However, only at a sufficiently large value of Da can it exceed $N_z Da$. Using Eq. (8.43), an asymptotic expression can be found for $N_z \to \infty$:

$$\frac{1}{M_0} = \frac{1}{2} \exp \sqrt{N_z Da} \left(1 + O\left(\frac{1}{\sqrt{N_z}} \right) \right) \qquad (8.55)$$

where $1 + O\left(\frac{1}{\sqrt{N_z}} \right)$ is the order of error.

The expression $\frac{1}{M_0} = \frac{1}{2} \exp \sqrt{N_z Da}$ can be related to the result for the one-zone reactor in which a single irreversible reaction (adsorption) takes place (see Chapter 5). For this reactor, $\mathcal{L}\{J_{\text{outlet}}\}$ is given by Eq. (5.56) with $\varepsilon_b s$ replaced with $\varepsilon_b s + k_a$:

$$\mathcal{L}\{J_{\text{outlet}}\} = \frac{1}{\cosh \left(L \sqrt{\frac{\varepsilon_b s + k_a}{D_{\text{eff}}}} \right)} \qquad (8.56)$$

and

$$\frac{1}{M_0} = \cosh \left(L \sqrt{\frac{k_a}{D_{\text{eff}}}} \right) \approx \frac{1}{2} \exp \left(L \sqrt{\frac{k_a}{D_{\text{eff}}}} \right) \qquad (8.57)$$

So, an infinite series of equidistant active zones merges into a single continuous one-zone reactor.

8.5.3 Activity for One Zone: Shadowing or Cooperation?

Let us now relate the "apparent coefficient" $\frac{X}{1-X}$ to the number of active zones to determine the activity per zone, that is

$$\bar{k}_a = \frac{1}{N_z} \frac{X}{1-X} = \frac{1}{N_z} \left(\frac{1}{M_0} - 1 \right) \qquad (8.58)$$

Eq. (8.43) gives a general expression for $1/M_0$. From this expression, it follows that for $N_z = 1$, $\bar{k}_a = Da$. A thorough analysis shows that there are only three scenarios:

(1) When $Da > 2$, the value of $\bar{k}_a$ increases with increasing N_z.

(2)   When $Da = 2$, $\bar{k}_a = Da$ for both $N_z = 1$ and $N_z = 2$ and then increases with increasing N_z.

(3)   When $Da < 2$, the value of $\bar{k}_a$ initially decreases with increasing N_z, until some minimum value. This domain can be considered as the domain of "shadowing," in which the apparent coefficient is less than the intrinsic coefficient. In the other domains—domains of cooperation—the apparent coefficient only increases.

In all cases, the following asymptotic relationship holds

$$\bar{k}_a \sim \frac{\exp\sqrt{N_z Da}}{2N_z} \tag{8.59}$$

8.5.4 Comparison of Model Reactors

We can refine this analysis comparing different cases with cascades of three model reactors, namely CSTRs, plug-flow reactors (PFRs), and TAP reactors (including thin-zone TAP reactors), for which the models are presented in Chapters 3 and 5. All results of our analysis are given in Table 8.1.

In Case A, the catalyst mass of every reactor (or every zone for the multizone TAP reactor) is fixed. As the number of reactors (N_r) or zones (N_z) increases, $N \to +\infty$, $\frac{1}{M_0} = \frac{1}{1-X}$ approaches $+\infty$ as well, but in different ways for the three types of reactors: for the CSTR as an exponential function of N with base $(1+Da)$ and for the PFR as an exponential function of N with base $\exp(Da)$. For the TAP reactor, it is a rather complicated function of the Damköhler number, but still of exponential growth in N.

Obviously, the PFR dependence on Da increases faster than the CSTR dependence and the conversion in the PFR is larger than that in the CSTR: "forced propagation is more efficient than local mixing only." In particular, our rigorous analysis shows that the TAP dependence lies in between those of the CSTR and the PFR.

In Case B, the total residence time of the cascade of reactors is kept constant. Let us distribute this total over our reactors (see Fig. 8.3, Case B). For the CSTR, the residence time of every reactor is τ/N, so $Da_N = Da/N$, where τ and Da relate to the residence time and Damköhler number of the cascade of reactors/zones and Da_N is the Damköhler number for reactor/zone N. Hence, in the second column of Table 8.1, Da is replaced with Da/N. Then, for the CSTR in the limit of $N \to +\infty$, $\left(1 + \dfrac{Da}{N}\right)^N$ becomes equal to $\exp(Da)$, which is identical to the result obtained for the PFR in Case B. This could be expected from the fact that a PFR can be conceived as a cascade of infinitesimal CSTRs. Finally, in Case B the asymptotic analysis of

Table 8.1 Expression for $\dfrac{1}{M_0} = \dfrac{1}{1-X}$ for three types of model reactors

	Case A	Case B	Case C
CSTR	$(1+\text{Da})^N \to \infty$ as $N \to \infty$	$\left(1+\dfrac{\text{Da}}{N}\right)^N \to \exp(\text{Da})$	$\left(1+\dfrac{\text{Da}}{N^2}\right)^2 \to 1$ as $N \to \infty$
PFR	$\exp(\text{Da})^N \to \infty$ as $N \to \infty$	$\exp(\text{Da})$	$\exp\left(\dfrac{\text{Da}}{N}\right) \to 1$ as $N \to \infty$
TAP	$\dfrac{1}{2}\left(1+\dfrac{\text{Da}}{2}+a\right)^N \times (1+b) +$ $\dfrac{1}{2}\left(1+\dfrac{\text{Da}}{2}-a\right)^N \times (1-b)$ with $a = \sqrt{\left(2+\dfrac{\text{Da}}{2}\right)\dfrac{\text{Da}}{2}};\ b = \dfrac{1}{\sqrt{1+\dfrac{4}{\text{Da}}}}$	$\dfrac{1}{2}\left(1+\dfrac{\text{Da}}{2N}+a\right)^N \times (1+b) +$ $\dfrac{1}{2}\left(1+\dfrac{\text{Da}}{2N}-a\right)^N \times (1-b)$ $\approx \dfrac{1}{2}\exp\sqrt{N\text{Da}} \to \infty$ as $N \to \infty$ with $a = \sqrt{\left(2+\dfrac{\text{Da}}{2N}\right)\dfrac{\text{Da}}{2N}},\ b = \dfrac{1}{\sqrt{1+\dfrac{4N}{\text{Da}}}}$	$\dfrac{1}{2}\left(1+\dfrac{\text{Da}}{2N^2}+a\right)^N \times (1+b) +$ $\dfrac{1}{2}\left(1+\dfrac{\text{Da}}{2N^2}-a\right)^N \times (1-b) \to \cosh\sqrt{\text{Da}}$ as $N \to \infty$ with $a = \sqrt{\left(2+\dfrac{\text{Da}}{2N^2}\right)\dfrac{\text{Da}}{2N^2}};\ b = \dfrac{1}{\sqrt{1+\dfrac{4N^2}{\text{Da}}}}$

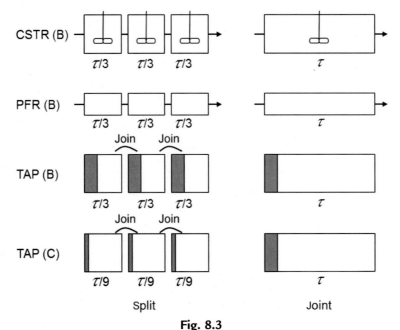

Fig. 8.3

Distribution of residence times for different types of reactors (Table 8.1, columns B and C).

the expressions for the TAP reactor shows that it tends to infinity as $\frac{1}{2}\exp\sqrt{N\mathrm{Da}}$. Put in a simple way, one can say that the PFR is a limit case of a cascade of CSTRs, but the TAP reactor is essentially different. The limit conversion (as $N \to +\infty$) in the TAP reactor will always be larger than the conversion in both the CSTR and the PFR due to the backmixing, which is of diffusional origin. This is an important conceptual statement, which is formulated under the assumption of the same total residence time in all three reactor configurations.

For Case C, the distribution of residence times over the reactors is different (see Fig. 8.3, Case C). We will treat this case based on the concept of the TAP reactor. The residence time in an active zone for the TAP reactor is $\dfrac{\Delta L_z L_i}{D_{\mathrm{eff}}}$, with ΔL_z the thickness of the active catalyst zone and L_i the thickness of the inert zone, or rather the distance between two active zones. We now replace ΔL_z with $\Delta L_z/N$ and L_i with L_i/N so

$$\mathrm{Da}_{\mathrm{zone}} = \frac{\bar{k}_a}{D_{\mathrm{eff}}}\left(\frac{\Delta L_z}{N}\right)\left(\frac{L_i}{N}\right) = \frac{\mathrm{Da}}{N^2} \tag{8.60}$$

For the CSTR and the PFR we could understand our virtual procedure in Case C as catalyst "diluting" down to "nothing" at the limit $N \to +\infty$. Consequently, in this limit, the conversion vanishes and $\dfrac{1}{M_0} = \dfrac{1}{1-X} \to 1$.

For the TAP reactor in Case C, the phenomenon of diffusion provides backmixing, so that the conversion at the limit $N \to +\infty$ will be finite and nonzero:

$$\frac{1}{M_0} = \frac{1}{1-X} = \cosh\sqrt{Da} \tag{8.61}$$

so

$$X = 1 - M_0 = 1 - \frac{1}{\cosh\sqrt{Da}} \tag{8.62}$$

Nomenclature

A	matrix for active zone
a	constant in Eqs. (8.20)–(8.22)
B	matrix for inactive zone
D_{eff}	effective diffusion coefficient (m^2 s^{-1})
Da	Damköhler number
J	flux (mol^{-2} s^{-1})
k_a	adsorption rate coefficient (s^{-1})
$\bar{k}_a$	adsorption rate coefficient per zone (s^{-1})
L	length of reactor (m)
ΔL_z	thickness of zone (m)
L_i	distance between zones $i-1$ and i (m)
M	matrix
M_0	zeroth moment
N	total number of reactors or thin zones
N_r	total number of reactors
N_z	total number of active zones
Q	matrix
s	variable of Laplace transform (s^{-1})
$T_n(x)$	Chebyshev polynomial of the first kind of order n
$U_n(x)$	Chebyshev polynomial of the second kind of order n
X	conversion

Greek symbols

α	parameter in matrix **Q**
β	parameter in matrix **Q**
Δ	determinant
ε_b	bed voidage
κ	$k_a \Delta L$ (m s^{-1})
τ	residence time (s)

Subscripts

0	initial
outlet	at reactor outlet
z	active thin zone

Reference

Constales, D., Shekhtman, S.O., Yablonsky, G.S., Marin, G.B., Gleaves, J.T., 2006. Multi-zone TAP-reactors theory and application IV. Ideal and non-ideal boundary conditions. Chem. Eng. Sci. 61, 1878–1891.

## CHAPTER 9

Experimental Data Analysis: Data Processing and Regression

9.1 The Least-Squares Criterion

The least-squares criterion for estimating the parameters in mathematical models,

$$S(\boldsymbol{\beta}) = \boldsymbol{\varepsilon}^T \boldsymbol{\varepsilon} \xrightarrow{\beta} \min \tag{9.1}$$

for models that are linear in the parameters can be easily mathematically elaborated toward the following parameter estimate vector (Stewart and Caracotsios, 2008):

$$\mathbf{b} = \left(\mathbf{X}^T \mathbf{X}\right)^{-1} \mathbf{X}^T \mathbf{y} \tag{9.2}$$

The criterion as defined in Eq. (9.1) is, evidently, not limited to linear models. For nonlinear models, however, in general form it is written as

$$\mathbf{y} = f(\mathbf{x}, \boldsymbol{\beta}) + \boldsymbol{\varepsilon} \tag{9.3}$$

The sum of the squares of the experimental errors:

$$S(\boldsymbol{\beta}) = \sum_{i=1}^{n} (y_i - f(\mathbf{x}_i, \boldsymbol{\beta}))^2 \tag{9.4}$$

or

$$S(\boldsymbol{\beta}) = \|y - f(\boldsymbol{\beta})\|^2 \tag{9.5}$$

with vector function

$$f : \boldsymbol{\beta} \mapsto \begin{bmatrix} f(\mathbf{x}_1, \boldsymbol{\beta}) \\ f(\mathbf{x}_2, \boldsymbol{\beta}) \\ \vdots \\ f(\mathbf{x}_n, \boldsymbol{\beta}) \end{bmatrix} \tag{9.6}$$

is not a quadratic function in the parameters. Consequently, the set of normal equations is not linear and thus possibly has more than one solution. One of these solutions

Advanced Data Analysis and Modelling in Chemical Engineering. http://dx.doi.org/10.1016/B978-0-444-59485-3.00009-6

corresponds to the global minimum of the objective function in Eq. (9.4), while possible other solutions correspond to local minima. One could attempt solving the set of normal equations numerically, but this meets with great technical difficulties. Therefore, one usually opts for direct minimization of the sum of squares, Eq. (9.4), with respect to $\boldsymbol{\beta}$.

9.2 The Newton-Gauss Algorithm

The Newton-Gauss method consists of linearizing the model equation using a Taylor series expansion around a set of initial parameter values $\mathbf{b}_0$, also called *preliminary estimates*, whereby only the first-order partial derivatives are consider

$$f(\mathbf{b}_0 + \boldsymbol{\Delta}\mathbf{b}) \approx f(\mathbf{b}_0) + \mathbf{J}_0 \cdot \boldsymbol{\Delta}\mathbf{b} \tag{9.7}$$

with $\mathbf{J}_0$ the $n \times p$ Jacobian matrix:

$$\mathbf{J}_0 = \begin{bmatrix} \dfrac{\partial f}{\partial \beta_1}(\mathbf{x}_1, \mathbf{b}_0) & \cdots & \dfrac{\partial f}{\partial \beta_p}(\mathbf{x}_1, \mathbf{b}_0) \\ \vdots & \ddots & \vdots \\ \dfrac{\partial f}{\partial \beta_1}(\mathbf{x}_n, \mathbf{b}_0) & \cdots & \dfrac{\partial f}{\partial \beta_p}(\mathbf{x}_n, \mathbf{b}_0) \end{bmatrix} \tag{9.8}$$

The linear theory is applicable to this linearized model. The set of linearized observation equations is

$$\mathbf{y} - f(\mathbf{b}_0) = \mathbf{J}_0 \cdot \boldsymbol{\Delta}\mathbf{b} + \boldsymbol{\varepsilon} \tag{9.9}$$

As the elements in the Jacobian $\mathbf{J}_0$ depend on the location $\mathbf{b}$ in the parameter space and therefore reflect local properties, a single application of the linear algorithm does not yield the optimal parameter estimates. These are determined by successive approximations:

$$\boldsymbol{\Delta}\mathbf{b}_{k+1} = \left(\mathbf{J}_k^T \mathbf{J}_k\right)^{-1} \mathbf{J}_k^T (\mathbf{y} - f(\mathbf{b}_k)) \tag{9.10}$$

$$\mathbf{b}_{k+1} = \mathbf{b}_k + \boldsymbol{\Delta}\mathbf{b}_{k+1} \tag{9.11}$$

In each step of the Newton-Gauss procedure, the model function f is approximated by its first-order Taylor series around a tentative set of parameter estimates. The linear regression theory then yields a new set of parameter estimates. The Newton-Gauss procedure assumes that these stay within the region in which the first-order Taylor series gives a sufficiently good approximation of f. If this is the case every time, then it is proved that the Newton-Gauss method converges quadratically. If not, then neglecting the higher-order Taylor terms can lead to an undesired systematic increase of the residual sum of squares, so that the algorithm diverges.

9.3 Search Methods From Optimization Theory

The Newton-Gauss method assumes that with good approximation, f can be assumed to be linear over the full length of every parameter step. It is also possible, however, to search for the minimum of the objective function, Eq. (9.4), numerically without having to make such an assumption by using algorithms derived from optimization theory. The price that these methods pay for their better robustness, compared to the Newton-Gauss method, is a slower rate of convergence. In what follows, the reader can briefly get acquainted with some methods from optimization theory. The methods discussed in Sections 9.3.1 and 9.3.2 are *direct methods*: they require only the calculation of objective-function values. The method discussed in Section 9.3.3 is a *gradient method*, which in addition requires the calculation of the partial derivatives of the objective function with respect to the parameters.

9.3.1 Univariate Search

In each step of the univariate search method, $p - 1$ parameters of the previous estimate vector remain fixed, while a minimum of the objective function is searched for by varying only the remaining parameter over a predetermined range. Starting from the initial parameter vector $\mathbf{b}_0$, in p steps all parameters are optimized in a fixed order, after which this process is repeated until each additional decrease of the objective function becomes negligible. Fig. 9.1 illustrates this for the case of two parameters. Through a univariate search, the

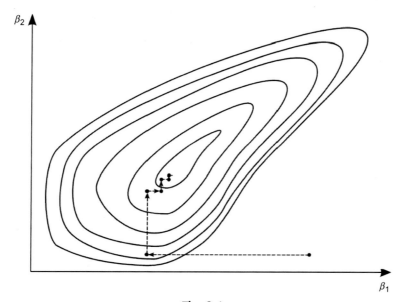

Fig. 9.1

Univariate search in the case of two parameters, β_1 and β_2. The contours of the objective function and the successive parameter estimates are shown. As every time one parameter is kept constant, one always searches parallel to one of the parameter axes.

multivariable optimization problem is split up into optimization problems with only one parameter, for which one has the disposal of efficient numerical methods from optimization theory. In view of the desired short total calculation time, it is advantageous to choose relatively inaccurate but fast methods. A disadvantage is that toward the end the algorithm has the propensity to only slowly oscillate toward the minimum. This is illustrated well in Fig. 9.1.

9.3.2 The Rosenbrock Method

In contrast with algorithms using univariate search, the Rosenbrock method is a so-called acceleration method, which makes the direction and/or the distance (in this case both) of the parameter jumps dependent on the degree of success of the previous parameter jumps. With p parameters, the algorithm proceeds as follows:

(1) In the Euclidean parameter space choose an orthonormal basis $\{\mathbf{e}_1^{(1)}, \mathbf{e}_2^{(1)}, \ldots, \mathbf{e}_p^{(1)}\}$ and a set of step sizes $\{s_1, s_2, \ldots, s_p\}$. Usually, for $\mathbf{e}_i^{(j)}$ the unit vectors according to the parameter axes are chosen. Tentatively, define $\mathbf{b}$ as the initial parameter vector $\mathbf{b}_0$. Further choose two numbers α and β, with $\alpha > 1$ and $0 < \beta < 1$. Usually, $\alpha = 3$ and $\beta = 0.5$ are chosen. Go to step (2) with $j = 1$.

(2) (At stage j). For i going from 1 to p, take steps of size s_i in the directions $\mathbf{e}_i^{(j)}$. Every time a step is successful, in other words if $S\left(\mathbf{b} + s_i \mathbf{e}_i^{(j)}\right) < S(\mathbf{b})$, take $\mathbf{b} + s_i \mathbf{e}_i^{(j)}$ as new parameter estimate $\mathbf{b}$. Now, in a second cycle, again search all directions $\mathbf{e}_i^{(j)}$, but with adjusted step sizes. If during the last cycle the last step in direction $\mathbf{e}_i^{(j)}$ was successful, then the corresponding step size is taken a factor α larger. If not, then the step size is multiplied by $-\beta$. Thus, the step is taken in the other direction and is smaller. Repeat the search cycle until every search direction $\mathbf{e}_i^{(j)}$ has yielded at least one success and one failure. Then calculate the numbers d_k as the sum of all successful steps in the direction of $\mathbf{e}_i^{(j)}$. If $\sum_{k=1}^{p} d_k^2$ is below a predetermined limit, then consider the optimization as finished with $\mathbf{b}$ as the optimum. If not, go to step (3).

(3) (At stage j). Define vectors $\{\mathbf{v}_1^{(j)}, \mathbf{v}_2^{(j)}, \ldots, \mathbf{v}_p^{(j)}\}$ as follows:

$$\mathbf{v}_i^{(j)} = \sum_{k=i}^{p} d_k \mathbf{e}_k^{(j)} \tag{9.12}$$

$\mathbf{v}_1^{(j)}$ defines the first search direction at the next stage $j+1$. According to the definition, of all vectors $\mathbf{v}_i^{(j)}$, $\mathbf{v}_1^{(j)}$ is influenced most by the previously successfully taken steps. $\mathbf{v}_1^{(j)}$ thus represents the future search direction with the greatest probability. For the vectors $\mathbf{v}_1^{(j)}$, this probability decreases fast as a function of i.

Define the orthonormal basis $\left\{\mathbf{e}_1^{(j+1)}, \mathbf{e}_2^{(j+1)}, \ldots, \mathbf{e}_p^{(j+1)}\right\}$ for the next search stage as

$$\mathbf{e}_i^{(j+1)} = \frac{\mathbf{d}_i^{(j)}}{\left|\mathbf{d}_i^{(j)}\right|} \tag{9.13}$$

with the vectors $\mathbf{d}_i^{(j)}$ found from the $\mathbf{v}_i^{(j)}$ through a Gram-Schmidt orthogonalization procedure:

$$\mathbf{d}_1^{(j)} = \mathbf{v}_1^{(j)}$$
$$\mathbf{d}_2^{(j)} = \mathbf{v}_2^{(j)} - \left(\mathbf{v}_2^{(j)} \cdot \mathbf{e}_1^{(j+1)}\right)\mathbf{e}_1^{(j+1)}$$
$$\mathbf{d}_3^{(j)} = \mathbf{v}_3^{(j)} - \left(\mathbf{v}_3^{(j)} \cdot \mathbf{e}_1^{(j+1)}\right)\mathbf{e}_1^{(j+1)} - \left(\mathbf{v}_3^{(j)} \cdot \mathbf{e}_2^{(j+1)}\right)\mathbf{e}_2^{(j+1)} \qquad (9.14)$$
$$\vdots$$
$$\mathbf{d}_p^{(j)} = \mathbf{v}_p^{(j)} - \sum_{i=1}^{p-1}\left(\mathbf{v}_3^{(j)} \cdot \mathbf{e}_i^{(j+1)}\right)\mathbf{e}_i^{(j+1)}$$

As the orthogonal directions are derived from those of $\mathbf{v}_1^{(j)}, \mathbf{v}_2^{(j)}, \ldots, \mathbf{v}_p^{(j)}$ in order of succession, $\mathbf{e}_1^{(j+1)}$ represents the probable most interesting search direction, $\mathbf{e}_2^{(j+1)}$ the second most interesting search direction, and so on. Now return to step (2), and raise j by 1.

Fig. 9.2 illustrates the Rosenbrock method for a model with two parameters. Because the algorithm attempts searching along the axis of a valley of the objective function S, many evaluations of S can be omitted compared with, for example, the algorithm with univariate

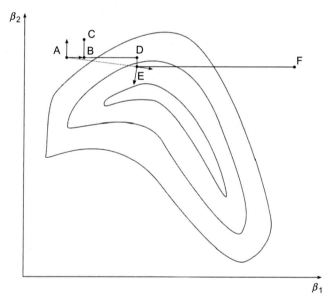

Fig. 9.2

The Rosenbrock method applied to the objective function of a regression analysis with a model consisting of two parameters, β_1 and β_2. The figure shows the first stage of the optimization. This stage goes from A to E through B and D, with C and F as rejected points. The step sizes are equal and the optimization directions for this first stage have been taken parallel to the parameter axes. This is indicated with the small coordinate system in A. The small coordinate system in E shows the optimization directions for the second stage. The first direction is a continuation of $\overrightarrow{\text{AE}}$, the second is perpendicular to this direction.

search. A problem with this algorithm is that it may become very slow if one or more step sizes s_i are chosen much too small or much too large, so that these must first be increased or decreased several times in each stage.

9.3.3 The Gradient Method

The gradient method, also called steepest descent or steepest ascent method, depending on whether one searches for a minimum or a maximum, is based on the following observation: if it is possible to calculate the partial derivatives of the objective function S with respect to the parameters, or discrete approximations thereof, then for each parameter vector $\boldsymbol{\beta}$, it can be calculated along which direction $S(\mathbf{b})$ changes fastest. This direction is given by the reverse of the gradient of S in $\mathbf{b}$:

$$-\nabla S(\boldsymbol{\beta}) = - \begin{bmatrix} \dfrac{\partial S}{\partial \beta_1}(\boldsymbol{\beta}) \\[2mm] \dfrac{\partial S}{\partial \beta_2}(\boldsymbol{\beta}) \\[2mm] \vdots \\[2mm] \dfrac{\partial S}{\partial \beta_p}(\boldsymbol{\beta}) \end{bmatrix} \qquad (9.15)$$

From the initial estimate $\mathbf{b}_0$ of the parameter vector, this direction is searched for a minimum. In this minimum, a new direction of steepest decent is chosen, and so on. In symbols

$$\mathbf{b}_{k+1} = \mathbf{b}_k - s_{\min} \nabla S(\mathbf{b}_k) \qquad (9.16)$$

with $s_{\min}$ defined as the positive parameter s for which the minimum of $S(\mathbf{b}_k - s\nabla S(\mathbf{b}_k))$ is attained. As already mentioned in Section 9.3.1, efficient numerical methods exist for solving such a one-dimensional optimization problem. If this is done with high accuracy, then it is clear that successive directions of greatest decrease will be quasiperpendicular to each other. The gradient method seems optimal in the sense that each time one searches in the direction of steepest descent. One should realize, however, that this steepest descent often is only very locally valid. Consider, for example, the situation in Fig. 9.3, where S has a long, narrow valley. The gradient method will zigzag down, whereby many evaluations of S and its gradient are required. In order to prevent this problem, Booth (1955) suggested shortening every four successive steps by one-tenth and only take every fifth step completely. Since the angles between the successive optimization directions need no longer be perpendicular to each other, the axes of a narrow valley of S can be followed better, and thus convergence to the solution is achieved faster. Fig. 9.3 illustrates this.

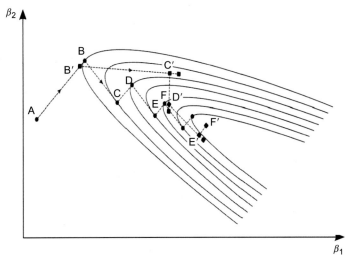

Fig. 9.3

The gradient method applied to the objective function of a regression analysis with a model consisting of two parameters, β_1 and β_2. The initial estimates are represented by point A. From there, it goes zigzag to B, C, D, E, F, and further *(indicated by dots)*. The method using Booth's modification evolves faster to the minimum: from A to B', C', D', E', F', and further *(indicated by filled squares)*.

9.4 The Levenberg-Marquardt Compromise

In Section 9.2, the method of Newton-Gauss has been presented as a method with fast convergence, but low robustness; convergence is ensured only if f behaves quasilinearly as a function of the parameter vector $\boldsymbol{\beta}$ over the full length of each parameter step. If one is sufficiently close to the solution, then the parameter steps still to be taken will be sufficiently short, so that the previously mentioned condition is fulfilled. In this case, the Newton-Gauss method can be employed advantageously to rapidly reach the solution. In practice, the initial parameter vector $\mathbf{b}_0$ will usually be too far removed from the solution. At the beginning, one should resort to the more robust, but more slowly converging methods from optimization theory presented in Section 9.3. Levenberg (1944) suggested an algorithm that, at the start of the regression, searches through the gradient of the objective function S, but later gradually changes over to the Newton-Gauss algorithm. This algorithm was rediscovered and popularized by Marquardt (1963), working as a statistician at DuPont. Today, the Levenberg-Marquardt algorithm is the algorithm of choice for nonlinear regression.

As with the Newton-Gauss method, f is linearized around a tentative parameter vector $\mathbf{b}_0$:

$$f(\mathbf{b}_0 + \boldsymbol{\Delta b}) \approx f(\mathbf{b}_0) + \mathbf{J}_0 \boldsymbol{\Delta b} \tag{9.17}$$

A special feature of the Levenberg-Marquardt algorithm is that the step size is determined beforehand. If it is chosen sufficiently small, then the risk of leaving the region in which the linear approximation holds is eliminated a priori. It is attempted to minimize the (estimated) residual sum of squares

$$\|\mathbf{y} - f(\mathbf{b}_k + \Delta\mathbf{b}_{k+1})\|^2 \approx \|\mathbf{y} - f(\mathbf{b}_k) - \mathbf{J}_k\Delta\mathbf{b}_{k+1}\|^2 \tag{9.18}$$

under the condition

$$\|\Delta\mathbf{b}_{k+1}\|^2 = s_{k+1}^2 \tag{9.19}$$

with step size s_{k+1} sufficiently small. This is a constrained-extreme problem, which can be solved with the Lagrange multiplier method. A minimum of the Lagrange function L_{k+1}, defined as

$$L_{k+1}: \ (\Delta\mathbf{b}_{k+1}, \lambda) \mapsto \|\mathbf{y} - f(\mathbf{b}_k) - \mathbf{J}_k\Delta\mathbf{b}_{k+1}\|^2 + \lambda_{k+1}\left(\|\Delta\mathbf{b}_{k+1}\|^2 - s_{k+1}^2\right) \tag{9.20}$$

is searched for.

The function value can be rewritten as

$$\begin{aligned} L_{k+1}(\Delta\mathbf{b}_{k+1}, \lambda_{k+1}) &= (\mathbf{y} - f(\mathbf{b}_k))^T(\mathbf{y} - f(\mathbf{b}_k)) - 2\Delta\mathbf{b}_{k+1}^T\mathbf{J}_k^T(\mathbf{y} - f(\mathbf{b}_k)) \\ &\quad + \Delta\mathbf{b}_{k+1}^T\mathbf{J}_k^T\mathbf{J}_k\Delta\mathbf{b}_{k+1} + \lambda_{k+1}\left(\Delta\mathbf{b}_{k+1}^T\Delta\mathbf{b}_{k+1} - s_{k+1}^2\right) \end{aligned} \tag{9.21}$$

Setting the partial derivatives with respect to the parameter increments and the multiplier λ equal to zero gives, in matrix form,

$$\mathbf{0} = -\mathbf{J}_k^T(\mathbf{y} - f(\mathbf{b}_k)) + \mathbf{J}_k^T\mathbf{J}_k\Delta\mathbf{b}_{k+1} + \lambda_{k+1}\Delta\mathbf{b}_{k+1} \tag{9.22}$$

and

$$0 = \|\Delta\mathbf{b}_{k+1}\|^2 - s_{k+1}^2 \tag{9.23}$$

The first equation can be rewritten as

$$\left(\mathbf{J}_k^T\mathbf{J}_k + \lambda_{k+1}\mathbf{I}_p\right)\Delta\mathbf{b}_{k+1} = \mathbf{J}_k^T(\mathbf{y} - f(\mathbf{b}_k)) \tag{9.24}$$

Solved for $\Delta\mathbf{b}_{k+1}$ this gives

$$\Delta\mathbf{b}_{k+1} = \left(\mathbf{J}_k^T\mathbf{J}_k + \lambda_{k+1}\mathbf{I}_p\right)^{-1}\mathbf{J}_k^T(\mathbf{y} - f(\mathbf{b}_k)) \tag{9.25}$$

Because the matrix $\mathbf{J}_k^T\mathbf{J}_k$ is symmetric, it can be diagonalized by an orthonormal basis transformation:

$$\mathbf{S}_k^T\mathbf{J}_k^T\mathbf{J}_k\mathbf{S}_k = \mathbf{D}_k \tag{9.26}$$

with $\mathbf{D}_k$ a diagonal matrix and

$$\mathbf{S}_k^T \mathbf{S}_k = \mathbf{I}_p \tag{9.27}$$

so that

$$\mathbf{J}_k^T \mathbf{J}_k = \mathbf{S}_k \mathbf{D}_k \mathbf{S}_k^T \tag{9.28}$$

The diagonal elements of $\mathbf{D}_k$ are the eigenvalues of $\mathbf{J}_k^T \mathbf{J}_k$, which are strictly positive because $\mathbf{J}_k^T \mathbf{J}_k$ is positive definite. Substitution of Eq. (9.28) into Eq. (9.24) and premultiplication with $\mathbf{S}_k^T$ gives

$$\left(\mathbf{D}_k + \lambda_{k+1} \mathbf{I}_p \right) \mathbf{S}_k^T \Delta \mathbf{b}_{k+1} = v_k \tag{9.29}$$

with

$$\mathbf{v}_k = \mathbf{S}_k^T \mathbf{J}_k^T (\mathbf{y} - f(\mathbf{b}_k)) \tag{9.30}$$

Solving for $\Delta \mathbf{b}_{k+1}$, taking into account Eq. (9.27), gives

$$\Delta \mathbf{b}_{k+1} = \mathbf{S}_k \left(\mathbf{D}_k + \lambda_{k+1} \mathbf{I}_p \right)^{-1} \mathbf{v}_k \tag{9.31}$$

This allows finding an expression for s_{k+1}^2 from Eq. (9.23):

$$s_{k+1}^2 = \mathbf{v}_k^T \left[\left(\mathbf{D}_k + \lambda_{k+1} \mathbf{I}_p \right)^2 \right]^{-1} \mathbf{v}_k \tag{9.32}$$

or, with $\eta_{k,1}, \eta_{k,2}, \ldots, \eta_{k,p}$ the strictly positive eigenvalues of $\mathbf{J}_k^T \mathbf{J}_k$,

$$s_{k+1}^2 = \sum_{i=1}^{p} \frac{v_{k,i}^2}{\left(\eta_{k,i} + \lambda_{k+1} \right)^2} \tag{9.33}$$

From this equation, it appears that s_{k+1} evolves strictly descending as a function of the multiplier λ_{k+1}, provided that this is kept positive. It is preferred to preserve λ_{k+1} at the expense of the step size s_{k+1}, knowing that the latter, although a priori unknown, is inversely proportional to λ_{k+1}. For the same tentative set of parameter estimates, one thus limits the step size more when λ_{k+1} is taken larger. Therefore, the larger λ_{k+1}, the more cautiously one proceeds during the optimization. At the start of a regression analysis, when one is still far from the solution, it is thus necessary to work with a large λ_{k+1}. After that, however, λ_{k+1} is systematically reduced as one approaches the solution more closely. How this is done exactly is somewhat arbitrary. Not only the size but also the direction of the parameter steps is changed. In the immediate vicinity of the solution, finally λ_{k+1} is set equal to zero. It is interesting to look at the two limiting cases, $\lambda_{k+1} \to +\infty$ and $\lambda_{k+1} = 0$.

With $\lambda_{k+1} \to +\infty$, the parameter step $\Delta \mathbf{b}_{k+1}$, according to Eq. (9.25), becomes proportional to $\mathbf{J}_k^T (\mathbf{y} - f(\mathbf{b}_k))$. One can easily verify that the direction of this vector is opposite to the gradient of the objective function S in $\mathbf{b}_k$. At the start of the regression, the parameter estimate thus moves in the direction of the fastest decrease of the residual sum of squares. With $\lambda_{k+1} = 0$, Eq. (9.25) is reduced to Eq. (9.10). At the end of the regression, one thus applies the Newton-Gauss algorithm. This explains a popular interpretation of the Levenberg-Marquardt algorithm as a compromise between the robust gradient method and the fast Newton-Gauss method.

9.5 Initial Estimates

As mentioned previously, the Newton-Gauss method converges very fast if the initial values b_0 can be chosen close to the optimum. Now the problem is that of obtaining good initial estimates. Naturally, no general techniques exist for this, for the location of the optimum is precisely what one is going to search for. Inspection of some types of current models can nevertheless indicate in which direction one can look for solving the problem of the initial estimates.

9.5.1 Power-Law Models for the Reaction Rate Equation

The rate equations of catalytic reactions of the type

$$A + B \rightarrow products \tag{9.34}$$

may be expressed in the form of power-law equations:

$$r = k p_A^m p_B^n \tag{9.35}$$

Incidentally, this type of kinetic equation also acts as an empirical model of order (m, n) for a homogeneous noncatalytic reaction. Taking the natural logarithm of both sides of Eq. (9.35) gives

$$\ln r = \ln k + m \ln p_A + n \ln p_B \tag{9.36}$$

which is a linear relation in $\ln k$, m, and n and thus is of the type

$$E(y) = \alpha_0 + \alpha_1 x_1 + \alpha_2 x_2 \tag{9.37}$$

to which the theory of linear regression is applicable, resulting in initial estimates for $\ln k$, m, and n.

9.5.2 Arrhenius Equation

The temperature dependence of the rate coefficient is of the well-known exponential type

$$k = k_0 \exp\left(-\frac{E_a}{R_g T}\right) \tag{9.38}$$

in which E_a is the activation energy and k_0 the preexponential factor. For adsorption equilibrium coefficients, an analogous relation holds

$$K_i = \exp\left(-\frac{\Delta H_i}{R_g T}\right) \exp\left(\frac{\Delta S_i}{R_g}\right) \tag{9.39}$$

in which ΔH_i and ΔS_i are the enthalpy and the entropy of adsorption of reaction component i.

Taking the natural logarithm of both sides of the Arrhenius equation, Eq. (9.38) gives

$$\ln k = \ln k_0 - \frac{E_a}{R_g}\frac{1}{T} \tag{9.40}$$

which again is a relation of the type

$$E(y) = \alpha_0 + \alpha_1 x_1 \tag{9.41}$$

A graph in which $\ln k$ is plotted as a function of $1/T$ (see Fig. 9.4) should give a linear relation if the model is correct. From this relation, estimates follow for the slope of the best fitting straight line, $-E_a/R_g$, and its intercept, $\ln k_0$.

Naturally, estimates for k_0 and E_a (or for ΔH_i and ΔS_i) can be obtained more accurately by linear regression based on Eq. (9.40):

$$\sum_{j=1}^{N_T}\left(\ln k_j - \left(\ln k_0 - \frac{E_a}{R_g}\frac{1}{T_j}\right)\right)^2 \xrightarrow{k_0, E_a} \min \tag{9.42}$$

in which N_T represents the number of temperature levels at which experiments were performed. Drawing of the graph as suggested in Fig. 9.4 is nevertheless a preceding necessity: representing the data in a graph in such a way that a linear relation must be obtained if the assumptions made are correct, cannot be matched by any other technique for the verification of these assumptions. In the present domain, it concerns the assumptions that underlay the reaction mechanism giving rise to the rate equation used.

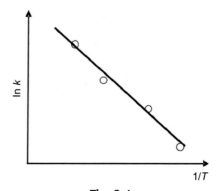

Fig. 9.4
Arrhenius diagram displaying the linear dependence of the rate coefficient with respect to the inverse temperature.

9.5.3 Rate Laws for Heterogeneously Catalyzed Reactions

A typical Langmuir-Hinshelwood-Hougen-Watson model for the reversible gas-phase reaction $A \rightleftarrows R+S$ taking place on a solid catalyst is given by

$$r = \frac{k\,K_A\left(p_A - \dfrac{p_R p_S}{K_{eq}}\right)}{\left(1 + K_A p_A + K_R p_R + K_S p_S\right)^2} \tag{9.43}$$

where $K_{eq} = K_A/K_R K_S$. This model can be transformed into:

$$\sqrt{\frac{\left(p_A - \dfrac{p_R p_S}{K_{eq}}\right)}{r}} = \frac{1}{\sqrt{k K_A}} + \frac{K_A}{\sqrt{k K_A}}p_A + \frac{K_R}{\sqrt{k K_A}}p_R + \frac{K_S}{\sqrt{k K_A}}p_S\left(+\varepsilon'\right) \tag{9.44}$$

which is a linear equation of the type

$$E(y) = \alpha_0 + \alpha_1 x_1 + \alpha_2 x_2 + \alpha_3 x_3 \tag{9.45}$$

On the basis of Eq. (9.44), very accurate preliminary estimates for the parameters k, K_A, K_R, and K_S can be obtained by simple linear regression. These parameters can be used with great advantage for nonlinear regression on the basis of the reaction rate equation as such.

Another possible transformation is given by

$$\sqrt{\frac{1}{r}} = \frac{1}{\sqrt{k\,K_A}}\frac{1}{\sqrt{DF}} + \frac{K_A}{\sqrt{k\,K_A}}\frac{1}{\sqrt{DF}}p_A + \frac{K_R}{\sqrt{k\,K_A}}\frac{1}{\sqrt{DF}}p_R + \frac{K_S}{\sqrt{k\,K_A}}\frac{1}{\sqrt{DF}}p_S\left(+\varepsilon'\right) \tag{9.46}$$

whereby DF represents the driving force $\left(p_A - \dfrac{p_R p_S}{K_{eq}}\right)$. Eq. (9.46) is a linear equation without constant term

$$E(y) = \alpha_1 x_1 + \alpha_2 x_2 + \alpha_3 x_3 + \alpha_4 x_4 \tag{9.47}$$

It is important to notice that the estimates for the groups $\dfrac{1}{\sqrt{kK_A}}, \dfrac{K_A}{\sqrt{kK_A}}, \dfrac{K_R}{\sqrt{kK_A}}$, and $\dfrac{K_S}{\sqrt{kK_A}}$,

that are obtained from the rearranged forms, Eqs. (9.44) and (9.46), by means of linear regression are different: the least-squares criterion is different for both. In general form, this can be written as

$$\sum_{i=1}^{n}\left(\phi(y_i; \mathbf{x}_i) - \sum_{j=1}^{p}(\varsigma_j(\mathbf{x}_i)\alpha_j)\right)^2 \xrightarrow{\alpha} \min \tag{9.48}$$

Here $\varsigma_j(\mathbf{x}_i)$ are the "new independent variables" and α_j are the combinations of the model parameters $\boldsymbol{\beta}$ on the right-hand side of the rearranged equations. The new independent variables $\varsigma_j(\mathbf{x}_i)$ are the independent variables $\mathbf{x}_i$ themselves, as in the case of rearrangement Eq. (9.44), or somewhat more complicated combinations in several to all independent variables, such as in rearrangement Eq. (9.46). The parameter ϕ groups the operations that must be performed to obtain the numerical value of the new dependent variable. In the first rearrangement, this new dependent variable is a combination of the original dependent variable, namely the reaction rate, and the input values of the independent variables, namely the partial pressures. In the second rearrangement, the new dependent variable is a function only of the original dependent variable. Consequently, it is clear that both forms of the rearranged rate equation give rise to different numerical values for the estimates.

The described—and obvious—modifications of the nonlinear rate equation that are used to arrive at a form that is linear in the groups $\boldsymbol{\alpha}$ of the true parameters $\boldsymbol{\beta}$ and on the basis of which estimates for the groups $\boldsymbol{\alpha}$ are obtained using simple linear regression, should always be considered as a preamble to a regression of the real, measured dependent variable; two important objections can be raised against the procedures described above.

The first objection is purely from the point of view of the nonoptimal fit of the model to the measured data: as not the dependent variable y as such was confronted with the model, but only a function $\phi(y_i; x_i)$ derived from y, the obtained parameter estimates make the best possible model predictions for the function ϕ and not for the dependent variable y. This is obvious from the previous discussion with respect to the rearrangements $\phi(y_i; x_i)$ and $\varsigma_j(x_i)\alpha_j$ that produce numerically different estimates. The best possible estimates, that is, those estimates that fit the model to the measured data in the strictest way, can therefore only be obtained on the basis of the criterion:

$$\sum (r_i - \hat{r}_i)^2 \to \min \tag{9.49}$$

supposing that the reaction rate indeed is the true dependent variable that is observed as such.

The second objection is from a statistical point of view and is of a more fundamental nature. The rearrangement of the nonlinear model $f(x, \beta)$ to $\sum \varsigma_j(x_i)\alpha_j$ is possible only if the additively occurring experimental error ε is first neglected and afterward, artificially a new error ε' is added to the rearranged equation in an additive way. The questions that now arise, and cannot easily be answered, are first whether this operation is allowed, and second whether the new, transformed error ε' obeys the general preconceived assumptions, namely:

$$\varepsilon' \sim N\left(\mathbf{0}, \mathbf{I}_n, \sigma'^2\right) \tag{9.50}$$

If this were not the case, the reliability of all tests developed would be questionable. Even worse, the tests would be worthless!

Suppose that the condition of Eq. (9.50) still would be properly fulfilled. The test results are thus reliable, but, in view of the rearranged form of the model on which they are based (eg, Eq. 9.46), relate to $\dfrac{1}{\sqrt{k}\,K_A}$, $\dfrac{K_A}{\sqrt{k}\,K_A}$, $\dfrac{K_R}{\sqrt{k}\,K_A}$, and $\dfrac{K_S}{\sqrt{k}\,K_A}$, and not to the estimates for k, K_A, K_R, and K_S themselves! The latter can be calculated from the former by application of the rule with respect to the propagation of uncertainty. The uncertainty associated with the component factors a of a function naturally propagates itself on this function. The resulting uncertainty on this function can be approximated by using a Taylor series expansion whereby all partial derivatives of second and higher order are neglected:

$$b_i = f(a_1, a_2, \ldots, a_p) \cong f(a_{10}, a_{20}, \ldots, a_{p0}) + \sum_{j=1}^{p} \frac{\partial f}{\partial \alpha_j}\bigg|_{\alpha = a} (a_j - a_{j0}) \qquad (9.51)$$

and applying the propagation rule to the variance on a linear function of multiple random variables:

$$\text{var}(b_i) \cong \sum_{j=1}^{p} \left(\frac{\partial f}{\partial \alpha_j}\right)^2 \text{var}(a_j) + 2 \sum_{i=1}^{p-1} \sum_{j=i+1}^{p} \frac{\partial f}{\partial \alpha_i} \frac{\partial f}{\partial \alpha_j} \text{cov}(a_i, a_j) \qquad (9.52)$$

For rearrangements of Hougen-Watson models, that is, rearranged forms of type (9.44) or (9.46), the functions $f(a_1, a_2, \ldots, a_p)$ can rather generally be represented by simple power-law functions of usually only two component factors, for example:

$$k = f(a_i, a_j) = a_i^m a_j^n \qquad (9.53)$$

Similar equations can be written for K_A, K_R, and K_S. The a_j always are the groups of parameters that figure on the right-hand side of the rearranged equation. Moreover, the powers m and n in Eq. (9.53) are often limited to the values 1 and -1. For example, for the rearrangement of Eq. (9.44):

$$k = f\left(\frac{1}{\sqrt{k\,K_A}}, \frac{K_A}{\sqrt{k\,K_A}}\right) = \frac{1}{\dfrac{1}{\sqrt{k\,K_A}} \dfrac{K_A}{\sqrt{k\,K_A}}} = a_1^{-1} a_2^{-1} \qquad (9.54)$$

and

$$K_A = f'\left(\frac{1}{\sqrt{k\,K_A}}, \frac{K_A}{\sqrt{k\,K_A}}\right) = \frac{\dfrac{K_A}{\sqrt{k\,K_A}}}{\dfrac{1}{\sqrt{k\,K_A}}} = a_1^{-1} a_2 \qquad (9.55)$$

It then follows that

$$\text{var}(K_A) = \text{var}(a_1^{-1} a_2) = \sum_{j=1}^{2} \left(\frac{\partial f}{\partial \alpha_j}\right)^2 \text{var}(a_j) + 2 \frac{\partial (a_1^{-1} a_2)}{\partial a_1} \frac{\partial (a_1^{-1} a_2)}{\partial a_2} \text{cov}(a_1, a_2) \qquad (9.56)$$

or

$$\text{var}(K_A) = \left(\frac{a_2}{a_1^2}\right)^2 \text{var}(a_1) + \left(\frac{1}{a_1}\right)^2 \text{var}(a_2) - 2 \frac{a_2}{a_1^3} \text{cov}(a_1, a_2) \qquad (9.57)$$

and analogously:

$$\text{var}(k) = \left(\frac{1}{a_1^2 a_2}\right)^2 \text{var}(a_1) + \left(\frac{1}{a_1 a_2^2}\right)^2 \text{var}(a_2) + 2\frac{1}{a_1^3 a_2^3}\text{cov}(a_1, a_2) \tag{9.58}$$

9.6 Properties of the Estimating Vector

The properties of the parameter estimates of linear models are no longer valid exactly for the estimates of nonlinear parameters, for the optimal estimates cannot be expressed as linear combinations of the observation outcomes. The estimating vector is not multidimensional normally distributed. Its distribution is not even known! The true covariance matrix and the joint confidence region of the parameter estimates are approximated by the expression obtained in the case of linear model regression:

$$V(\mathbf{b}) = \left(\mathbf{X}^T\mathbf{X}\right)^{-1}\sigma^2 \tag{9.59}$$

and

$$(\mathbf{b} - \boldsymbol{\beta})^T\mathbf{X}^T\mathbf{X}(\mathbf{b} - \boldsymbol{\beta}) = ps^2 F(p, n-p; 1-\alpha) \tag{9.60}$$

where the matrix $\mathbf{X}$ is replaced by the matrix $\mathbf{J}$, defined in Eq. (9.8) but evaluated at the converged optimal estimates $\mathbf{b}$ with s^2 given by

$$s^2 = \frac{\sum_{i=1}^{n}(y_i - f(\mathbf{x}_i, \mathbf{b}))^2}{n-p} \tag{9.61}$$

Thus

$$\hat{V}(\mathbf{b}) \cong \left(\mathbf{J}^T\mathbf{J}\right)^{-1}s^2 \tag{9.62}$$

and

$$(\mathbf{b} - \boldsymbol{\beta})^T\mathbf{J}^T\mathbf{J}(\mathbf{b} - \boldsymbol{\beta}) = ps^2 F(p, n-p; 1-\alpha) = \delta \tag{9.63}$$

The proofs of Eqs. (9.59) and (9.60) explicitly rely on the linear character of the model. The above relations are thus only correct under the same conditions. Therefore, one speaks of the joint confidence region under linear assumptions. If the model is not linear in its parameters, the surface in the parameter space defined by Eq. (9.63) no longer is a contour of constant residual sums of squares. Although the probability level is correct, the contour itself has only been approximated. This property provides a qualitative measure of the degree of nonlinearity of the model; it is rather simple to determine the coordinates of the points of intersection of the ellipsoid surface, Eq. (9.63), with its principal axes by canonical reduction. These points of intersection represent the most important combinations of parameter values, as these are the combinations that are closest to or farthest away from the optimum set **b**, but still just qualify as plausible estimates, see Figure 9.5. In these points, denoted as b_{1l}, b_{1h}, b_{2l} and b_{2h}, with l referring to the lower end of the interval and h to the higher end, $S(\boldsymbol{\beta})$ can be calculated and compared with the value resulting from the definition of the joint confidence region:

$$S(\boldsymbol{\beta}) = S(\mathbf{b}) \left(1 + \frac{p}{n-p} F(p, n-p; 1-\alpha) \right) \tag{9.64}$$

If Eq. (9.63) approximates the joint confidence region well, in other words, if the model is only weakly nonlinear, the deviations will not be too large. Instead of determining the approximating joint confidence region at an exact probability level using Eq. (9.63), the exact joint confidence region can also be evaluated using Eq. (9.64). This construction consists of determining parameter combinations $\boldsymbol{\beta}$ that obey Eq. (9.64) and that must be sufficiently large in number and must be spread over the entire region in order to be able to draw the contours sufficiently precise. Now, however, the probability level is only approximating because the distribution of the parameter estimates is not known. The shape of this exact confidence region again is a qualitative indication of the degree of nonlinearity of the model. The contour will no longer take on the purely ellipsoid shape but will deviate from this shape more strongly according as the model shows stronger nonlinearities (Fig. 9.5). The contours $S(\boldsymbol{\beta})$ show obliqueness and usually take the shape of a banana. They are asymmetric with respect to **b**.

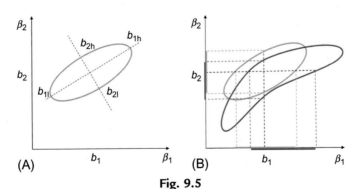

Fig. 9.5

Verification of linear behavior. (A) Differences between the values of the objective function calculated directly with the nonlinear model at the points of intersection with the ellipsoid indicate how pronounced the nonlinearity of the model is; (B) Oblique, exact contour calculated directly with the nonlinear model versus ellipsoid-shaped approximating contour calculated based on the linear approximation of the model.

9.7 Temperature Dependence of the Kinetic Parameters k and K_i

The dependence on the temperature of the rate coefficients and adsorption equilibrium coefficients is well known and given by Eqs. (9.38) and (9.39). A technique for obtaining initial, though rough, estimates for the parameters k_0 and E_a or ΔH and ΔS has already been explained in Section 9.5.2. Of course, it is better to perform the regression formally on the basis of criterion (9.42), or even better, because the logarithmic transformation is not applied, on the basis of

$$\sum_{j=1}^{N_T} \left(k_j - k_0 \exp\left(\frac{E_a}{R_g} \frac{1}{T_j} \right) \right)^2 \xrightarrow{k_0, E_a} \min \tag{9.65}$$

Nevertheless, the above approach again does not yield the best possible values for the parameters k_0 and E_a or ΔH and ΔS and thus the model does not fit the data in the best possible way. The determination of k_0 and E_a or ΔH and ΔS has not been carried out on the basis of criterion (9.49) with the real measured variable, but on the basis of criterion (9.65), which leans on derived data, in this case the point estimates for k_j obtained by processing of the data for each temperature separately. A correct analysis consists in *simultaneously* fitting *all* data to the model. To this end, the temperature dependence of the parameters should be incorporated in the model explicitly. Parameters k and K_i are then replaced by their functional temperature dependence, Eq. (9.38) or (9.39). The temperature is included in the set of independent variables. In other words, the correct criterion that was used in the analysis of the data for each temperature, Eq. (9.66), is extended to Eq. (9.67), whereby each parameter k, $K_i, \dots$ gives rise to two parameters in the criterion (9.67):

$$\sum_{i=1}^{n} \left(r_i - \hat{r}_i\left(k, K_i, \dots; \frac{W}{F}, p_t \right) \right)^2 \xrightarrow{k, K_i, \dots} \min \tag{9.66}$$

$$\sum_{i=1}^{n} \left(r_i - \hat{r}_i\left(k_0, E_a, \Delta H, \Delta S, \dots; \frac{W}{F}, p_t, T \right) \right)^2 \xrightarrow{k_0, E_a, \Delta H, \Delta S, \dots} \min \tag{9.67}$$

It should be mentioned that, however useful, highly desirable, and necessary this analysis may be, it forces the parameters to obey the Arrhenius relation even though this relation may not be valid in the case studied. Deviations from the Arrhenius relation can be caused by the occurrence of significant internal diffusion limitations in part of the temperature range investigated or by the use of an inadequate model. An analysis for each temperature, for example, can reveal these internal diffusion limitations through a kink in the Arrhenius plot. By forcing the relations (9.38) or (9.39) upon the parameters, however, the possible occurrence of such a bend is naturally prevented. Therefore, it is absolutely necessary to *always* perform an analysis of the data for each temperature before carrying out the nonisothermal analysis. The estimates following from this analysis then also allow calculating accurate preliminary estimates for k_0 and E_a or ΔH en ΔS using Eq. (9.65) and analogous criteria for the adsorption equilibrium coefficients. This can result in a fast convergence of the nonisothermal analysis.

The analysis for each temperature does not always yield handsome linear Arrhenius relations in a graph of $\ln k$ versus $1/T$, especially not for the adsorption equilibrium coefficients. One of the causes for this often is the high correlation between the rate coefficient k and the adsorption equilibrium coefficients K_i, resulting in small deviations in k being compensated by much larger deviations in K_i. Furthermore, the model also appears to be much less sensitive to K_i than to k, which seems rather logical in view of the structure of a Langmuir-Hinshelwood rate equation. This is all the more reason for performing the complete nonisothermal analysis. However, introducing the exponential relations (9.38) and (9.39) causes an even greater correlation than the one already existing between k and K_i, namely the one between k_0 and E_a or between ΔH and ΔS. This correlation frequently is 0.99 or higher, which gives rise to convergence difficulties or at least very slow progress in the optimization process. The fundamental reason is the very limited range over which the temperature can be varied in practice in a chemical process: a range of 100 K can rarely be realized, because at the lower limit the reaction does not proceed sufficiently fast to be measured, while at the upper limit complete conversion is achieved, irrespective of the values of the other independent variables. In order to eliminate these convergence difficulties as much as possible, in all circumstances it is recommended to perform the following substitution, also called reparametrization:

$$k = k_0 \exp\left(-\frac{E_a}{R_g T}\right) = k_0 \exp\left(-\frac{E_a}{R_g T_{ave}}\right) \exp\left(-\frac{E_a}{R_g}\left(\frac{1}{T} - \frac{1}{T_{ave}}\right)\right) \qquad (9.68)$$

or

$$k = k_0 \exp\left(-\frac{E_a}{R_g}\left(\frac{1}{T} - \frac{1}{T_{ave}}\right)\right) \qquad (9.69)$$

9.8 Genetic Algorithms

Genetic algorithms form a modern alternative to the regression methods described in the previous sections. They minimize the residual sum of squares in a heuristic way that is inspired by Darwin's theory of evolution and genetics. They differ from the classical regression methods in three respects. Genetic algorithms:

- operate with a population of estimating vectors instead of a single estimating vector
- do not operate with these estimating vectors as such, but with codes thereof, called *chromosomes*
- use probabilistic transition rules instead of deterministic ones

Genetic algorithms are not based on initial estimates for each parameter, but on intervals in which these parameters are supposed to lie. At the start of the algorithm, for each parameter an entirely random selection is made from its interval. Each selection is then converted into a code of fixed length: a sequence of *genes*. These genes all have a certain number of possible values, called *alleles*. Usually, coding consists of nothing more than the binary notation of all parameter values. In this binary code, each bit position then is a gene with "0" and "1" as alleles. The codes of the separate parameters are arranged in a fixed order. The result is a *chromosome*. The whole procedure, starting from a random selection for each of the parameters, is subsequently repeated $2N - 1$ times, so that finally a *population* of $2N$ chromosomes is obtained. Fig. 9.6 shows an example.

Each chromosome in the population is given a score, which reflects how successful the corresponding set of parameter values is with respect to the optimization problem. For the present optimization problem, the minimization of the objective function S, an obvious choice is to take as score the opposite of the function value below S. Subsequently, $2N$ chromosomes are randomly selected from the population. The probability that a chromosome is selected is taken proportional to the score of the chromosome. On average, successful chromosomes will be selected multiple times at the expense of less-successful chromosomes, which are not retained. This is in accordance with the Darwinist principle of survival of the fittest. In other words, the selected chromosomes will now, as it were, reproduce themselves. To this end, the chromosomes are randomly paired. Each pair will give rise to two new chromosomes. Hereby, the chromosomes will be crossed with a probability that is to be determined beforehand. Again randomly, a location of crossing is chosen. This location divides both chromosomes of the pair into two parts: one part that will be exchanged during the propagation and one part that will not. If no crossing occurs, then the two new chromosomes will be copies of the original ones. Fig. 9.7 illustrates a crossing for two pairs of chromosomes from Fig. 9.6.

Once the new generation of chromosomes has been formed, a mutation mechanism comes into operation; in the chromosomes, a number of genes are modified in a completely arbitrary manner. Typically, however, this modification concerns only one gene to a thousand, so

	β_1	β_2
$C_1 =$	0011011010	0010110101
$C_2 =$	0101100100	1001001100
$C_3 =$	1001001110	0010100111
$C_4 =$	1101100101	0110111011

Fig. 9.6

A population of four chromosomes, C_1, C_2, C_3, and C_4, each consisting of binary coded values for two parameters, β_1 and β_2.

$$C_2 = 0101100100\ 1001001100$$
$$C_4 = 1101100101\ 0110111011$$
$$\longrightarrow$$
$$C_1' = 0101100101\ 0110111011$$
$$C_2' = 1101100100\ 1001001100$$

$$C_3 = 1001001110\ 0010100111$$
$$C_4 = 1101100101\ 0110111011$$
$$\longrightarrow$$
$$C_3' = 1001001110\ 0010111011$$
$$C_4' = 1101100101\ 0110100111$$

Fig. 9.7

From a population of four, the chromosomes C_2, C_3, and C_4 were chosen to propagate, the last of which two times. C_2 and C_4, and C_3 and C_4 were randomly paired. For both pairs, a location of crossing was chosen arbitrarily. The result is the new generation of chromosomes C_1', C_2', C_3', and C_4'.

mutation is of secondary importance compared to selection and crossing. In contrast, in nature mutation is important to avoid certain potentially useful genetic material being lost. Analogously, mutation is applied in genetic algorithms in order not to "forget" certain regions of the parameter space during optimization. Besides selection, crossing, and mutation, sometimes other biologically inspired evolutionary mechanisms are applied in genetic algorithms. Nevertheless, it is already possible to build an effective program with the three mechanisms mentioned above. After mutation, the propagation cycle can start again, beginning with the selection of the chromosomes that will propagate. The cycle is repeated a large number of times, whereby the population is becoming increasingly successful with respect to the optimization problem. Hereby, the chromosomes will approximate the optimal chromosome increasingly better. The corresponding estimating vector then is the solution of the optimization problem.

Nomenclature

$\mathbf{b}$	vector of parameter estimates
$\mathbf{b}_0$	set of initial parameter values or preliminary estimates
C_i	chromosome i
$\mathbf{D}_k$	diagonal matrix
d_k	elements of diagonal matrix
$\mathbf{d}_i^{(j)}$	vector of optimal search direction in orthonormal basis
DF	driving force
$E(y)$	dependent variable in independent equation
E_a	activation energy (J mol^{-1})
$\mathbf{e}_i^{(j)}$	vector of orthonormal basis
$F(m,n;\alpha)$	F value at the α probability level with m degrees of freedom in the numerator and n degrees of freedom in the denominator
f	model function
ΔH_i	adsorption enthalpy of component i (J mol^{-1})

$\mathbf{J}_0$	Jacobian matrix containing the derivatives of the nonlinear model with respect to the model parameters
j	counter
K_{eq}	equilibrium coefficient for reaction converting m reactants into n products (Pa^{n-m})
K_i	adsorption equilibrium coefficient of component i (Pa^{-1})
k	reaction rate coefficient for reaction of order n ($(mol\ m^{-3})^{n-1}\ s^{-1}$ or $(mol\ kg_{cat}^{-1})^{n-\&}\ s^{-1}$)
k_0	preexponential factor, same as k
$\mathbf{I}$	unit matrix
L	Lagrange function
$N(\mathbf{0}, \mathbf{I}_n\sigma'^2)$	normal distribution with zero average and constant variance
N_T	number of temperature levels
n	total number of experiments
p	number of parameters (and steps in univariate method)
p_i	parameter i
p_i	partial pressure of component i (Pa)
p_t	total pressure (Pa)
R_g	universal gas constant (J mol^{-1} K^{-1})
r	reaction rate (mol m^{-3} s^{-1})
r_i	actual reaction rate (mol m^{-3} s^{-1})
$\hat{r}_i$	estimated reaction rate (mol m^{-3} s^{-1})
S	objective function
$S(\boldsymbol{\beta})$	sum of squares of the residuals between the experimental observations and the model simulations
s	step size
ΔS_i	adsorption entropy of component i (J mol^{-1} K^{-1})
T	temperature (K)
$V(\mathbf{b})$	true covariance matrix
$\hat{\mathbf{V}}(\mathbf{b})$	approximate covariance matrix
$\mathbf{v}_i^{(j)}$	vector of optimal search direction
W/F	ratio of catalyst weight to molar flow rate (kg mol^{-1} s^{-1})
$\mathbf{X}$	matrix of the independent variables
x_i	independent variable
y_i	dependent variable
$\mathbf{y}$	vector of the dependent variables

Greek symbols

α	positive number in Rosenbrock method
α	parameter group vector

β	negative number in Rosenbrock method
$\boldsymbol{\beta}$	parameter vector
$\Delta\mathbf{b}$	estimated adjustment of the parameter estimates in a single iteration
δ	contour value of the objective function
ε	experimental error
$\boldsymbol{\varepsilon}$	vector of experimental errors
$\eta_{k,i}$	eigenvalues of $\mathbf{J}_k^T\mathbf{J}_k$
λ	multiplier
σ	experimental standard deviation
$\varsigma_j(\mathbf{x}_i)$	new independent variables
ϕ	group of operations

Subscripts

0	initial
$'$	transformed

Superscripts

m, n	exponents in power-law equation

References

Booth, A.D., 1955. Numerical Methods. Butterworth, London.

Levenberg, K., 1944. A method for the solution of certain non-linear problems in least squares. Q. Appl. Math. 2, 164–168.

Marquardt, D.W., 1963. An algorithm for least-squares estimation of nonlinear parameters. J. Soc. Ind. Appl. Math. 11, 431–441.

Stewart, W.E., Caracotsios, M., 2008. Computer-Aided Modeling of Reactive Systems. Wiley, New York, NY.

## CHAPTER 10

Polymers: Design and Production

10.1 Introduction

This chapter provides an overview of the most frequently applied numerical methods for the simulation of polymerization processes, that is, the calculation of the polymer microstructure as a function of monomer conversion and process conditions such as the temperature and initial concentrations. It is important to note that such simulations allow one to optimize the macroscopic polymer properties and to influence the polymer processability and final polymer product application range. Both deterministic and stochastic modeling techniques are discussed. In deterministic modeling techniques, time variation is seen as a continuous and predictable process, whereas in stochastic modeling techniques, a random-walk process is assumed instead.

It should be emphasized that from a mathematical point of view, polymerization processes are inherently complicated, since many species of various *chain lengths* not only are involved but also are characterized by a wide range of intrinsic and/or apparent rate coefficients on the microscale ($k_{chem/app} = 10^{-4}$–$10^6 \, \text{m}^3 \, \text{mol}^{-1} \, \text{s}^{-1}$; Asua, 2007; D'hooge et al., 2013). During a typical polymerization process, the viscosity of the reaction mixture may increase by several orders of magnitude (Achilias, 2007) and thus reactions characterized by a sufficiently high intrinsic chemical reactivity can become diffusion controlled. In such cases, the diffusivity of the reactants determines the observed reaction rate, leading to apparent polymerization kinetics. A large number of model parameters have thus to be determined to ensure a reliable simulation of the polymer microstructure.

The numerical methods are illustrated for chain-growth polymerizations, which are mechanistically the most important group of polymerization processes. According to the chain-growth polymerization mechanism, polymer chains are formed via three successive steps: (1) chain initiation, in which the first active center (radical, anion, or cation) is formed and the growth of the chain is initiated; (2) propagation, during which the chain grows by the successive addition of monomers containing carbon-carbon double bonds to the active center at the end of the chain; (3) termination, in which the growth is stopped by either neutralization or transfer of the active center. Via these three steps, macrospecies with a chain length of up to 10^5 (or even 10^6) are typically formed on the timescale of (milli) seconds, leading to the necessity to calculate chain length distributions (CLDs) or at least average CLD

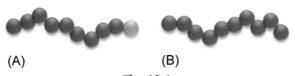

Fig. 10.1

Definition of (A) "living" and (B) "dead" macromolecules in chain-growth polymerizations. *Dark gray sphere*, monomer unit; *light gray sphere*, monomer unit with active center.

characteristics. A first distinction has to be made between "living" and "dead" macromolecules (Fig. 10.1). The former group represents macrospecies having an active center (eg, radical or organometallic species; light gray sphere in Fig. 10.1), whereas the latter group corresponds to macrospecies without an active center forming the polymer product (only dark gray spheres).

Moreover, in chain-growth copolymerizations, which involve the polymerization of two or more comonomers, different *comonomer sequences* have to be tracked by chain length, as illustrated in Fig. 10.2, for simplicity considering only well-defined monomer sequences (Matyaszewski et al., 2012). This clearly complicates the mathematical description of the polymerization kinetics. In addition, if different CLDs can be obtained by type of active center leading to an observed CLD that is a superposition of individual CLDs, the computational cost increases further (Soares and Hamielec, 1995).

Fig. 10.2

Examples of well-defined monomer sequences (Matyaszewski et al., 2012): (A) block copolymer chain, (B) linear gradient copolymer chain, (C) graft copolymer chain with all arms having the same chain length, different shades of gray (*light* or *dark*) correspond to different comonomer types.

Fig. 10.3

Increasing degree of branching of polymer molecules leads to greater model complexity; only homopolymers are shown.

Another increase in modeling complexity is obtained when it is also desired to track the presence, type, and location of *branches* in the polymer molecules (Fig. 10.3). Accounting for such topological aspects is important as these codetermine the macroscopic properties, such as the density and crystallinity, and the processability of the polymer (Konkolewicz et al., 2011).

Further mathematical complications arise depending on the *polymerization method* applied, which can be either homogeneous or heterogeneous. For homogeneous polymerization, that is, bulk and solution polymerization, simulations of large reactors require accounting for mixing and/or heat transport phenomena on the macroscale (Meyer and Keurentjes, 2005; Roudsari et al., 2013). In particular, local hot spots, which are unwanted for process control, may be invoked. For heterogeneous polymerization, a distinction can be made between polymerizations conducted in a dispersed medium (eg, suspension or emulsion polymerization) and polymerizations performed using solid catalysts (eg, Ziegler-Natta catalysts or supported metallocenes). In both cases, not only the CLD has to be tracked but also the particle size distribution (PSD). Control over this PSD at the mesoscale is important for downstream operations as it is directly related to the polymer rheology and thus the final polymer application. In particular, for Ziegler-Natta catalyzed polymerizations, it has been indicated that radial concentration and temperature gradients along the growing polymer particles cannot be ignored, as mass and heat transfer limitations can be very pronounced (McKenna and Soares, 2001).

Hence, for a given polymerization reactor configuration ideally every species present should be tracked spatially as a function of the polymerization time taking into account its diffusivity, chemical reactivity, and the possible segregation of polymer particles. In practice, this approach is characterized by a too high computational cost, explaining the development of mathematical models in which a simplification with respect to the description of the hydrodynamics, the multiphase character, and/or the involved chemical reactions is made. The validity of such model reductions is strongly related to the studied polymerization process and the purpose of the mathematical model. For example, for pure on-line reactor control, it can suffice to consider only the main polymerization reactions. This allows a reliable calculation of the polymerization rate and the number/mass average chain length ($x_{n/m}$) as a function of the polymerization time (Richards and Congalidis, 2006). Another example of model simplification relates to the modeling of suspension polymerization processes in which the polymerization reaction occurs in monomer droplets of different sizes (50–500 μm; Asua, 2007). For the simulation of the polymerization kinetics, it can be assumed that such droplets behave as "bulk reactors" and one droplet of average size can be considered for the calculation of the overall conversion, average chain length, and content of structural defects as a function of the polymerization time (De Roo et al., 2005; Kotoulas and Kiparissides, 2006). In parallel, the evolution of the monomer droplet size distribution (MDSD), which after monomer removal eventually results in a PSD, can be obtained using coalescence and breaking coefficients that are linked only with the polymerization kinetics through overall polymerization characteristics, such as the overall monomer conversion (Kotoulas and Kiparissides, 2006). A third example of simplification is the use of pseudohomogeneous models to describe polymerizations with solid catalysts, that is, the homogeneous equations are used but with adapted coefficients to mimic the heterogeneous aspect (Asua, 2007).

In what follows, first attention is focused on deterministic and stochastic modeling techniques applied at the microscale. Such models can be selected in case the aforementioned meso- and macroscale phenomena can be neglected or in case the local concentrations and local temperature are known. It is shown that the highest level of detail is obtained with more advanced numerical techniques albeit at the expense of an increase in computational cost. Furthermore, it is explained how macroscale effects can be included in the kinetic modeling technique and it is indicated to which extent the mathematical treatment can be adapted to accurately describe heterogeneous polymerizations in dispersed media, including a brief explanation of the relevance of the interaction of micro- and mesoscale effects. Finally, the main mathematical techniques for modeling heterogeneously catalyzed polymerizations are briefly highlighted, focusing again on the interaction between the micro- and mesoscale. The discussion is complementary to earlier work presented by Asua (2007) and Meyer and Keurentjes (2005) and includes a significant update with respect to the calculation of the polymer microstructure and PSD, made possible by the availability of more advanced computer technologies.

10.2 Microscale Modeling Techniques

This section is an overview of the most important deterministic and stochastic modeling techniques to obtain the polymer microstructure as a function of monomer conversion and polymerization conditions at the microscale. It is assumed that, for this scale, the bulk concentrations and temperature are known. The simplest case is the simulation of a batch polymerization reactor on laboratory scale with perfect macromixing and isothermicity implying a reactor with spatial homogeneity of the bulk concentrations and temperature.

10.2.1 Deterministic Modeling Techniques

Within the deterministic modeling techniques, a first distinction can be made between numerical methods enabling the simulation of only averages of the CLD as a function of monomer conversion, that is, the *method of moments*, and methods enabling the simulation of the full CLD, that is, the *full CLD methods*. In what follows, the main aspects of these methods are discussed in detail. For illustration purposes, it is assumed that only two populations of macrospecies, namely the living and dead polymer molecules (Fig. 10.1), exist and that only a limited number of chain-growth polymerization reactions can take place.

10.2.1.1 Method of moments

In the method of moments, the CLD is represented by a discrete number of sth ($s > 0$) order averages (Achilias and Kiparissides, 1988; Zhu, 1999):

$$x_s = \frac{\sum_i i^s ([R_i] + [P_i])}{\sum_i i^{s-1} ([R_i] + [P_i])} \qquad (10.1)$$

in which $[R_i]$ is the concentration of living polymer molecules with chain length i, and $[P_i]$ is the concentration of dead polymer molecules with chain length i. The calculation is typically limited to the first, second, and third average (x_1, x_2, and x_3). These first three averages are also known as the number-, mass-, and z-average chain length ($x_{n/m/z}$), as illustrated in Fig. 10.4.

The relative position of these three averages allows assessment of the broadness of the CLD without its explicit calculation:

$$D_n = \frac{x_m}{x_n} \qquad (10.2)$$

$$D_m = \frac{x_z}{x_m} \qquad (10.3)$$

For values of D_n and D_m close to one, the CLD is narrow, whereas for values higher than 1.5, the CLD is classified as broad. Typically only D_n, referred to as the polydispersity index or dispersity, is used as a measure of the broadness of the CLD (Gilbert et al., 2009).

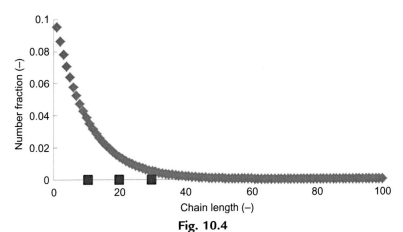

Fig. 10.4

Representation of the number CLD (*diamonds*) via three averages (Eq. 10.1); $s=1, 2, 3$ or n, m, z (*squares*; left to right); simplified case of a maximum chain length of 100.

From Eq. (10.1), it follows that to calculate x_s ($s=1, 2, 3$), the following moments ($0 \leq s' < 3$) have to be known:

$$\lambda_{s'} = \sum_i i^{s'} [R_i] \qquad (10.4)$$

$$\mu_{s'} = \sum_i i^{s'} [P_i] \tag{10.5}$$

in which $\lambda_{s'}$ and $\mu_{s'}$ represent the s'th moment of the population of living and dead polymer molecules. Note that for s' equal to zero, the total population concentrations are obtained. The moments $\lambda_{s'}$ and $\mu_{s'}$ $(0 \leq s' < 3)$ at a given polymerization time t can be obtained by integration of the following equations:

$$\frac{d\lambda_{s'}}{dt} = \sum_i i^{s'} \frac{d[R_i]}{dt} \tag{10.6}$$

$$\frac{d\mu_{s'}}{dt} = \sum_i i^{s'} \frac{d[P_i]}{dt} \tag{10.7}$$

These moment equations are typically integrated by linear multistep methods, such as the Adams method and the backward differentiation formula (Petzold, 1983). The right-hand sides of Eqs. (10.6), (10.7) are directly related to the selected reaction scheme. As an example, Scheme 10.1 shows the main reaction steps for a simplified chain-growth radical polymerization.

$$I_2 \xrightarrow{k_1} 2R_0$$

$$R_0 + M \xrightarrow{k_2} R_1$$

$$R_i + M \xrightarrow{k_2} R_{i+1}$$

$$R_i + R_j \xrightarrow{k_{3a}} P_i^H + P_j^=$$

$$R_i + R_j \xrightarrow{k_{3b}} P_i^= + P_i^H \quad (i \neq j)$$

Scheme 10.1

Basic reaction scheme to illustrate the derivation of the moment equations (Eqs. 10.8–10.11). I_2, initiator; R_0, initiator fragment; M, monomer; R_i, living polymer molecule with chain length i; P_i, dead polymer molecule with chain length i; $k_{1/2/3a/3b}$, rate coefficient for dissociation/propagation/termination (with intrinsic or apparent chain length dependencies ignored). For termination by disproportionation of polymer molecules with different chain lengths two reaction paths (a and b) are shown as chemically two end-groups are possible for P.

As termination mode, termination by disproportionation has been selected. A perfect chain initiation is also considered, which implies no use of an initiator efficiency (Moad and Solomon, 2006). The dependency of the rate coefficients on the chain length is neglected as well. For this scheme, the following equations can be derived, assuming that all dead polymer products can be lumped $(k_{3a} = k_{3b} = k_3)$ and volume effects due to a difference in density between monomer and polymer can be ignored:

$$\frac{d\lambda_{s'}}{dt} = -k_3\lambda_0\lambda_{s'} + k_2[M][R_0] + \begin{cases} k_2[M]\lambda_0 & (s'=1) \\ k_2[M]\lambda_0 + 2k_2[M]\lambda_1 & (s'=2) \\ k_2[M]\lambda_0 + 3k_2[M]\lambda_1 + 3k_2[M]\lambda_2 & (s'=3) \end{cases} \tag{10.8}$$

$$\frac{d\mu_{s'}}{dt} = k_3\lambda_0\lambda_{s'} \tag{10.9}$$

$$\frac{d[M]}{dt} = -k_2[M][R_0] - k_2[M]\lambda_0 \tag{10.10}$$

$$\frac{d[R_0]}{dt} = 2k_1[I_2] - k_2[M][R_0] \tag{10.11}$$

In these equations, only $[M]$ and $[I_2]$ have a nonzero value at $t=0$. In particular, Eq. (10.10) allows for the calculation of the monomer conversion profile without the interference of higher order moments.

In case the reactions in Scheme 10.1 are assumed to be dependent on the chain length, k_2 and k_3 have to be replaced by population-weighted rate coefficients defined as

$$<k_{2,s}> = \frac{\sum_i i^s k_{2,i}[R_i]}{\sum_i i^s[R_i]} \tag{10.12}$$

$$<k_{3,s}> = \frac{\sum_{i,j} i^s k_{3,ij}[R_i][R_j]}{\sum_{i,j} i^s[R_i][R_j]} \tag{10.13}$$

in which $k_{2,i}$ and $k_{3,ij}$ are the corresponding (apparent) chain length dependent rate coefficients (with $k_{3,ij}=2k_{3a,ij}$). However, in the classical method of moments, no assessment of the individual concentrations of the living macromolecules is performed and the dependency on the chain length is either ignored or the population-weighted rate coefficients are calculated based on overall polymerization characteristics (eg, the polymer mass fraction).

It should be noted that for termination by recombination (Fig. 10.5A), inherently an error is introduced regarding its contribution in the classical moment equations. For the individual living polymer continuity equations, it can be derived that the contribution of termination by recombination is

$$\frac{d[R_i]}{dt} = \cdots - 2k_{tc,ii}[R_i]^2 - \sum_{j(i \neq j)} k_{tc,ij}[R_i][R_j] \tag{10.14}$$

in which $k_{tc,ii}$ and $k_{tc,ij}$ $(i \neq j)$ are the homo- and cross-termination rate coefficients.

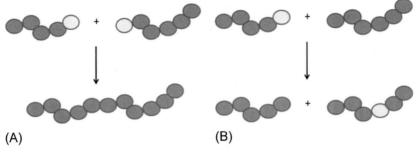

Fig. 10.5

Termination by (A) recombination and (B) chain transfer to dead polymer. *Dark gray spheres, monomer unit; light gray sphere*, monomer unit with active center.

Note that due to factor two in Eq. (10.14), additional rate coefficients have to be formally introduced to derive the moment equations:

$$\frac{d[R_i]}{dt} = \cdots - \sum_j k^*_{tc,ij}[R_i][R_j] \quad \left(k^*_{tc,ii} = 2k_{tc,ii}; \ k^*_{tc,ij} = k_{tc,ij} \ \text{for } i \neq j\right) \tag{10.15}$$

This manipulation allows one to write:

$$\frac{d\lambda_{s'}}{dt} = \cdots - <k_{tc}, s'> \lambda_0 \lambda_{s'} - \cdots \tag{10.16}$$

in which the population-weighted rate coefficient is defined as

$$<k_{tc}, s'> = \frac{\displaystyle\sum_{i,j} i^{s'} k^*_{tc,ij}[R_i][R_j]}{\displaystyle\sum_{i,j} i^{s'}[R_i][R_j]} \tag{10.17}$$

Since in the classical method of moments, no explicit calculation of the individual concentrations is performed, the manipulation introduced previously disappears when writing down the final classical moment equation, explaining the inherent error for this reaction:

$$\frac{d\lambda_{s'}}{dt} = \cdots - k_{tc}\lambda_0 \lambda_{s'} - \cdots \tag{10.18}$$

Furthermore, closure terms have to be included for the moment equations in case chain transfer to dead polymer (Fig. 10.5B) is included as reaction possibility (rate coefficient: k_{tp}). Such terms are needed as this reaction leads to a contribution in the sth order moment equation that depends on the $(s+1)$th order moment of the dead polymer molecules, as the reactivity is linked to the number of monomer units in the dead polymer chain and not to the concentration of this chain:

$$\frac{d[P_i]}{dt} = \cdots - \sum_j k_{tp,ij}[R_j] \, i[P_i] + \cdots \tag{10.19}$$

$$\frac{d\mu_{s'}}{dt} = \cdots - <k_{tp}, s'> \lambda_0 \mu_{s'+1} - \cdots \tag{10.20}$$

in which the population-weighted rate coefficient is defined as

$$<k_{tp}, s'> = \frac{\sum_{i,j} i^{s'+1} k_{tp,ij} [P_i] [R_j]}{\sum_{i,j} i^{s'+1} [P_i] [R_j]} \tag{10.21}$$

In many kinetic studies, the Hulburt and Katz (Baltsas et al., 1996; Hulburt and Katz, 1964; Pladis and Kiparissides, 1998) approximation is selected as closure method while considering only moment equations up to the second order:

$$\mu_3 = \frac{\mu_2}{\mu_0 \mu_1} \left(2\mu_2 \mu_0 - \mu_1^2 \right) \tag{10.22}$$

10.2.1.2 Full CLD methods

Within the full CLD methods, a distinction can be made based on the discretization of the chain length range, whether the quasi-steady-state approximation (QSSA) is applied for intermediate reactive species at some stage in the numerical procedure or whether moments are used for benchmark purposes. A discretization of the chain length range to simulate CLDs is necessary as otherwise too stiff systems of differential equations, and too high storage capacities and computational times result. Linear multistep methods (Petzold, 1983) are frequently used to integrate the obtained continuity equations after discretization of the chain length range.

The most important full CLD methods are the extended methods of moments, such as the probability generating function method and coarse-graining-based techniques; the fixed pivot method; and the discrete weighted Galerkin formulation. In what follows, the most important aspects of these methods are highlighted.

10.2.1.2.1 Extended method of moments

In the extended method of moments, a CLD is reconstructed based on calculated moments, as introduced above. The simplest way is a predefined mass CLD, such as the two-parameter Wesslau distribution (Pladis and Kiparissides, 1988):

$$m_i = \frac{1}{\sqrt{2\pi\sigma^2 i^2}} \exp\left(-\frac{(\ln i - \ln p_1)^2}{2\sigma^2} \right) \tag{10.23}$$

in which m_i is the mass fraction of polymer molecules with chain length i and the parameters p_1 and σ are directly linked with the sth order moments:

$$p_1 = \left(\frac{\mu_2}{\mu_0}\right)^{0.5} \quad \exp\left(\sigma^2\right) = \frac{\mu_0 \mu_2}{\mu_1^2} \tag{10.24}$$

A more advanced method is the inverse (discrete) Laplace transformation (or probability generating function [pgf]) method (Asteasuain et al., 2002a,b, 2004). In this method, the CLD is reconstructed from the integrated moment equations, as illustrated in Fig. 10.6. A number and mass probability are first introduced for the living polymer molecules:

$$P_{0,R}(i) = \frac{[R_i]}{\lambda_0} \tag{10.25}$$

$$P_{1,R}(i) = \frac{i[R_i]}{\lambda_1} \tag{10.26}$$

Similarly, for the dead polymer molecules, a number and mass probability are defined

$$P_{0,P}(i) = \frac{[P_i]}{\mu_0} \tag{10.27}$$

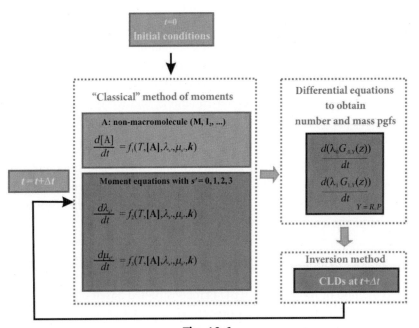

Fig. 10.6

Principle of Laplace transformations (or probability generating function (pgf) method) to reconstruct the CLD based on the moment equations. **k**, vector with rate coefficients; **[A]**, vector with concentrations of nonmacromolecules; Δt, time step for integration; *CLD*, chain length distribution; G, transformed z-variable.

$$P_{1,P}(i) = \frac{i[P_i]}{\mu_1} \tag{10.28}$$

These four probabilities are subsequently used in number/mass pgfs ($a=0,1$ and $Y=R, P$) of the transformed (complex) z-variable:

$$G_{a,Y}(z) = \sum_i z^i P_{a,Y}(i) \tag{10.29}$$

Hence, in case these pgfs are obtained for a given range of (complex) z-values, the CLD can be reconstructed via an appropriate Laplace inversion method. The z-based pgfs needed for this inversion are obtained by deriving the right-hand sides of

$$\frac{d(\lambda_0 G_{0,Y}(z))}{dt} \quad \text{and} \quad \frac{d(\lambda_1 G_{1,Y}(z))}{dt} \tag{10.30}$$

followed by their integration using the moments as obtained by integration of the moment equations in parallel. For example, the contribution of termination by recombination (Fig. 10.5A) to the mass pgf can be derived as follows:

$$\frac{d[R_i]}{dt} = \cdots - \sum_j k^*_{tc,ij}[R_i][R_j] - \cdots \tag{10.31}$$

$$\frac{d\lambda_1\left(\sum_i z^i i[R_i]/\lambda_1\right)}{dt} = \cdots - \lambda_1 \frac{\sum_j [R_j] \sum_i k^*_{tc,ij} z^i i [R_i]}{\lambda_1} \tag{10.32}$$

$$\frac{d(\lambda_1 G_{1,R})}{dt} = \cdots - k_{tc}\lambda_0(\lambda_1 G_{1,R})$$

where in the last step the chain length dependency of the population-weighted rate coefficient is ignored, which means a classical method of moments is applied. Similarly, a z-dependent term results for the chain initiation step (Step 2 in Scheme 10.1):

$$\frac{d[R_1]}{dt} = \cdots + k_2[M][R_0] - \cdots \tag{10.33}$$

$$\frac{d(\lambda_1 G_{1,R})}{dt} = \cdots + zk_2[M][R_0] - \cdots \tag{10.34}$$

Alternatively, a reconstruction can be performed using the correct moment equations, that is, including explicit population-weighted rate coefficients (cf. Eqs. (10.8)–(10.11) corrected with Eqs. (10.12), (10.13); Bentein et al., 2011). Each time, step convergence is required

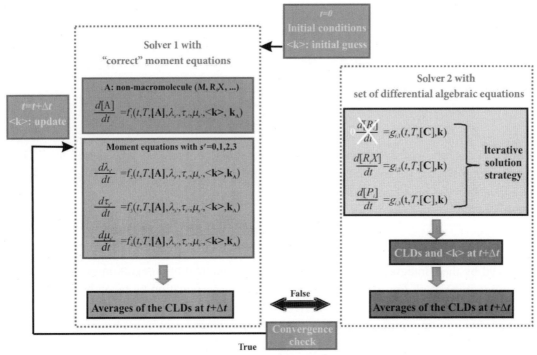

Fig. 10.7

Principle of the extended method of moments applied for NMP using "correct" moment equations (solver 1) and requiring convergence for the resulting CLD averages and those obtained via iterative integration of the set of differential algebraic equations using the QSSA for the living species and keeping the continuity equations for the dormant macrospecies (solver 2). [A], vector with concentrations of nonmacromolecules; [C]: vector with all concentrations; <k>, vector with population-weighted (apparent) rate coefficients; k, vector with all individual (apparent) rate coefficients; k_A, vector with all (apparent) rate coefficients for reactions involving only nonmacromolecules; f and g, functions; Δt, time step for the integration (Bentein et al., 2011).

between the averages of the CLD as obtained with these "correct" moment equations using fixed but updated population-weighted apparent rate coefficients per time step and the averages as obtained based on the reconstructed CLD in a parallel solver. In this parallel solver, for the calculation of the concentrations of part of the individual species, the QSSA is applied and an iterative solution strategy is selected. In particular, for controlled radical polymerizations (CRP), this method has proven to be successful. Fig. 10.7 shows the principle of this solution strategy for an important CRP process, namely nitroxide-mediated polymerization (NMP; Scheme 10.2; Malmström and Hawker, 1998).

$$R_0X \quad \underset{k_{deact}}{\overset{k_{act}}{\rightleftarrows}} \quad R_0 + X$$

$$R_i + M \quad \overset{k_2}{\longrightarrow} \quad R_{i+1}$$

$$R_iX \quad \underset{k_{deact}}{\overset{k_{act}}{\rightleftarrows}} \quad R_i + X$$

$$R_i + R_j \quad \overset{k_{tc}}{\longrightarrow} \quad P_{i+j}$$

Scheme 10.2

Principle of controlled radical polymerization (CRP) exemplified via nitroxide-mediated polymerization (NMP). X, nitroxide; R_i, living polymer molecule; P_{i+j}, dead polymer molecule; R_iX, dormant polymer molecule; i, chain length; R_0X, NMP initiator; k_{act}, activation; k_{deact}, deactivation; k_2, propagation; k_{tc}, termination by recombination (Fig. 10.5A) (Malmström and Hawker, 1998); for simplicity the activation/deactivation rate coefficients of the initiator species are assumed the same as those for the macrospecies.

In NMP, living macrospecies can be temporarily trapped by a nitroxide species X resulting in the formation of dormant macrospecies (R_iX), which are the targeted polymer molecules for CRP, in contrast with the typical dead polymer product P in other chain-growth polymerizations. For a sufficiently fast deactivation (k_{deact}; Scheme 10.2), this dormant state is favored and the contribution of dead polymer molecules is minimized. This favoritism is enhanced as X does not undergo self-termination.

With the additional s'th ($s' \geq 0$) order moments for the dormant species defined as

$$\tau_{s'} = \sum_i i^{s'} [R_iX] \tag{10.35}$$

it can be seen in Fig. 10.7 that for the iterative solver (solver 2) the QSSA is applied for the living species but not for the dormant species for which the original continuity equations are kept. Convergence can be obtained for the averages as typically a quasi-equilibrium is obtained between the dormant and living species strongly favoring the dormant state (Bentein et al., 2011; Matyjaszwski and Xia, 2001). For a very broad chain length domain, coarse graining (Russell, 1994) can be considered to reduce the number of equations for the iterative solution strategy, explaining why this alternative approach is often classified as a coarse-graining based extended method of moments. However, for an accurate simulation of the polymerization kinetics, still individual contributions are recommended for small chain lengths (< 200) (Bentein et al., 2011).

10.2.1.2.2 Fixed pivot method

In the fixed pivot method, which is a population balance method, the chain length domain is divided into N_{max} (typically 50) intervals, which are fixed during the integration but can be

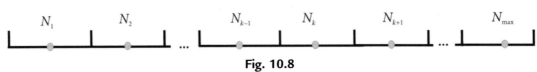

Fig. 10.8
Division of the chain length range for the fixed pivot method; each interval can be of a different size (Δ_k) but the sizes are fixed during the integration; typically a logarithmic scale is used; *gray spheres* represent the pivot elements.

chosen to be of a different size (Butté et al., 2002; Kumar and Ramkrishna, 1996). Typically the size variation is of a logarithmic nature in order to accurately capture the contributions of macrospecies with a small chain length.

Each interval N_k ($k=1,\ldots, N_{max}$) is characterized by only one chain length belonging to that interval, that is, the characteristic chain length or pivot i_k (see Fig. 10.8). In other words, the total concentration of, for example, dead polymer molecules in N_k is the product of the concentration of dead polymer molecules with that particular chain length and the interval size Δ_k. The continuity equations are integrated only per interval leading to an overall reduction of the number of equations to be considered. For the living and dead polymer molecules, the symbols $[R_k^*]$ and $[P_k^*]$ ($k=1,\ldots, N_{max}$) are typically used to denote the corresponding overall concentrations.

$$R_i + M \xrightarrow{k_{trM}} P_i + R_0$$

Scheme 10.3
Chain transfer to monomer.

In the simplest case, chain transfer to the monomer (Scheme 10.3, rate coefficient: k_{trM}) is considered and a contribution to the continuity equation for $[R_k^*]$ can be derived directly, as the reactants and products belong to the same interval:

$$\frac{d[R_k^*]}{dt} = -k_{trM}[R_k^*][M] + \ldots \tag{10.36}$$

However, for certain reactions only particular chain lengths belonging to other intervals contribute to the population balance for a given interval. For example, for a product of termination by recombination belonging to the interval k, the macroradical chain lengths i_m and i_l (with m and l the corresponding intervals) have to be identified and a reverse lever rule has to be applied (cf. Fig. 10.9):

$$\frac{d[P_k^*]}{dt} = \ldots + k_{tc} \sum_{i=1}^{i_{max,k}} \sum_{i_l=1}^{i_l+i_m=i} A_{pivot}(i)\frac{[R_m^*]}{\Delta_m}\frac{[R_l^*]}{\Delta_l} \tag{10.37}$$

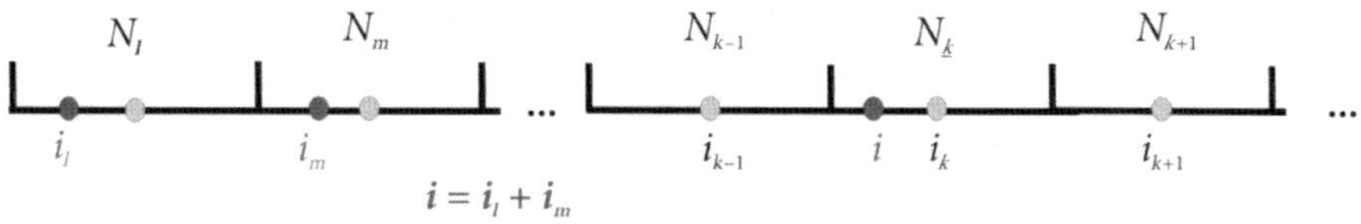

Fig. 10.9

Representation of termination by recombination leading to product with a chain length i belonging to the interval k of the fixed pivot method chain length domain range; the intervals of the reactants are l and m with pivot chain lengths i_l and i_m. As the considered chain length i is closer to the pivot i_k than to the pivot i_{k-1} the contribution of the termination product to the interval k is larger (Eq. 10.38); *light gray spheres* represent pivot elements; *dark gray spheres* represent chain lengths related to terminating radicals.

in which $i_{\mathrm{max},k}$ is the longest chain length for the interval k and $A_{\mathrm{pivot}}(i)$ is a function defined as

$$A_{\mathrm{pivot}}(i) = \begin{cases} \dfrac{i - i_{k-1}}{i_k - i_{k-1}} & i_{k-1} \leq i \leq i_k \\[2mm] \dfrac{i_{k+1} - i}{i_{k+1} - i_k} & i_k \leq i \leq i_{k+1} \\[2mm] 0 & \text{otherwise} \end{cases} \tag{10.38}$$

10.2.1.2.3 Discrete weighted Galerkin formulation

In the discrete weighted Galerkin formulation, the CLD is represented by an expansion in basis functions until a truncated value (N_{tr}) (Budde and Wulkow, 1991; Iedema et al., 2003; Wulkow, 1996). For example, for the dead polymer this implies that the CLD at time t is given by

$$P_i(t) = \psi(i; q) \sum_k^{N_{\mathrm{tr}}} a_k(t; q) l_k(i; q) \quad \text{with } i > 0 \tag{10.39}$$

in which ψ is a positive weight function, q is a parameter a_k ($k = 1, \ldots, N_{\mathrm{tr}}$) are expansion coefficients and l_k ($k = 1, \ldots, N_{\mathrm{tr}}$) are the basis functions, which are orthogonal with respect to a weight function ψ:

$$\sum_i l_j(i; q) \, l_k(i; q) \psi(i; q) = \gamma_k \delta_{jk} \tag{10.40}$$

in which δ_{jk} is the Dirac delta function and γ_k an orthogonality factor.

Typically the Schulz-Flory weight function (Eq. 10.41) is used and discrete Laguerre polynomials (Eqs. 10.42, 10.43) are used as basis functions.

$$\psi(i; q) = (1 - q) q^{i-1} \quad 0 < q < 1 \quad i > 0 \tag{10.41}$$

$$(k + 1) l_{k+1}(i; q) = [(k + 1)q + k - (1 - q)(i - 1)] l_k(i; q) - k \rho l_{k-1}(i; q) \tag{10.42}$$

$$l_{-1} = 0, \quad l_0 = 1, \quad \text{and} \quad \gamma_k = q^k \tag{10.43}$$

The expansion coefficients are determined via substitution and application of the orthogonality relation. For example, if the time dependency of the polymerization kinetics (denoted here with ') can be described by

$$P_i'(t) = \mathbf{B}' P_i(t) \tag{10.44}$$

with $\mathbf{B}'$ a representative matrix, it can be derived (Budde and Wulkow, 1991) that

$$\gamma_j a_j'(t) = \Psi(i) \sum_k a_{ki}(t)(l_j, \mathbf{B}' l_k) \tag{10.45}$$

with the inner product of two functions f and g defined as

$$(f, g) = \sum_i f(i)g(i)\psi(i) \tag{10.46}$$

10.2.2 Stochastic Modeling Techniques

In contrast to deterministic modeling techniques, stochastic modeling techniques do not require the numerical integration of a set of coupled differential equations to simulate chemical kinetics but only require stochastic executions of discrete events. Two important stochastic techniques are the *Gillespie-based kinetic Monte Carlo* (kMC) and the *Tobita-based Monte Carlo* technique, which are both addressed in this section. In general, full CLDs are directly obtained with stochastic modeling approaches as long as a sufficiently high resolution, that is, a sufficiently large initial number of molecules, is selected.

10.2.2.1 Gillespie-based kMC technique

The most frequently applied kMC algorithm to study polymerization processes is that developed by Gillespie (1977). In this algorithm, different species are tracked in a representative microscopic-scale homogeneous volume V (eg, 10^{-20} m^3) and reactions are selected in a discrete manner via stochastic time steps. For each reaction ($\nu = 1,..., N_r$ with N_r the total number of reactions), a Monte Carlo (MC) reaction probability P_ν^{MC} is first defined:

$$P_\nu^{MC} = \frac{r_\nu^{MC}}{\displaystyle\sum_{\nu=1}^{N_r} r_\nu^{MC}} \tag{10.47}$$

in which r_ν^{MC} is the "MC rate of reaction ν" expressed per second.

As in the kMC technique the reactions are represented discretely the reaction rates must thus be converted from macroscopic values, which are in general on a per volume basis, to stochastic rates on the basis of the total number of molecules within the scaled reaction volume. In order to do so, the macroscopic concentrations ([C_m]; $m = 1,..., N_{sp}$; with N_{sp} the number of different types of species) must be converted into a total number of molecules N_m within V:

$$N_m = [C_m]N_A V \tag{10.48}$$

in which N_A is Avogadro's number. Depending on the order of the reaction, either first or second, the rate coefficients (k_ν; $\nu = 1,\ldots, N_r$) are converted accordingly:

$$\text{first order}: \quad k_\nu^{MC} = k_\nu \tag{10.49}$$

$$\text{second order}: \quad k_\nu^{MC} = \frac{bk_\nu}{N_A V} \tag{10.50}$$

in which the value of b is one for a reaction between different species and two for a reaction between identical species.

The stochastic time interval between reactions (τ^{MC}) is determined as follows (with r_1 a random number uniformly distributed between 0 and 1):

$$\tau^{MC} = \frac{-\ln(r_1)}{\sum_{\nu=1}^{N_r} r_\nu^{MC}} \tag{10.51}$$

The reaction to be executed, μ^{MC}, is chosen by selecting a second random number, r_2, uniformly distributed between 0 and 1, for which the following equation holds

$$\sum_{\nu=1}^{\mu^{MC}-1} r_\nu^{MC} < r_2 < \sum_{\nu=1}^{\mu^{MC}} r_\nu^{MC} \tag{10.52}$$

The principle of this cumulative-based selection is illustrated in Fig. 10.10. It can be clearly seen that, via this criterion, reactions characterized by a high reaction probability are favored to be executed more intensively. For polymerization reactions, however, N_r is very

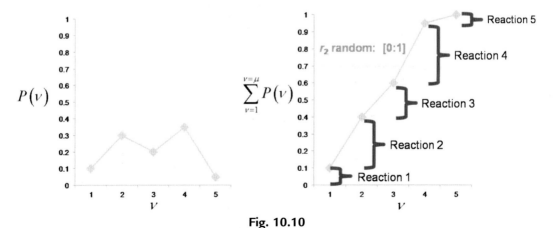

Fig. 10.10

Principle of the random selection of a reaction based on cumulative reaction probabilities (Eq. 10.52); for macromolecular reactions in an additional step the involved chain lengths are selected; here the random number r_2 is 0.7: reaction 4 is selected.

large ($> 10^5$) as apparent chain length dependent rate coefficients result, mainly due to diffusional limitations. Therefore "lumped" kMC reaction rates are used in practice so that a limited number of reaction types is obtained. In an additional step, the macrospecies involved are selected by generating extra numbers.

For example, for Scheme 10.1, only four reaction channels remain with the following "lumped" kMC reaction rates:

$$r_1^{MC} = k_1^{MC} N_{I_2} \tag{10.53}$$

$$r_2^{MC} = k_2^{MC} N_{R_0} N_M \tag{10.54}$$

$$r_3^{MC} = <k_2^{MC}> N_{\lambda_0} N_M \tag{10.55}$$

$$r_4^{MC} = <k_3^{MC}> N_{\lambda_0} N_{\lambda_0} \tag{10.56}$$

in which $<k_2^{MC}>$ and $<k_3^{MC}>$ are MC population-weighted rate coefficients, which are typically updated only at distinct monomer conversions.

It is important to realize that the total simulation time is related to the selected volume V. Depending on the complexity of the reaction scheme and the requirement to accurately calculate not only the average properties but also the concentrations of nonabundant species, a high computational cost may result. The simulation time can be reduced, however, by implementing search and reaction execution algorithms based on binary tree structures and using the overall kMC rates and not the reaction probabilities, that is, skipping the normalization (Van Steenberge et al., 2011).

The principle of a binary tree operation, as originally explored by Chaffey-Millar et al. (2007), is shown in Fig. 10.11 considering for simplicity only oligomeric species with a maximum

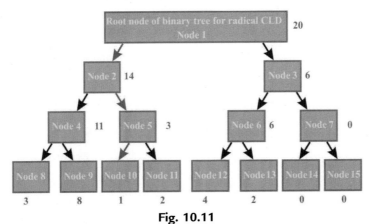

Fig. 10.11

Example of the use of a binary tree for the random selection of the chain length i^* (here 3) of the nth (here 12th) macrospecies of a radical CLD when the radicals are ranked in order of increasing chain length from left to right. Random number is 0.60; 0.60*20 = 12; from bottom row the CLD can be constructed.

chain length of eight. The top of the binary tree is denoted as the "root" node, whereas the bottom of the binary tree consists of the so-called leaf nodes. In every layer the nodes are connected in pairs. In Fig. 10.11, the *i*th leaf node contains the number of living macrospecies with chain length *i*. It is further supposed that random number generation requires that the 12th macrospecies (if ranked according to increasing chain length) reacts. Clearly, due to the binary character of the tree, a fast identification of this species is obtained by following the light gray arrows. For completeness it is mentioned here that for the calculation of chemical composition CLDs so-called composite binary trees have been introduced (Van Steenberge et al., 2011) in which, for a given chain length (thus a given leaf node in Fig. 10.11), a subtree is introduced of which the root node is the previous leaf node and information is stored for the chemical composition (eg, the number of comonomer units of a certain type or the number of branches).

It should be further stressed that a more detailed level of the calculation of the polymer structure results in case the kMC reaction event history is also stored. For growing macrospecies, this implies the use of matrices to show comonomer sequences and branching points. The principle of this matrix-based method is illustrated in Fig. 10.12 for a copolymerization leading to a linear copolymer. The complete polymer microstructure can be obtained after removal of nonpropagating reaction events. Based on this information, derivative properties can be calculated, which allows one to obtain the quality of the polymer microstructure with respect to a targeted microstructure (TM). In particular, such calculations have been performed to determine the linear gradient (Van Steenberge et al., 2012) and diblock (Toloza Porras et al., 2013) quality of linear copolymers obtained through CRP via an average linear gradient deviation and an average block deviation value. In general form, an average structural deviation, $<SD>$, can be calculated

$$<SD> = \sum_{z=1}^{N_c} SD(z) \qquad (10.57)$$

in which $SD(z)$ is the deviation of the *z*th chain of the N_c chains tracked with respect to the TM in case the same chain length is considered and N_c is sufficiently large so that a representative kMC sample is obtained. For an $<SD>$ of zero, a perfect match is obtained.

The individual $SD(z)$ ($z = 1, \dots, N_c$) terms are obtained by evaluating the minimum of the following four values (Van Steenberge et al., 2012):

$$SD_1 = \sum_{z=1}^{N_c} \frac{1}{2} \frac{|S_A(y,z) - S_{A,TM,B \text{ to } A}(y,z)| + |S_B(y,z) - S_{B,TM,B \text{ to } A}(y,z)|}{i^2} \qquad (10.58)$$

$$SD_2 = \sum_{z=1}^{N_c} \frac{1}{2} \frac{|S_A(y,z) - S_{A,TM,A \text{ to } B}(y,z)| + |S_B(y,z) - S_{B,TM,A \text{ to } B}(y,z)|}{i^2} \qquad (10.59)$$

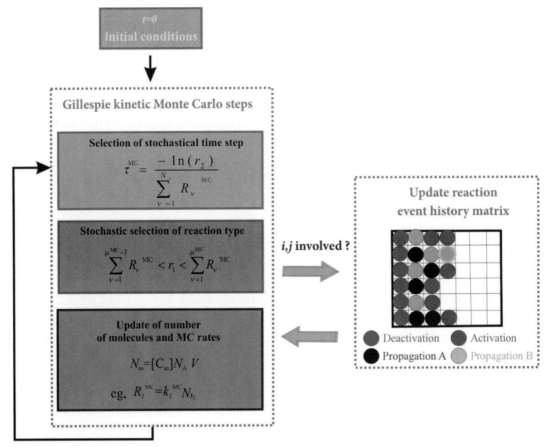

Fig. 10.12

Principe of Gillespie based kinetic Monte Carlo technique for the simulation of the reaction event history shown (Van Steenberge et al., 2012) for an NMP copolymerization (Scheme 10.2); removal of nonpropagation events allows retrieval of the comonomer sequences per chain; i,j, chain lengths if macrospecies are involved; for simplicity no termination reaction events are shown.

$$SD'_1 = \sum_{z=1}^{N_c} \frac{1}{2} \frac{\left|S'_A(y,z) - S'_{A,TM,B \text{ to } A}(y,z)\right| + \left|S'_B(y,z) - S'_{B,TM,B \text{ to } A}(y,z)\right|}{i^2} \qquad (10.60)$$

$$SD'_2 = \sum_{z=1}^{N_c} \frac{1}{2} \frac{\left|S'_A(y,z) - S'_{A,TM,A \text{ to } B}(y,z)\right| + \left|S'_B(y,z) - S'_{B,TM,A \text{ to } B}(y,z)\right|}{i^2} \qquad (10.61)$$

in which $S_x(y,z)$ is the cumulative number of monomer units of types x at position y in chain z (chain length: i) of the kMC sample as counted from left to right, $S_{x,TM,r}(y,z)$ is the cumulative number of monomer units of type x at position y of the targeted chain in the direction r (always defined from left to right) with the same length as the considered chain z, and $S_x'(y,z)$ and $S_{x,\mathrm{TM},r}'(y,z)$ are the corresponding cumulative functions when counting from right to left. Four evaluations are needed to treat both monomer types on the same basis and to account for the possibility that the chains are not ordered a priori.

10.2.2.2 Tobita-based MC technique

For branched polymers, the matrix-based kMC approach discussed previously can be extended by adding additional rows for the composition of the individual branches and storing the connection points. However, for highly branched polymers this requires very high storage capacities and intensive coding efforts. Alternatively, the Tobita-based MC technique (Meyer and Keurentjes, 2005; Tobita, 1993) can be applied to assess the topology of the polymer. With this technique, the polymer topology is reconstructed a posteriori from building blocks that are generated via different random numbers. In this section, the basic steps of the calculation are discussed. This technique is reliable, however, only in case apparent kinetics are of minor importance.

In a first step, the so-called birth conversion x^* is sampled at which an initial zeroth-order building block (label 0) is generated taking into account the maximum conversion, $x_{\max}$, reached. In a second step, the chain length i^* of this block is obtained by random selection, typically following the Flory mass CLD (Tobita, 1993):

$$m_i = \frac{i}{x_n} \exp\left(-\frac{i}{x_n}\right) \tag{10.62}$$

in which m_i is the mass fraction of polymer chains with chain length i. The average chain length x_n is determined via the ratio of the reaction rates leading to chain growth and those leading to termination. The concentrations involved are assessed based on the conversion selected (eg, via moment equations or pseudoanalytical expressions).

In a third step, for the selected building block, the number of branching points (N_{br}) is determined via sampling based on a binominal distribution (Meyer and Keurentjes, 2005):

$$p(N_{\mathrm{br}}) = \binom{i}{N_{\mathrm{br}}} \rho_{\mathrm{br}}^{N_{\mathrm{br}}} \rho_{\mathrm{br}}^{i-N_{\mathrm{br}}} \tag{10.63}$$

in which ρ_{br} is a predefined branching probability. Assuming chain transfer to polymer as the dominant branching point creator, the following expression has been proposed for ρ_{br}:

$$\rho_{\mathrm{br}} = \frac{k_{\mathrm{tp}}}{k_{\mathrm{p}}} \ln\left(\frac{1-x^*}{1-i}\right) \tag{10.64}$$

in which k_{tp} and k_{p} are the transfer to polymer and propagation rate coefficients.

In a fourth step, the birth conversions as well as the lengths have to be selected for these branches. To obey physical boundaries, the conversion interval $[x^*, x_{\max}]$ has to be considered for the selection of each additional birth conversion x^{**}. The following conditional probability distribution function, which inherently reflects a higher branching density at higher conversions, has been proposed to retrieve x^{**}:

$$p(xx^*|x^*) = \frac{\ln\left(\dfrac{1-x^*}{1-x^{**}}\right)}{\ln\left(\dfrac{1-x^*}{1-i}\right)} \tag{10.65}$$

The corresponding length l^{**} (still label 0) is obtained via sampling based on

$$p(l^{**}) = \frac{l^{**}}{l_n} \exp\left(\frac{-l^{**}}{l_n}\right) \tag{10.66}$$

with l_n a characteristic length. In a fifth step, it is determined whether this building block (with its possible branches) is connected to an earlier building block (at conversion x^{***} between 0 and x^*), which relates to the relative importance of the chain transfer to polymer reaction. If random selection based on the corresponding probability reveals that a connection has to be made, the birth conversion and the length (Eq. 10.64) of this additional branch, which is still given a label 0, are determined.

Subsequently, the number of branches (Eq. 10.63) and lengths (Eq. 10.66) of this "older" building block are sampled. For these branches, the label is changed from 0 to 1 to emphasize the transition to another building block. A similar approach is followed as for the branches of the zeroth-building block. The algorithm stops if no older building blocks and new branches are selected.

10.3 Macroscale Modeling Techniques

In general, the continuity equations of the reaction components depend on the selected reactor configuration. Industrial-scale polymerizations are commonly carried out in batch, tubular, or continuous stirred-tank reactors. Usually, relatively simple reactor models are used and thus ideal flow patterns in the reactor (eg, without backmixing) and isothermal temperature profiles are assumed. Deviations from the ideal flow pattern and from spatial isothermicity can lead, however, to significantly different local kinetics resulting in different local apparent rates and thus in different reactor averaged apparent rates as compared to those predicted assuming an ideal flow pattern and isothermicity.

The simplest way to model nonideal flow patterns and temperature profiles is to divide the reactor into a discrete number of compartments and include exchange/recycle streams between the compartments (Topalos et al., 1996; Zhang and Ray, 1997). For each compartment, a kinetic scheme is applied and the conversion profile and the polymer properties are obtained from averaging the properties of each compartment with respect to their size.

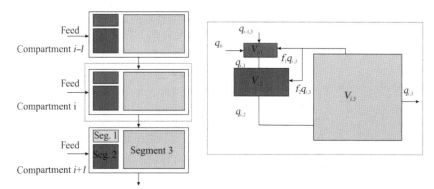

Fig. 10.13

Example of a compartment model with three perfectly mixed segments (two small and one large) to account for macromixing; $V_{i,k}$, volume of segment k in compartment i; $q_{Vi,k}$, exit volumetric flow rate for segment k in compartment i; q_{V0}, volumetric feed flow rate; f_l, recycle ratio ($l = 1, 2$); subscript V in the flow rates is not shown in order not to overload the figure.

An important first "compartment model" is based on the division of each compartment into a discrete number of perfectly mixed segments. For example, in Fig. 10.13, for each compartment of an industrial-scale reactor for a polymerization process still in nondispersed medium, two small perfectly mixed segments and one large perfectly mixed segment are considered. The small segments reflect imperfect mixing (eg, of the initiator) and temperature gradients (eg, hot spots) at the inlet, whereas the large segment reflects the bulk zone of the compartment with an internal recycle to both small segments.

Denoting the volume of the kth segment as $V_{i,k}$, $q_{Vi,k}$ as its exit volumetric flow rate, $R_{Mi,k}$ as the corresponding net monomer production rate, $[M]_{i,k}$ as the monomer concentration, f_l as the lth recycle rate from the third segment, and q_{V0} as the volumetric feed flow rate with monomer concentration $[M]_0$, the following continuity equations can be written down for the monomer in each segment k of a compartment i:

$$V_{i,1}\frac{d[M]_{i,1}}{dt} = [M]_0 q_{V0} + [M]_{i-1,3} q_{Vi-1,3} + f_1 [M]_{i,3} q_{Vi,3} + R_{Mi,1} V_{i,1} - [M]_{i,1} q_{Vi,1} \tag{10.67}$$

$$V_{i,2}\frac{d[M]_{i,2}}{dt} = [M]_{i,1} q_{Vi,1} + f_2 [M]_{i,3} q_{Vi,3} + R_{Mi,2} V_{i,2} - [M]_{i,2} q_{Vi,2} \tag{10.68}$$

$$V_{i,3}\frac{d[M]_{i,3}}{dt} = [M]_{i,2} q_{Vi,2} + R_{Mi,3} V_{i,3} - f_1 [M]_{i,3} q_{Vi,3} - f_2 [M]_{i,3} q_{Vi,3} \tag{10.69}$$

in which it is assumed that the volumetric flow rates are balanced:

$$q_{Vi,1} = q_{V0} + q_{Vi-1,3} + f_1 q_{Vi,3} \tag{10.70}$$

$$q_{Vi,2} = q_{Vi,1} + f_2 q_{Vi,3} \tag{10.71}$$

$$(1 + f_1 + f_2) q_{Vi,3} = q_{Vi,2} \tag{10.72}$$

Analogously, overall sth order moment equations can be written down. For example, assuming that no reaction occurs in the feed, the total concentration of dead polymer in the first segment of compartment i follows from

$$V_{i,1}\frac{d\mu_{0i,1}}{dt} = \mu_{0i-1,3}q_{Vi-1,3} + f_1 q_{Vi,3}\mu_{0i,3} + R_{\mu_{0i,1}}V_{i,1} - \mu_{0i,1}q_{Vi,1} \qquad (10.73)$$

in which $\mu_{0i,1}$ and $R_{\mu_{0i,1}}$ are the zeroth-order moment for the dead polymer molecules and the corresponding net production rate in segment 1 of compartment i. The net production rates can be calculated with the method of moments, as explained previously, assuming a given reaction scheme per segment. In addition, a temperature variation can be accounted for. Taking again the first segment, from an energy balance it follows that

$$\rho c_p \frac{dT_{i,1}}{dt}V_{i,1} = Q_0 + Q_{i-1,3} + f_1 Q_{i,3} + Q_{ri,1} - Q_{i,1} \qquad (10.74)$$

in which $T_{i,1}$ is the temperature of the first segment of compartment i, ρ, and c_p are the corresponding density and specific heat capacity at constant pressure, and the subscript "r" is used to distinguish the net heat production by reaction from the flow contribution terms. The reaction term is given by

$$Q'_{i,1,r} = (-\Delta_r H)k_p[M]_{i,1}\lambda_{0,i,1}V_{i,1} \qquad (10.75)$$

in which $\Delta_r H$ is the propagation reaction enthalpy and $\lambda_{0i,1}$ is the zeroth-order moment of the living polymer molecules. Note that it is assumed that the generated heat is only due to propagation. The energy flow terms are proportional to the temperature of the flow while taking into account the contributions of the monomer and the polymer to obtain the correct density and specific heat capacity.

A second important compartment model comprises compartments consisting of a perfectly mixed segment and a plug-flow segment, which in turn can be represented by a series of perfectly mixed segments (see Fig. 10.14). This compartment model is selected in case segregation is important. Equations analogous to those for the compartment model of Fig. 10.13 can be derived.

Instead of using compartment models, the flow pattern in the reactor also can be calculated via computational fluid dynamics (CFD). However, when using CFD, relatively small reaction networks are often used to reduce the computational cost. An exception is gas-phase polymerization, such as the production of low-density polyethylene). For more details on the application of CFD calculations for polymerization processes, the reader is referred to Asua and De La Cal (1991), Fox (1996), Kolhapure and Fox (1999), and Pope (2000).

10.4 Extension Toward Heterogeneous Polymerization in Dispersed Media

This section pays attention to the standard modeling tools to simulate the polymer micro structure for polymerizations in dispersed media for which mesoscale phenomena are also relevant. In such polymerizations, a surfactant is present so that a dispersed phase can be

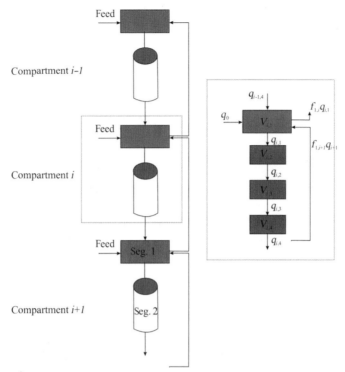

Fig. 10.14

Example of a compartment model with one perfectly mixed segment and one plug-flow segment to account for macromixing; the plug-flow segment can be represented by a series of perfectly mixed segments (here three); $V_{i,k}$, volume of segment k in compartment i; $q_{Vi,k}$, exit volumetric flow rate for segment k in compartment i; q_{V0}, volumetric feed flow rate; $f_{1,k}$, recycle ratio; in the figure subscript V has been omitted.

formed and the polymer product consists of polymer particles. A distinction can be made between suspension and emulsion polymerization. In suspension polymerization, large monomer droplets (50–500 μm; Asua, 2007) are the reaction locus, whereas in emulsion polymerization, different reaction loci can be identified and the segregated entities are much smaller (50–1000 nm; Asua, 2007).

10.4.1 Suspension Polymerization

For suspension polymerization, the polymerizing droplets are so large that it can be safely assumed that the kinetics are the same for every droplet and thus the microkinetic modeling tools introduced previously can be directly applied for one droplet to obtain the evolution of the CLD and its averages with conversion.

In addition, the PSD (or MDSD) can be calculated analogously by replacing the variable i by its diameter d_p or volume V_p. Most commonly, the fixed pivot technique has been applied while dividing the reactor into two compartments, an impeller and a circulation compartment,

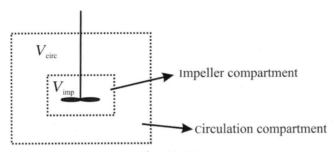

Fig. 10.15

Compartment model to calculate the evolution of the monomer droplet size distribution (MDSD) in a suspension polymerization; $V_{imp/circ}$, volume of impeller/circulation compartment.

as shown in Fig. 10.15 (Alexopoulos and Kiparissides, 2005). This distinction is made to account for the different turbulence intensity in the direct environment of the impeller and the region away from the impeller, that is, to account for macroscale effects. Such a simplification is justified, at least as a first approximation, based on CFD simulations of the flow pattern in batch reactors with an impeller (Alexopoulos et al., 2002).

For each compartment, a population balance is derived as basis for the further discretization according to the selected pivot elements along the V_p axis and using number density functions $n_{imp}(V,t)$ and $n_{circ}(V,t)$ for the impeller and circulation compartments. These density functions are defined such that $n_{imp}(V,t)dV$ and $n_{circ}(V,t)dV$ are the numbers of droplets per unit volume of the impeller and circulation compartments with a volume in the interval $[V, V+dV]$.

For the impeller compartment with a volume, V_{imp}, $n_{imp}(V,t)$ can be obtained based on the following nonlinear integrodifferential equation:

$$\frac{\partial n_{imp}(V,t)}{\delta t} = \int_V^{Vmax} \beta(U,V)u(U)g(U)n_{imp}(U,t)dU$$
$$+ \frac{1}{2}\int_0^V k(V-U,U)n_{imp}(V-U,t)n_{imp}(U,t)dU$$
$$- n_{imp}(V,t)\int_0^{Vmax} k(V,U)n_{imp}(U,t)dU - g(V)n_{imp}(V,t)$$
$$+ \frac{q_{V,e}}{V_{imp}}\left(n_{circ}(V,t) - n_{imp}(V,t)\right) = PB_{imp}^{rhs}$$

(10.76)

The first term of the population balance, PB_{imp}^{rhs}, relates to the formation of droplets with a volume in the interval $[V, V+dV]$ by breakage of droplets with a larger volume (maximally V_{max}). In this term, the mesoscale parameter $g(U)$ is the breakage coefficient for a droplet with a volume U, $u(U)$ is the number of droplets formed upon breakage of a droplet with a volume U (typically two), and $\beta(U,V)$ reflects the probability that a droplet with a volume U breaks into a droplet with a volume V. The second term represents the formation of droplets in the volume interval $[V,$

$V + dV$] by coalescence of two smaller droplets, with the mesoscale parameter $k(U,V)$ the coalescence coefficient of droplets with a volume of U and V. The third and fourth terms represent the disappearance of droplets by, respectively, coalescence and breakage, while the last term expresses the exchange of droplets between the two compartments with $q_{V,e}$ as exchange flow rate. The analogous nonlinear integrodifferential equation for the circulation compartment can be obtained by interchanging the subscripts "imp" and "circ" in Eq. (10.76). The mesoscale parameters are typically obtained by correlations that can depend on the microscale properties via a dependence on, for instance, the monomer conversion (Asua, 2007).

In practice, the continuous $n_{imp}(V,t)$ function is represented via delta functions:

$$n_{imp}(V,t) = \sum_{k=1}^{N_{max}} N_{imp,k}^*(t)\delta(V - x_k) \tag{10.77}$$

in which $N_{imp,k}^*$ is the contribution to the density function for the droplets in the interval k. Integration per grid interval of Eq. (10.76) gives

$$\int_{V_i}^{V_{i+1}} \frac{\partial n_{imp}(V,t)}{\partial t}dV = \int_{V_i}^{V_{i+1}} PB_{imp}^{rhs} dV \tag{10.78}$$

For the left-hand side of this equation, it follows that

$$\int_{V_i}^{V_{i+1}} \frac{\partial n_{imp}(V,t)}{\partial t}dV = \frac{\partial}{\partial t}\int_{V_i}^{V_{i+1}} \sum_{k=1}^{N_{max}} N_{imp,k}^*(t)\delta(V - x_k)dV$$

$$= \frac{\partial}{\partial t}\sum_{k=1}^{N_{max}} \int_{V_i}^{V_{i+1}} N_{imp,k}^*(t)\delta(V - x_k)dV = \frac{d}{dt}N_{imp,k}^*(t) \tag{10.79}$$

Similarly, for each coalescence/breakage term of the right-hand side mathematical manipulations can be performed. For example, for the positive coalescence contribution we can write

$$\int_{V_i}^{V_{i+1}} \frac{1}{2}\int_0^{V_x} k(V - U,U)n_r(V - U,t)n_{imp}(U,t)dU\,dV = \frac{1}{2}\int_{V_i}^{V_{i+1}} arg\,dV \tag{10.80}$$

while demanding that this contribution can be related to neighboring pivot volumes:

$$\int_{V_i}^{V_{i+1}} arg\,dV = \int_{x_{i-1}}^{x_i} b(V,x_i)\,arg\,dV + \int_{x_i}^{x_{i+1}} a(V,x_i)\,arg\,dV \tag{10.81}$$

in which the functions a and b follow again from the reverse lever rule (see also the CLD part of the fixed pivot technique):

$$a(V,x_i) = \frac{x_{i+1} - V}{x_{i+1} - x_i} \tag{10.82}$$

$$b(V, x_i) = \frac{V - x_{i-1}}{x_i - x_{i-1}} \tag{10.83}$$

For the first term of the right-hand side of Eq. (10.81), substitution of Eq. (10.77) leads to

$$\int_{x_{i-1}}^{x_i} b(V, x_i) \arg dV = \int_{x_{i-1}}^{x_i} b(V, x_i) \int_0^{V_x} k(V - U, U) \sum_{j=1}^{N_{max}} N_{imp,j}^*(t) \delta\left(V - U - x_j\right) \sum_{l=1}^{N_{max}} N_{imp,j}^*(t) \delta(U - x_l) dU dV \tag{10.84}$$

Taking into account the definition of the delta function for the integration of U and the maximum limit of V following simplification can be made

$$\int_{x_{i-1}}^{x_i} b(V, x_i) \arg dV = \int_{x_{i-1}}^{x_i} b(V, x_i) \sum_{l=1}^{i} k(V - x_l, x_l) \sum_{j=1}^{N_{max}} \delta\left(V - x_l - x_j\right) N_{imp,j}^*(t) N_{imp,l}^*(t) dV \tag{10.85}$$

Subsequently, considering the definition of the delta function for the integration of V and the integration limits, it can be derived that

$$\int_{x_i}^{x_{i+1}} b(V, x_i) \arg dV = \sum_{x_{i-1} \leq x_j + x_l \leq x_i}^{j \geq l} b(x_j + x_l, x_i) k(x_j, x_l) N_{imp,j}^*(t) N_{imp,l}^*(t) \tag{10.86}$$

Analogously, for the second term of the right-hand side of Eq. (10.81), it follows that

$$\int_{x_i}^{x_{i+1}} a(V, x_i) \arg dV = \sum_{x_i \leq x_j + x_l \leq x_{i+1}}^{j \geq l} a(x_j + x_l, x_i) k(x_j, x_l) N_{imp,j}^*(t) N_{imp,l}^*(t) \tag{10.87}$$

In general, Eq. (10.76) can thus be transformed into

$$\frac{dN_{imp,i}^*(t)}{dt} = f\left(N_{imp,1}^*, \ldots, N_{circ,1}^*, \ldots\right) \tag{10.88}$$

which allows for an efficient calculation of the evolution of the MDSD also taking into account the analogous set of continuity equations for the impeller region.

10.4.2 Emulsion Polymerization

The main classes of emulsion polymerization are macro-, micro-, and miniemulsion polymerization (Asua and De La Cal, 1991; Cunningham, 2008; Zetterlund et al., 2008). In macroemulsion polymerization, a relatively high surfactant concentration is present so that many small empty (or monomer-swollen) micelles can be formed besides large monomer droplets (Fig. 10.16). Initiation typically occurs in the water phase and after a few propagation steps, the resulting living macromolecule enters a micelle, as the micelles have a much higher surface area than the monomer droplets. This entering process is known as

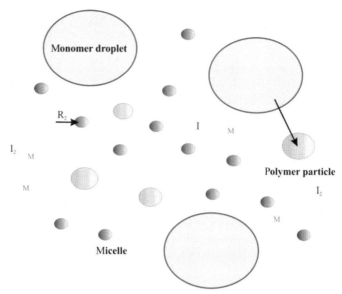

Fig. 10.16
The existence of three populations in a macroemulsion process: (i) (monomer-swollen) micelles, (ii) polymer particles, and (iii) large monomer droplets; only heterogeneous nucleation leading to transformation of micelles into polymer particles is considered (shown here with the entry of R_2 formed after two propagation steps with I); I_2, water soluble initiator; large arrow represents monomer diffusion towards polymer particle.

heterogeneous nucleation and transforms micelles into polymer particles, which are fed with fresh monomer from the large monomer droplets. In addition, new living macromolecules can enter the polymer particles or species can exit and reenter. This results in a very complex polymerization process. A further complication is that the monomer diffusivity in the aqueous phase also plays a role, and at one point the monomer reservoirs become depleted.

From a mathematical point of view, limiting cases of macroemulsion polymerization are mini- and microemulsion polymerizations. In miniemulsion polymerization, only small monomer droplets are present and these are also the main reaction locus. In microemulsion polymerization, the monomer droplets are also small and, in principle, reaction can take place in the monomer droplets as well as in the micelles and polymer particles. An important feature of a microemulsion is that it is thermodynamically stable, whereas the other emulsion types are only kinetically stable. However, if monomer is added very slowly and a small amount of surfactant is present, polymer particles gradually swell starting from a micelle population only. Thus, the emulsion polymerizations differ with respect to the populations present, but in all cases the latex obtained consists of segregated entities with a size at least one order of magnitude smaller than in suspension polymerization. In the most complicated situation, monomer droplets, monomer-swollen micelles, and polymer particles are present simultaneously, and initiation and propagation can occur in the continuous water phase and in the oil phase.

The kinetic behavior in each segregated entity can be different in view of the random nature of exit and entry phenomena and the nanometer scale of these identities, that is, a deviation from bulk kinetics ("one big droplet") is to be expected. Hence, for emulsion polymerization, it is crucial to track the number of low-abundant (radical) species per segregated entity, as compartmentalization of radical species may influence the overall kinetics and thus the development of the polymer microstructure. If u radical types are present, this implies the calculation of the number of segregated entities characterized by u indices, with each index reflecting the discrete presence of one radical type.

In this section, the concept of compartmentalization is illustrated by assuming the presence of one segregated entity, which for simplicity is referred to as a polymer particle. Macroscale effects are also neglected for simplicity. The number of polymer particles N_p is taken to be constant, as well as the particle diameter d_p and particle volume V_p. A distinction is made between the calculation of the polymerization rate and the CLD characteristics as a function of polymerization time, on the one hand, and a free radical polymerization (FRP) and CRP (NMP) reaction scheme on the other hand.

$$I_{2,w} \xrightarrow{\ k_1\ } 2R_{0,w}$$

$$R_{0,w} \xrightarrow{\ k_{entry}\ } R_{0,p}$$

$$R_{i,p} \xrightarrow{\ k_{exit}\ } R_{0,w}$$

$$R_{i,p} + M_p \xrightarrow{\ k_2\ } R_{i+1,p}$$

$$R_{i,p} + R_{j,p} \xrightarrow{\ k_3\ } P_{i,p} + P_{j,p}$$

Scheme 10.4

Simplified FRP reaction scheme for the calculation of the polymerization rate in a simplified macroemulsion process consisting of polymer particles only; the initiator efficiency is assumed to be 100%; w, water phase; p, polymer particle.

10.4.2.1 Calculation of polymerization rate

A crucial distribution for the polymerization rate is the radical number distribution, which describes the number of polymer particles with a given number of radicals ($N_{p(n)}$, $n > 0$). The average number of free radicals per polymer particle is defined by (Asua, 2007)

$$\bar{n} = \frac{\sum_{n=0}^{\infty} n N_{p(n)}}{\sum_{n=0}^{\infty} N_{p(n)}} \tag{10.89}$$

The polymerization rate with respect to the total volume of the reactor V_r is subsequently given by

$$r_p = k_2 [M]_p \frac{\bar{n} N_p}{N_A V_r} \tag{10.90}$$

in which $[M]_p$ is the monomer concentration per particle, which is assumed to be known, and k_2 is the propagation rate coefficient (Scheme 10.4).

In case it is assumed that radical initiation occurs in the water phase and radicals can only enter (absorb), exit (desorb), propagate, and terminate by disproportionation (Scheme 10.4), and chain length dependencies can be neglected, $N_{p(n)}$ can be calculated using the so-called Smith-Ewart (Smith and Ewart, 1948) equations ($n \geq 0$):

$$\frac{dN_{p(n)}}{dt} = (1 - \delta(n)) k_{entry} [R_w] N_{p(n+1)} + k_{exit} (n+1) N_{p(n+1)}$$
$$+ \frac{k_{to}}{N_A V_p} (n+2)(n+1) N_{p(n+2)} - k_{entry} [R_w] N_{p(n)} \tag{10.91}$$
$$- k_{exit} n N_{p(n)} - \frac{k_{to}}{N_A V_p} n(n-1) N_{p(n)}$$

$$\frac{d[R_w]}{dt} = 2 k_{dis} [I_{2w}] + k_{exit} \bar{n} \frac{N_p}{N_A V_w} - 2 k_{tw} [R_w]^2 - k_{entry} [R_w] \frac{N_p}{N_A V_w} \tag{10.92}$$

$$\frac{d[I_{2w}]}{dt} = -k_{dis} [I_{2w}] \tag{10.93}$$

where k_{entry} is the entry or absorption coefficient, k_{exit} is the exit or desorption coefficient, k_{dis} is the dissociation rate coefficient, and $k_{to/w}$ is the termination rate coefficient in the oil/water phase. Furthermore, V_w is the volume of the water phase, $[I_{2w}]$ is the concentration of conventional radical initiator I_2 in the water phase, and $[R_w]$ is the total concentration of radicals in the water phase.

Note that in order to solve Eqs. (10.91)–(10.93), the concentrations of nonradical components in the individual particles do not have to be known. In contrast, for CRP these concentrations do have to be known and multidimensional Smith-Ewart equations are needed to accurately describe the polymerization kinetics. Based on Scheme 10.2 with an oil-soluble NMP initiator R_0X, it follows that if a differentiation is made between the number of NMP initiator radicals n_0, the number of macroradicals n (as before), and the number of nitroxide persistent species n_X per polymer particle, the following contribution results for the activation of the NMP initiator (Bentein et al., 2012):

$$\frac{dN_{p(n_0, n, n_X)}}{dt} = N_A V_p k_{act}[R_0X]_p \left(N_{p(n_0-1, n, n_X-1)} - N_{p(n_0, n, n_X)} \right) \qquad (10.94)$$

in which the NMP initiator radical (R_0X) concentration is assumed to be the same for each polymer particle. The actual value of this concentration is adapted per time step based on all polymer particles and subsequent uniformization over all polymer particles:

$$\frac{d[R_0X]_p}{dt} = \sum_{n_0, n, n_X} \frac{k_{deact}}{N_p (N_A V_p)^2} n_0 n_X N_{p(n_0, n, n_X)} - k_{act}[R_0X]_p \qquad (10.95)$$

As for the microscale models, typically linear multistep methods are used for the numerical integration of the obtained set of equations. Alternatively, a kMC approach can be selected.

10.4.2.2 Calculation of average chain length characteristics

Theoretically, the description of the evolution of the CLD for an emulsion polymerization process is complex as not only the number of radicals has to be tracked per particle but also the chain lengths of these radicals. For FRP, however, the computational effort can be reduced in case a so-called zero-one system is obtained, in which the polymer particles either contain no radicals or only one radical at a given time.

In general, for emulsion polymerizations, mostly only average characteristics of the CLDs are assessed implying the approximate calculation of higher order moments starting from the zeroth-order moments as obtained based on the calculation of the polymerization rate (Eq. 10.90). Typically the moments are expressed as the total amount of moles for all polymer particles together and not as a concentration (per particle). In this chapter, the superscript $'$ is used to stress this change in dimension to moles. For example, for the NMP case (Scheme 10.2 with oil-soluble NMP initiator), the zeroth-order moment on an overall molar basis for the macroradical population is directly given by (Smith and Ewart, 1948)

$$\lambda'_0 = \bar{n} \frac{N_p}{N_A} \qquad (10.96)$$

Assuming semibulk kinetics, the following equation contributions result for the change of the overall number of moles of radicals with a chain length i considering activation and termination by disproportionation as reaction possibilities (Smith and Ewart, 1948):

$$\frac{dn_{R_i}}{dt} = k_{act} n_{R_iX} - k_{td}[R_i]_p \lambda'_0 + \dots \qquad (10.97)$$

in which n_{R_i} and n_{R_iX} are the total number of moles of radicals and dormant species with chain length i. Applying a QSSA, the higher order moment equations (on an overall molar basis) can be rewritten as follows:

$$\lambda_1' = \frac{k_{\text{act}} \tau_1' + \cdots}{k_{\text{td}} \lambda_0' + \cdots} \tag{10.98}$$

$$\lambda_2' = \frac{k_{\text{act}} \tau_2' + \cdots}{k_{\text{td}} \lambda_0' + \cdots} \tag{10.99}$$

For the dormant and dead polymer molecules, the higher moment equations also have to be written on an overall molar basis but no QSSA can be applied.

10.5 Extension Toward Heterogeneous Polymerization With Solid Catalysts

Besides polymerization in dispersed media as described in Section 10.4, another form of heterogeneous polymerization involves the use of a solid catalyst. The most commonly known examples are Ziegler-Natta and Phillips catalysts for the production of polyethylene products (Tobita and Yanase, 2007). In this section, the most frequently applied modeling approaches for the calculation of the polymer microstructure in such polymerization processes are highlighted, neglecting macroscale effects for simplicity. For a more detailed description, the reader is referred to Asua (2007) and Tobita and Yanase (2007).

In the simplest case, the mass CLD of a heterogeneous polymerization, which is defined as the mass fraction of polymer molecules with a given chain length i, is represented by a weighted sum over different Flory distributions (Flory, 1953), each corresponding to one type of catalyst site:

$$m_i = \sum_{j=1}^{N_d} m_{j,\text{CLD}} \frac{i}{x_{n,j}^2} \exp\left(-\frac{i}{x_{n,j}}\right) \tag{10.100}$$

in which N_d is the number of types of catalyst sites and $m_{j,\text{CLD}}$ is the mass fraction of the CLD for the jth catalyst site and with number average chain length $x_{n,j}$. In practice, the individual average chain lengths are determined via regression to experimental data of the total CLD. Note that this method can be seen as an extended method of moments in which the CLD is reconstructed based on the individual number average chain lengths.

For copolymerizations, the Flory distribution is extended to the so-called Stockmayer distribution (Soares and McKenna, 2012; Stockmayer, 1945). For a single CLD for a copolymerization involving the comonomers A and B, the corresponding copolymer composition CLD is defined by

$$m_{i,y} = \frac{i}{x_n^2} \exp\left(-\frac{i}{x_n}\right) \sqrt{\frac{i}{2\pi\kappa}} \exp\left(-\frac{iy^2}{2\kappa}\right) \tag{10.101}$$

with

$$y = F_{\text{pA}} - F_{\text{pA}}^{\text{inst}} \tag{10.102}$$

$$\kappa = F_{pA}^{inst}\left(1 - F_{pA}^{inst}\right)\sqrt{1 - 4F_{pA}^{inst}\left(1 - F_{pA}^{inst}\right)(1 - r_A r_B)} \tag{10.103}$$

in which F_{pA}^{inst} is the instantaneous composition based on the propagation rates using the comonomer feed concentrations, F_{pA} is the composition of a given polymer chain (in terms of the mole fraction of A units in the copolymer), and r_A and r_B are the monomer reactivity ratios:

$$r_A = \frac{k_{pAA}}{k_{pAB}} \tag{10.104}$$

$$r_B = \frac{k_{pBB}}{k_{pBA}} \tag{10.105}$$

in which $k_{pM_1M_2}$ is the propagation rate coefficient for the addition of a living macrospecies ending in a monomer unit M_1 to a monomer of type M_2.

The following distribution has been proposed for a single CLD in case long-chain branches can be formed (Asua, 2007):

$$m(i, z_{br}) = \frac{1}{(2z_{br} + 1)!} i^{2z_{br} + 1} \frac{1}{(x_n)^{2z_{br} + 1}} \exp\left(-\frac{i}{x_n}\right) \tag{10.106}$$

with z_{br} the number of long-chain branches considered. For a z_{br} equal to zero, the Flory distribution is obtained.

More recently, so-called multigrain models have been developed (Asua, 2007; Tobita and Yanase, 2007) to describe polymerizations with solid catalysts in a more detailed manner, including a clear link between the micro- and mesoscale. In such models, the polymer growth and the possibility to form radial concentration and temperature gradients are accounted for. The catalyst particle is seen as an agglomeration of macrograins, which in turn are composed of micrograins, as depicted in Fig. 10.17. The catalyst sites are assumed to be present at the surface of the catalyst particle.

First, the monomer concentration profile in the macrograin ($M_a[r_{mac}, t]$) can be calculated based on

$$\varepsilon_p \frac{\partial[M_{mac}]}{\partial t} = \frac{1}{r_{mac}^2} \frac{\partial}{\partial r_{mac}}\left(D_{eff,M} r_{mac}^2 \frac{\partial[M_{mac}]}{\partial r_{mac}}\right) - R_{M,mic} \tag{10.107}$$

in which the particles are assumed to be spherical, ε_p is the porosity, r_{mac} is the radial position with respect to the center of the macrograin, $R_{M,mic}$ is the monomer disappearance rate toward the micrograins, and $D_{eff,M}$ is the effective monomer diffusion coefficient, as defined by

$$D_{eff,M} = \frac{D_{b,M}\varepsilon_p}{\tau_p} \tag{10.108}$$

in which $D_{b,M}$ is the bulk monomer diffusivity and τ_p is the tortuosity of the catalyst particle.

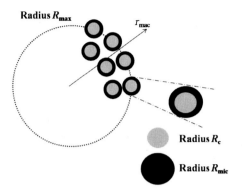

Fig. 10.17
Principle of a multigrain model for the description of radial temperature and concentration gradients in heterogeneous polymerizations with solid catalysts. A macrograin (radius R_{mac}) consists of micrograins (radius R_{mic} with polymer layer, R_c without polymer layer) in concentric layers surrounded by the gas phase.

As initial condition, it is typically assumed that the concentration profile is constant:

$$[M_{mac}](r_{mac}, 0) = [M_{mac}]_0 \qquad (10.109)$$

Furthermore, the following boundary conditions can be used:

$$\frac{\partial[M_{mac}]}{\partial r_{mac}}(0, t) = 0 \qquad (10.110)$$

$$D_{eff,M}\frac{\partial[M_{mac}]}{\partial r_{mac}}(R_{mac}, t) = k_{tf}([M_b] - [M_{mac}]) \qquad (10.111)$$

in which R_{mac} is the radius of the macrograin, k_{tf} is a mass transfer coefficient, and $[M_b]$ is the bulk monomer concentration.

For the micrograins, the analogous equation is

$$\frac{\partial[M_{mic}]}{\partial t} = \frac{1}{r_{mic}^2}\frac{\partial}{\partial r_{mic}}\left(D_{p,M}r_{mic}^2\frac{\partial[M_{mic}]}{\partial r_{mic}}\right) \qquad (10.112)$$

in which r_{mic} is the radial position with respect to the microparticle center, $D_{p,M}$ is the diffusivity of the monomer in the polymer layer around the microparticle of radius R_{mic} and original size R_c (surface location). $D_{p,M}$ is given by the ratio of the diffusivity of the monomer in the amorphous phase and a correction factor for the presence of crystallites. The following boundary conditions are used:

$$[M_{mic}](R_{mic}, t) = [M_{eq}] \qquad (10.113)$$

$$4\pi R_c^2 D_{p,M}\frac{\partial[M_{mic}]}{\partial r_{mic}}(R_c, t) = \frac{4}{3}\pi R_c^3 R_{M,mic}^V \qquad (10.114)$$

in which $R_{M,mic}^V$ is the monomer consumption rate per unit volume of micrograin particle, which is linked to the microkinetics, and $[M_{mic}]_{eq}$ is the equilibrium concentration, which can be linked to the macrograin concentration through a partition coefficient. As initial condition, again a constant profile is typically selected:

$$[M_{mic}](r_{mic}, 0) = [M_{mic}]_0 \qquad (10.115)$$

For the temperature profile, analogous equations can be derived (Asua, 2007; Tobita and Yanase, 2007).

10.6 Conclusions

Several numerical tools have been developed and successfully applied to model the polymer microstructure as a function of monomer conversion and process conditions. In this chapter, these tools have been reviewed and applied to model the polymer microstructure for both radical and catalytic chain-growth polymerizations, which are mechanistically the most important polymerization processes. Attention is focused both on the accurate simulation of the CLD and the PSD, following a multiscale modeling approach. The numerical tools discussed can be applied also for step-growth polymerizations provided that the model parameters are available or can be determined experimentally.

A main distinction has been made between deterministic and stochastic modeling techniques. A further distinction has been proposed based on the scale for which the mathematical model must be derived (eg, micro-, meso-, and/or macroscale). Notably, the complexity of the model approach depends on the desired model output. Detailed microstructural information is only accessible using advanced modeling tools but these are associated with an increase high in computational cost. The advanced models allow one to directly relate macroscopic properties to the polymer synthesis procedure and, thus, to broaden the application market for polymer products, based on a fundamental understanding of the polymerization kinetics and their link with polymer processing.

Nomenclature

A, B	comonomer type
A_{pivot}	function predefined by Eq. (10.38)
arg	short notation for part of population balance to obtain $n_{imp/circ}(V,t)$
a_k	kth expansion coefficient for Galerkin formulation
B'	representative matrix for polymerization kinetics
c_p	specific heat capacity at constant pressure ($J\,kg^{-1}\,K^{-1}$)
$D_{b,M}$	bulk monomer diffusivity ($m^2\,s^{-1}$)
$D_{eff,M}$	effective monomer diffusion coefficient ($m^2\,s^{-1}$)
$D_{p,M}$	diffusivity of the monomer in the polymer layer ($m^2\,s^{-1}$)

F_{pA}^{inst}	instantaneous copolymer composition (mole fraction of A units)
F_{pA}	composition of copolymer chain (mole fraction of A units)
f_l	lth recycle ratio
$G_{a,Y}$	pgf ($a = 0,1$: number, mass; Y: R, P: macromolecule type) of the transformed (complex) z variable
$g(U)$	breakage coefficient for a droplet with volume U (s^{-1})
i, j	chain length
i^*	randomly selected chain length
i_k	kth pivot chain length
k	vector with rate coefficients (dependent on reaction order)
$k(U,V)$	coalescence coefficient of droplets with volumes U and V (m^3 s^{-1})
k^*	adjusted rate coefficient to allow for derivation of moment equations involving termination by recombination (m^3 mol^{-1} s^{-1})
k_{act}	activation rate coefficient in NMP (s^{-1})
k_{app}	apparent rate coefficient (dependent on reaction order)
k_{chem}	intrinsic rate coefficient (dependent on reaction order)
k_r	rate coefficient for reaction r (dependent on reaction order)
k_{tf}	mass transfer coefficient (m s^{-1})
k_ν	"macroscopic" rate coefficient for reaction channel ν (dependent on reaction order)
k_ν^{MC}	"kinetic MC rate coefficient" for reaction channel ν (s^{-1})
l^{**}	length of branch (Tobita-based MC)
l_k	basis functions for Galerkin formulation
l_n	characteristic length of branch (Tobita-based MC)
m_i	mass fraction of polymer chains with chain length i; if also index y: relates to copolymer composition CLD
$m_{j,CLD}$	mass fraction of the CLD for the jth catalyst site
N_A	Avogadro's number (mol^{-1})
N_{br}	number of branching points/branches
N_c	number of tracked chains for matrix-based kinetic MC
$N_{imp,k}$	contribution to the density function for the interval k of the droplet size domain for the fixed pivot method (m^{-6})
N_k	kth chain length domain for the fixed pivot method
N_m	number of molecules for mth species type in kinetic MC simulations in V
N_p	number of polymer particles
$N_{p(n)}$	number of polymer particles with n radicals
$N_{p(n_0,n,n_X)}$	number of polymer particles with n_0 initiator radicals, n radicals, and n_X nitroxide radicals
N_r	number of reactions for kinetic MC simulations
N_{sp}	number of different species

N_{tr}	truncation number for Galerkin formulation
n	number of macroradicals per polymer particle
n_0	number of NMP initiator radicals per polymer particle
n_{R_i}	overall amount of radicals with chain length i (mol)
n_{R_iX}	overall amount of dormant species with chain length i (mol)
n_X	number of nitroxide persistent species per polymer particle
$n_{circ}(V,t)$	number density function for circulation compartment (m^{-6})
$n_{imp}(V,t)$	number density function for impeller compartment (m^{-6})
$\bar{n}$	average number of free radicals per polymer particle
P	dead species
$P_{a,Y}$	pgf ($a=0,1$: number, mass; Y: R, P: type of macromolecule)
P_ν^{MC}	MC reaction probability for reaction channel ν
PB_{imp}^{rhs}	right-hand side of population balance for impeller compartment
$p(xx^* \mid x)$	conditional probability for branching density (Tobita-based MC)
p_1	parameter in the Wesslau distribution
$Q_{ri,k}$	heat produced by reaction for segment k in compartment i (W)
$Q_{i,k}$	heat contribution via exit flow from segment k in compartment i (W)
Q_0	heat contribution via feed flow (W)
q	parameter in Schulz-Flory weight function for Galerkin formulation
$q_{V,e}$	exchange flow rate (m^3 s^{-1})
$q_{Vi,k}$	volumetric exit flow rate for segment k in compartment i (m^3 s^{-1})
q_{V0}	volumetric feed flow rate (m^3 s^{-1})
R_c	radius of microparticle without polymer layer (m)
$R_{Mi,k}$	net monomer production rate for segment k in compartment i (mol m^{-3} s^{-1})
R_{mac}	radius of macrograin (m)
R_{mic}	radius of micrograin with polymer layer (m)
$R_{\mu 0,i,k}$	net production rate of dead polymer for segment k in compartment i (mol m^{-3} s^{-1})
$R_{M,mic}$	monomer disappearance rate toward the micrograins per unit volume of macrograin (mol m^{-3} s^{-1})
$R_{M,mic}^V$	monomer consumption rate per unit volume of micrograin (mol m^{-3} s^{-1})
r_1, r_2	random numbers for kinetic MC simulations
r_A, r_B	reactivity ratios for propagation in copolymerization
r_{mac}	radial position with respect to the center of macrograin (m)
r_{mic}	radial position with respect to center of micrograin (m)
r_p	polymerization rate per unit volume of reactor (mol m^{-3} s^{-1})
r_ν^{MC}	MC reaction rate for reaction channel ν
SD(z)	structural deviation of chain z with respect to TM; minimum of four contributions SD$_k$ ($k=1,...,4$)
$S_x(y,z)$	cumulative number of monomer units of type x at position y in chain z (chain length: i) of the kMC sample as counted from "left to right"

$S_{x,\mathrm{TM},r}(y,z)$	cumulative number of monomer units of type x at position y of the targeted chain in the direction r (always defined from "left to right") with the same length as the considered chain z as counted from "left to right"
$S'_x(y,z)$	cumulative number of monomer units of type x at position y in chain z (chain length: i) of the kMC sample as counted from "right to left"
$S'_{x,\mathrm{TM},r}(y,z)$	cumulative number of monomer units of type x at position y of the targeted chain in the direction r (always defined from "left to right") with the same length as the considered chain z as counted from "right to left"
T	temperature (K)
U	volume of droplet (m^3)
$u(U)$	number of droplets formed upon breakage of a droplet with volume U
V	volume (m^3)
V_{max}	indicates border of size domain for droplets in fixed pivot method (m^3)
V_{p}	particle volume (m^3)
V_{r}	reactor volume (m^3)
X	nitroxide
x_s	sth order average chain length; $s = 1, 2, 3$ or n, m, z
x^*	birth conversion (Tobita-based MC)
x^{**}	birth conversion of branch (Tobita-based MC)
x^{***}	birth conversion for branch created in an earlier stage
y	parameter for Stockmayer distribution
z	complex variable (pgf method) or chain number (matrix-based kinetic MC)
z_{br}	number of long-chain branches
[A]	concentration of nonmacromolecule A (mol m^{-3})
[I$_2$]	concentration of radical initiator (mol m^{-3})
[M]	concentration of monomer (mol m^{-3})
[P$_i$]	concentration of dead polymer chain with chain length i (mol m^{-3})
[P$^*_k$]	concentration of dead polymer chains in the chain length domain k (mol m^{-3})
[R$_0$]	concentration of (NMP) initiator radical (mol m^{-3})
[R$_i$]	concentration of living polymer chain with chain length i (mol m^{-3})
[R$_i$X]	concentration of dormant polymer chain with chain length i (mol m^{-3})
[R$^*_k$]	concentration of living polymer chains in the chain length domain k (mol m^{-3})
[R$_{\mathrm{w}}$]	total concentration of radicals in the water phase (mol m^{-3})
$<k,s>$	sth order population-weighted rate coefficient (mol m^{-3} s^{-1})
$<\mathrm{SD}>$	average structural deviation

Greek symbols

$\beta(U,V)$	probability that a droplet with a volume U breaks into a droplet with a volume V
γ	orthogonality factor

$\Delta_r H$	enthalpy change of reaction (J mol^{-1})
Δ_k	size of kth chain length domain in fixed pivot method
Δt	time step (s)
δ	Dirac delta function
ε_p	porosity of catalyst particle
κ	parameter for Stockmayer distribution
μ,λ,τ	moments for dead, living, and dormant macrospecies (mol m^{-3})
μ',λ',τ'	rescaled moments for dead, living, and dormant macrospecies for all polymer particles (mol)
μ^{MC}	MC reaction channel
ρ	density (kg m^{-3})
ρ_{br}	predefined branching probability
σ	parameter in the Wesslau distribution
τ^{MC}	MC time step (s)
τ_p	tortuosity of catalyst particle

Subscripts

0	initial/feed (compartment model)
b	bulk
act	activation
circ	circulation
d	FRP in emulsion: desorption
deact	deactivation; 0 if NMP initiator related
dis	dissociation
entry	denotes entry of radicals in polymer particles
eq	equilibrium
exit	denotes exit of radicals from polymer particles
i,k	segment k of compartment i
imp	impeller
k	domain k
mac	macrograin
max	maximum
mic	micrograin
o	oil phase
p	particle
p	propagation
n,m,z	number, mass, z order for average of CLD
s'	order
tc	termination by recombination
td	termination by disproportionation

tp	(chain) transfer to polymer
trM	chain transfer to monomer
w	water phase

Superscripts

| MC | Monte Carlo |

Abbreviations

CFD	computational fluid dynamics
CLD	chain length distribution
CRP	controlled radical polymerization
FRP	free radical polymerization
NMP	nitroxide-mediated polymerization
MC	Monte Carlo
MDSD	monomer droplet size distribution
PSD	particle size distribution
pgf	probability generating function
QSSA	quasi-steady-state approximation
TM	targeted microstructure

Acknowledgments

This work is supported by the Long Term Structural Methusalem Funding by the Flemish government and the Interuniversity Attraction Poles Programme—Belgian State—Belgian Science Policy. D.R.D. acknowledges the Fund for Scientific Research Flanders (FWO) through a postdoctoral fellowship.

References

Achilias, D.S., 2007. A review of modeling of diffusion controlled polymerization reactions. Macromol. Theory Simul. 16, 319–347.

Achilias, D., Kiparissides, C., 1988. Modeling of diffusion-controlled free-radical polymerization reactions. J. Appl. Polym. Sci. 35, 1303–1323.

Alexopoulos, A.H., Kiparissides, C., 2005. Part II: dynamic evolution of the particle size distribution in particulate processes undergoing simultaneous particle nucleation, growth and aggregation. Chem. Eng. Sci. 60, 4157–4169.

Alexopoulos, A.H., Maggioris, D., Kiparissides, C., 2002. CFD analysis of turbulence non-homogeneity in mixing vessels: a two-compartment model. Chem. Eng. Sci. 57, 1735–1752.

Asteasuain, M., Sarmoria, C., Brandolin, A., 2002a. Recovery of molecular weight distributions from transformed domains. Part I. Application of pgf to mass balances describing reactions involving free radicals. Polymer 43, 2513–2527.

Asteasuain, M., Sarmoria, C., Brandolin, A., 2002b. Recovery of molecular weight distributions from transformed domains. Part II. Application of numerical inversion methods. Polymer 43, 2529–2541.

Asteasuain, M., Sarmoria, C., Brandolin, A., 2004. Molecular weight distributions in styrene polymerization with asymmetric bifunctional initiators. Polymer 45, 321–335.

Asua, J.M., 2007. Polymer Reaction Engineering. Blackwell Publishing, Oxford.

Asua, J.M., De La Cal, J.C., 1991. Entry and exit rate coefficients in emulsion polymerization of styrene. J. Appl. Polym. Sci. 42, 1869–1877.

Baltsas, A., Achilias, D.S., Kiparissides, C., 1996. A theoretical investigation of the production of branched copolymers in continuous stirred tank reactors. Macromol. Theory Simul. 5, 477–497.

Bentein, L., D'hooge, D.R., Reyniers, M.F., Marin, G.B., 2011. Kinetic modeling as a tool to understand and improve the nitroxide mediated polymerization of styrene. Macromol. Theory Simul. 20, 238–265.

Bentein, L., D'hooge, D.R., Reyniers, M.-F., Marin, G.B., 2012. Kinetic modeling of miniemulsion nitroxide mediated polymerization of styrene: effect of particle diameter and nitroxide partitioning up to high conversion. Polymer 53, 681–693.

Budde, U., Wulkow, M., 1991. Computation of molecular weight distributions for free-radical polymerization systems. Chem. Eng. Sci. 46, 497–508.

Butté, A., Storti, G., Morbidelli, M., 2002. Evaluation of the chain length distribution in free-radical polymerization, 1. Bulk polymerization. Macromol. Theory Simul. 11, 22–36.

Chaffey-Millar, H., Stewart, D., Chakravarty, M.M.T., Keller, G., Barner-Kowollik, C., 2007. A parallelised high performance Monte Carlo simulation approach for complex polymerisation kinetics. Macromol. Theory Simul. 16, 575–592.

Cunningham, M.F., 2008. Controlled/living radical polymerization in aqueous dispersed systems. Prog. Polym. Sci. 24, 365–398.

De Roo, T., Wieme, J., Heynderickx, G.J., Marin, G.B., 2005. Estimation of intrinsic rate coefficients in vinyl chloride suspension polymerization. Polymer 45, 8340–8354.

D'hooge, D.R., Reyniers, M.F., Marin, G.B., 2013. The crucial role of diffusional limitations in controlled radical polymerization. Macromol. React. Eng. 7, 362–379.

Flory, P.L., 1953. Principles of Polymer Chemistry. Cornell University Press, Ithaca, NY.

Fox, R.O., 1996. Computational methods for turbulent reacting flows in chemical process industry. Rev. Inst. Franç. Pétrole 51, 215–243.

Gilbert, R.G., Hess, M., Jenkins, A.D., Jones, R.G., Kratochvil, P., Stepto, R.F.T., 2009. Dispersity in polymer science. Pure Appl. Chem. 81, 351–353.

Gillespie, D.T., 1977. Exact stochastic simulation of coupled chemical reactions. J. Phys. Chem. 81, 2340–2361.

Hulburt, H.M., Katz, S., 1964. Some problems in particle technology. A statistical mechanical formulation. Chem. Eng. Sci. 19, 555–574.

Iedema, P.D., Grcev, S., Hoefsloot, H.C.J., 2003. Molecular weight distribution modeling of radical polymerization in a CSTR with long chain branching through transfer to polymer and terminal double bond (TDB) propagation. Macromolecules 36, 458–476.

Kolhapure, N.H., Fox, R.O., 1999. CFD analysis of micromixing effects on polymerization in tubular low-density polyethylene reactors. Chem. Eng. Sci. 54, 3233–3242.

Konkolewicz, D., Sosnowski, S., D'hooge, D.R., Szymanski, R., Reyniers, M.F., Marin, G.B., Matyjaszewski, K., 2011. Origin of the difference between branching in acrylates polymerization under controlled and free radical conditions: a computational study of competitive processes. Macromolecules 44, 8361–8373.

Kotoulas, C., Kiparissides, C., 2006. A generalized population balance model for the prediction of particle size distribution in suspension polymerization reactors. Chem. Eng. Sci. 61, 332–346.

Kumar, S., Ramkrishna, D., 1996. On the solution of population balance equations by discretization—I. A fixed pivot technique. Chem. Eng. Sci. 51, 1311–1332.

Malmström, E.E., Hawker, C.J., 1998. Macromolecular engineering via 'living' free radical polymerizations. Macromol. Chem. Phys. 199, 923–935.

Matyaszewski, K., Sumerlin, B.S., Tsarevsky, N.V., 2012. Progress in Controlled Radical Polymerization: Materials and Applications. American Chemical Society, Washington, DC.

Matyjaszwski, K., Xia, J., 2001. Atom transfer radical polymerization. Chem. Rev. 101, 2921–2990.

McKenna, T.F., Soares, J.P.B., 2001. Single particle modeling for olefin polymerization on supported catalysts: a review and proposals for future developments. Chem. Eng. Sci. 56, 3931–3949.

Meyer, T., Keurentjes, J., 2005. Handbook of Polymer Reaction Engineering, vol. 1. Wiley-VCH, Weinheim.

Moad, G., Solomon, D.H., 2006. The Chemistry of Radical Polymerization, second ed. Elsevier, Oxford.

Petzold, L.R., 1983. Automatic selection of methods for solving stiff and nonstiff systems of ordinary differential equations. SIAM J. Sci. Stat. Comput. 4, 136–148.

Pladis, P., Kiparissides, C., 1988. A comprehensive model for the calculation of molecular weight long-chain branching distribution in free-radical polymerizations. Chem. Eng. Sci. 53, 3315–3333.

Pladis, P., Kiparissides, C., 1999. Dynamic modeling of multizone, multifeed high-pressure LDPE autoclaves. J. Appl. Polym. Sci. 73, 2327–2348.

Pope, S.B., 2000. Turbulent Flows. Cambridge University Press, Cambridge.

Richards, J.R., Congalidis, J.P., 2006. Measurement and control of polymerization reactors. Comput. Chem. Eng. 30, 1447–1463.

Roudsari, S.F., Ein-Mozaffari, F., Dhib, R., 2013. Use of CFD in modeling MMA solution polymerization in a CSTR. Chem. Eng. J. 219, 429–442.

Russell, G.T., 1994. On exact and approximate methods of calculating an overall termination rate coefficient from chain length dependent termination rate coefficients. Macromol. Theory Simul. 3, 439–468.

Smith, W.V., Ewart, R.H., 1948. Kinetics of emulsion polymerization. Chem. Phys. 16, 592–599.

Soares, J.P.B., Hamielec, A.E., 1995. Deconvolution of chain-length distributions of linear polymers made by multiple-site-type catalysts. Polymer 36, 2257–22263.

Soares, J.P.B., McKenna, T.F., 2012. Polyolefin Reaction Engineering. Wiley-VCH, Weinheim.

Stockmayer, W.H.J., 1945. Distribution of chain lengths and compositions in copolymers. Chem. Phys. 13, 199–207.

Tobita, H., 1993. Molecular weight distribution in free radical polymerization with long-chain branching. J. Polym. Sci. B Polym. Phys. 31, 1363–1371.

Tobita, H., Yanase, F., 2007. Monte Carlo simulation of controlled/living radical polymerization in emulsified systems. Macromol. Theory Simul. 16, 476–488.

Toloza Porras, C., D'hooge, D.R., Van Steenberge, P.H.M., Reyniers, M.F., Marin, G.B., 2013. A theoretical exploration of the potential of ICAR ATRP for one- and two-pot synthesis of well-defined diblock copolymers. Macromol. React. Eng. 7, 311–326.

Topalos, E., Pladis, P., Kiparissides, C., Goossens, I., 1996. Dynamic modeling and steady-state multiplicity in high-pressure multizone LDPE autoclaves. Chem. Eng. Sci. 51, 2461–2470.

Van Steenberge, P.H.M., Vandenbergh, J., D'hooge, D.R., Reyniers, M.F., Adriaensens, P.J., Lutsen, L., Vanderzande, D.J.M., Marin, G.B., 2011. Kinetic Monte Carlo modeling of the sulfinyl precursor route for poly (p-phenylene vinylene) synthesis. Macromolecules 44, 8716–8726.

Van Steenberge, P.H.M., D'hooge, D.R., Wang, Y., Zhong, M., Reyniers, M.F., Konkolewicz, D., Matyjaszewski, K., Marin, G.B., 2012. Linear gradient quality of ATRP copolymers. Macromolecules 45, 8519–8531.

Wulkow, M., 1996. The simulation of molecular weight distributions in polyreaction kinetics by discrete Galerkin methods. Macromol. Theory Simul. 5, 393–416.

Zetterlund, P.B., Kagawa, Y., Okubo, M., 2008. Controlled/living radical polymerization in dispersed systems. Chem. Rev. 108, 3747–3794.

Zhang, S.X., Ray, W.H., 1997. Modeling of imperfect mixing and its effects on polymer properties. AIChE J. 43, 1265–1277.

Zhu, S., 1999. Modeling of molecular weight development in atom transfer radical polymerization. Macromol. Theory Simul. 8, 29–37.

## CHAPTER 11

Advanced Theoretical Analysis in Chemical Engineering

Computer Algebra and Symbolic Calculations

11.1 Critical Simplification

The principle of critical simplification was first explained by Yablonsky et al. (Yablonskii and Lazman, 1996; Yablonsky et al., 2003) using the catalytic oxidation reaction as an example. The authors presented a dramatic simplification of the kinetic model for this reaction at critical conditions relating to bifurcation points. In this section, results obtained by Gol'dshtein et al. (2015) are also used.

11.1.1 Model of the Adsorption Mechanism

The simplest mechanism for interpreting critical phenomena in heterogeneous catalysis is the Langmuir adsorption mechanism, also referred to as the Langmuir-Hinshelwood mechanism. This mechanism includes three elementary steps: (1) adsorption of one type of gas molecule on a catalyst active site; (2) adsorption of a different type of gas molecule on another active site; (3) reaction between these two adsorbed species. For the oxidation of carbon monoxide on platinum, this mechanism can be written as follows:

$$
\begin{array}{lllll}
(1) & 2Pt & + & O_2 & \rightleftarrows & 2PtO \\
(2) & Pt & + & CO & \rightleftarrows & PtCO \\
(3) & PtO & + & PtCO & \rightarrow & 2Pt + CO_2
\end{array}
\qquad (11.1)
$$

where Pt is a free active platinum site and PtO and PtCO are platinum sites with adsorbed oxygen and carbon monoxide. The adsorption steps, (1) and (2), are usually reversible, although oxygen adsorption can be considered to be irreversible at low to moderate temperatures.

Advanced Data Analysis and Modelling in Chemical Engineering. http://dx.doi.org/10.1016/B978-0-444-59485-3.00011-4

The kinetic model related to this adsorption mechanism with irreversible adsorption of oxygen can be presented as

$$
\begin{cases}
\dfrac{dx}{dt} = 2r_1^+ - r_3^+ \\[2mm]
\dfrac{dy}{dt} = r_2^+ - r_2^- - r_3^+ \\[2mm]
\dfrac{dz}{dt} = -2r_1^+ - r_3^+ + r_2^- + 2r_3^+
\end{cases}
\tag{11.2}
$$

where x, y, and z are the normalized surface concentrations of respectively adsorbed oxygen and carbon monoxide and of free active sites: $x = \theta_{PtO}, y = \theta_{PtCO}, z = \theta_{Pt}$; r_1^+, r_2^+ and r_3^+ are rates of the forward reactions and r_2^- is the rate of reverse reaction (2) in Eq. (11.1) ($r_1^- = r_3^- = 0$). The reaction rates correspond to the mass-action law: $r_1^+ = 2k_1^+ p_{O_2} z^2, r_2^+ = 2k_2^+ p_{CO} z, r_2^- = k_2^- y, r_3^+ = k_3^+ xy$, where $k_{1,2,3}^{+/-}$ are the kinetic coefficients of the reactions and p_{O_2} and p_{CO} are the partial pressures of oxygen and carbon monoxide. The third row in Eq. (11.2) is a linear combination of the first and second rows.

In addition, the law of mass conservation holds regarding the total normalized surface concentration of platinum sites:

$$
x + y + z = 1 \tag{11.3}
$$

Therefore, the model can be reduced to

$$
\begin{cases}
\dfrac{dx}{dt} = 2k_1^+ p_{O_2} (1 - x - y)^2 - k_3^+ xy \\[2mm]
\dfrac{dy}{dt} = k_2^+ p_{CO} (1 - x - y) - k_2^- y - k_3^+ xy
\end{cases}
\tag{11.4}
$$

From the corresponding steady-state model $\left(\dfrac{dx}{dt} = 0; \dfrac{dy}{dt} = 0\right)$, it follows that

$$
\begin{cases}
2k_1^+ p_{O_2} (1 - x - y)^2 - k_3^+ xy = 0 \\[2mm]
k_2^+ p_{CO} (1 - x - y) - k_2^- y - k_3^+ xy = 0
\end{cases}
\tag{11.5}
$$

The partial pressures p_{O_2} and p_{CO} are considered to be parameters of our model with p_{O_2} constant and $p_{O_2} \gg p_{CO}$ while the steady-state rate dependence on the gas-phase composition is assumed to be monoparametric: $r = f(p_{CO})$.

The set of equations, Eq. (11.5) has one obvious boundary steady-state solution: $x = 1; y = 0; z = 0$, that is, all active Pt sites are occupied by adsorbed oxygen. In order to determine other steady states, we can transform the equations as follows:

$$-2k_1^+ p_{O_2} \left(k_3^+\right)^2 x^3 + \left[2k_1^+ p_{O_2} \left(k_3^+\right)^2 - 4k_1^+ p_{O_2} k_2^- k_3^+ - \left(k_3^+\right)^2 k_2^+ p_{CO}\right] x^2$$
$$+ \left[4k_1^+ p_{O_2} k_2^- k_3^+ - 2k_1^+ p_{O_2} \left(k_2^-\right)^2 - k_2^- k_3^+ k_2^+ p_{CO}\right] x + 2k_1^+ p_{O_2} \left(k_2^-\right)^2 = 0 \tag{11.6}$$

This is a cubic equation in the variable x, the normalized surface concentration of adsorbed oxygen. The corresponding value of y, the normalized surface concentration of adsorbed carbon monoxide, can be calculated from Eq. (11.5).

In a typical Langmuir adsorption mechanism. the elementary reaction between two adsorbed species is very fast. In this case, such a fast reaction occurs between adsorbed oxygen and adsorbed carbon monoxide, that is, $k_3^+ \gg k_1^+ p_{O_2}, k_2^+ p_{CO}, k_2^-$. Consequently, we can introduce a small parameter $\varepsilon = 1/k_3^+$ and write the discriminant of Eq. (11.6) in the following form:

$$\Delta = \left(-2k_1^+ p_{O_2} + k_2^+ p_{CO}\right)^2 \left[\left(k_2^-\right)^2 - 8k_1^+ p_{O_2} k_2^- + 2k_2^+ p_{CO} k_2^- + \left(k_2^+\right)^2 p_{CO}^2\right]$$
$$- \varepsilon 4 k_1^+ p_{O_2} \left\{ 24 \left(k_1^+\right)^2 p_{O_2}^2 \left(k_2^-\right)^2 + k_2^+ p_{CO} \left(k_2^+ p_{CO} + k_2^-\right) \left[2\left(k_2^+\right)^2 p_{CO}^2 + 2k_2^+ p_{CO} k_2^- - \left(k_2^-\right)^2\right] \right.$$
$$\left. - 2k_1^+ p_{O_2} k_2^- \left[10 k_2^+ p_{CO}^2 + k_2^+ p_{CO} k_2^- + \left(k_2^-\right)^2\right] \right\}$$
$$- \varepsilon^2 4 k_1^+ p_{O_2}^2 \left(k_2^+\right)^2 p_{CO}^2 \left(k_2^-\right)^2 \left[-\left(k_2^-\right)^2 + 8\left(k_2^+\right)^2 p_{CO}^2 + 8k_2^- \left(3k_1^+ p_{O_2} + k_2^+ p_{CO}\right)\right]$$
$$- \varepsilon^3 32 \left(k_1^+\right)^3 p_{O_2}^3 \left(k_2^+\right)^2 p_{CO}^2 \left(k_2^-\right)^4. \tag{11.7}$$

In the zero approximation, $\varepsilon = 0$, the expression for the discriminant becomes a very simple product of two terms:

$$\Delta = \left(k_2^+ p_{CO} - 2k_1^+ p_{O_2}\right)^2 \left[\left(k_2^-\right)^2 - 8k_1^+ p_{O_2} k_2^- + 2k_2^+ p_{CO} k_2^- + \left(k_2^+\right)^2 p_{CO}^2\right] \tag{11.8}$$

The corresponding quantities

$$B_1 = 1 - \frac{k_2^+ p_{CO}}{2k_1^+ p_{O_2}} \tag{11.9}$$

and

$$B_2 = \left(k_2^-\right)^2 - 8k_1^+ p_{O_2} k_2^- + 2k_2^+ p_{CO} k_2^- + \left(k_2^+\right)^2 p_{CO}^2 \tag{11.10}$$

will be used as bifurcation parameters. A bifurcation occurs if $B_1 = 0$ or $B_2 = 0$. If $B_1 = B_2 = 0$ then the corresponding bifurcation has maximum complexity (Gol'dshtein et al., 2015).

11.1.2 Transformation of the Langmuir System to a Slow-Fast System

In the example of the oxidation of carbon monoxide, the rate of the fast reaction, $k_3^+ xy$, is present in both equations in Eq. (11.4). The original system does not have any fast-slow division in the (x,y) coordinate system. In accordance with the main concepts of the theory of singularly perturbed vector fields (Bykov et al., 2006, 2008), it is possible to construct another orthogonal coordinate system where the original model becomes a slow-fast one. In the case of the model of the adsorption mechanism, this transformation involves only a simple rotation. We use the following orthogonal coordinate transformation:

$$\begin{cases} u = \dfrac{x-y}{\sqrt{2}} \\ v = \dfrac{x+y}{\sqrt{2}} \end{cases} \tag{11.11}$$

Expressed in the new coordinates, the system has the form

$$\begin{cases} \dfrac{du}{dt} = \sqrt{2}k_1^+ p_{O_2}\left(1-\sqrt{2}v\right)^2 - \dfrac{k_2^+}{\sqrt{2}}p_{CO}\left(1-\sqrt{2}\right)v + \dfrac{k_2^-}{2}(v-u) \\ \varepsilon\dfrac{dv}{dt} = \dfrac{u^2-v^2}{\sqrt{2}} + \dfrac{\varepsilon}{2\sqrt{2}}\left[4k_1^+ p_{O_2}\left(1-\sqrt{2}u\right)^2 + 2k_2^+ p_{CO}\left(1-\sqrt{2}v\right) - \sqrt{2}k_2^-(v-u)\right] \end{cases} \tag{11.12}$$

with small parameter $\varepsilon = 1/k_3^+$.

This is a standard slow-fast system (singularly perturbed system) that combines singular and regular perturbations in the second equation; the singular perturbation induces the small parameter ε before the time derivative of v and this same small parameter, caused by the regular perturbation, appears before the second term on the right-hand side.

11.1.3 Zero Approximation for Singular Perturbation

Now, all the tools of the standard singular perturbation theory can be applied to the analysis of the system. These tools include its decomposition to slow and fast subsystems and reduction of the near steady-state dynamics to dynamics on invariant slow manifolds (slow curves in the present model). The zero approximation, $\varepsilon=0$ ($k_3^+ \to \infty$), of the slow invariant curve is given by setting $\varepsilon\dfrac{dv}{dt}=0$, yielding the following equation for the slow curve

$$u^2 - v^2 + \dfrac{\varepsilon}{2}\left[4k_1^+ p_{O_2}\left(1-\sqrt{2}u\right)^2 + 2k_2^+ p_{CO}\left(1-\sqrt{2}v\right) - \sqrt{2}k_2^-(v-u)\right] = 0 \tag{11.13}$$

Then, putting $\varepsilon=0$ for the regular perturbation simplifies the analysis of steady states dramatically. The approximation of the slow curve is now given by

$$u^2 - v^2 = 0 \tag{11.14}$$

This curve has two branches, $u + v = 0$ and $u - v = 0$, or in the original coordinates, $x = 0$ and $y = 0$, that is, there is no adsorbed oxygen or carbon monoxide present on the catalyst surface. Both branches of the slow invariant curve are stable. The two branches have an intersection point $(0,0)$, which is definitely not a steady-state point of the original system. Therefore, around this point a more accurate approximation is necessary.

11.1.4 First Approximation for Singular Perturbation

The zero approximation, $\varepsilon = 0$, is not informative at the intersection point $(0,0)$ of the two branches $u = v$ and $u = -v$ so the first approximation of the slow curve must be calculated (see Fig. 11.1):

$$u^2 - v^2 + \varepsilon \left[2k_1^+ p_{O_2} \left(1 - \sqrt{2} \right)^2 \frac{v - u}{v} + k_2^+ p_{CO} \left(1 - \sqrt{2}v \right)^2 \frac{u + v}{v} \right] = 0 \qquad (11.15)$$

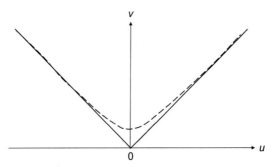

Fig. 11.1

Schematic of the system dynamics. *Solid line*: zero approximation of slow manifold; *dashed line*: first approximation of slow manifold.

We will now analyze the vicinity of the point of intersection $(0,0)$. The first bifurcation parameter in this approximation is

$$B_{1,1} = \left(1 - \frac{k_2^+ p_{CO}}{2k_1^+ p_{O_2}} \right)^2 - \varepsilon \frac{2\left(k_2^+\right)^2 p_{CO}^2}{k_1^+ p_{O_2}} \qquad (11.16)$$

The system has two, three, or four steady states depending on this bifurcation parameter. If $B_{1,1} = 0$, there are three steady states: $(0,1)$, $(1,0)$, and $\left(\frac{1}{2} \sqrt{ \varepsilon \frac{2\left(k_2^+\right)^2 p_{CO}^2}{k_1^+ p_{O_2}} }, \frac{1}{2} \sqrt{ \varepsilon \frac{2\left(k_2^+\right)^2 p_{CO}^2}{k_1^+ p_{O_2}} } \right)$.

The third steady state belongs to the ε-vicinity of $(0,0)$. As $\varepsilon \to 0$, this point disappears from the vicinity of $(0,0)$ and this steady state disappears for $B_{1,1} < 0$.

In the invariant domain, the following inequality holds:

$$2k_1^+ p_{O_2} \left(1 - \sqrt{2}\right)^2 \frac{v-u}{v} + k_2^+ p_{CO} \left(1 - \sqrt{2}v\right)^2 \frac{u+v}{v} > 0 \tag{11.17}$$

From this inequality, we can conclude that there exist nontrivial steady states inside the invariant domain. If $B_{1,1} = 0$, two nontrivial steady states merge, resulting in one steady state, a saddle node. If the initial conditions satisfy the inequality $x < y$, all trajectories attract to the point $(0,1)$. If, on the other hand, the initial condition satisfies the inequality $x > y$, all

trajectories attract to the point $\left(\dfrac{1}{2} \sqrt{\varepsilon \dfrac{2\left(k_2^+\right)^2 p_{CO}^2}{k_1^+ p_{O_2}}}, \dfrac{1}{2} \sqrt{\varepsilon \dfrac{2\left(k_2^+\right)^2 p_{CO}^2}{k_1^+ p_{O_2}}} \right)$.

If $B_1 = 0$ and $B_2 = 0$, the situation is more complex and the standard asymptotic analysis is not applicable at these conditions. Actually, this is a more interesting case because this bifurcation point is the point of maximum bifurcational complexity (MBC). At the same time, it is the point where critical simplification is observed and the following relationship between the kinetic parameters is obtained:

$$2k_1^+ p_{O_2} = k_2^+ p_{CO} = k_2^- \tag{11.18}$$

11.1.5 Bifurcation Parameters

For the branch $u - v = 0$ (or $y = 0$), the bifurcation parameter, B_1, is given by Eq. (11.9). There are only one or two steady states on this branch. The first is $(1,0)$ and does not depend on B_1. The second steady state is $\left(1 - \dfrac{k_2^+ p_{CO}}{2k_1^+ p_{O_2}}, 0 \right)$. If $B_1 > 0$, this steady state is stable. If $B_1 < 0$, the steady state does not belong to the invariant domain U, that is, it has no physical meaning there. At the bifurcation point $B_1 = 0$, so

$$k_2^+ p_{CO} = 2k_1^+ p_{O_2} \tag{11.19}$$

This equation is an example of critical simplification. At the bifurcation point, the kinetic parameters $k_1^+ p_{O_2}$ and $k_2^+ p_{CO}$ are interdependent. Knowing, for example, the partial pressures of carbon monoxide and oxygen at the bifurcation point and the adsorption coefficient of carbon monoxide (k_2^+), we can determine the adsorption coefficient of oxygen (k_1^+).

A change of sign of B_1 from positive to negative corresponds to extinction. For the branch $u + v = 0$ (or $x = 0$), the bifurcation parameter, B_2, is given by Eq. (11.10). The bifurcation parameters B_1 and B_2 are not independent. By a simple calculation, we obtain

$$\begin{aligned} B_2 &= \left(k_2^-\right)^2 + 2k_2^- \left(-k_2^+ p_{CO} - 4k_1^+ p_{O_2} B_1\right) + \left(k_2^+\right)^2 p_{CO}^2 \\ &= \left(k_2^+ p_{CO} - k_2^-\right)^2 - 8k_2^- k_1^+ p_{O_2} B_1 \end{aligned} \tag{11.20}$$

From this relation, it follows that if $B_1 < 0$, then $B_2 > 0$ and if $B_1 = 0$, then $B_2 = \left(k_2^+ p_{CO} - k_2^-\right)^2 \geq 0$.

The situation $B_1 < 0$, $B_2 < 0$ is not possible.

There are zero, one, or two steady states on this branch, depending on the sign of B_2. If $B_2 < 0$, there are no steady states; if $B_2 = 0$, there is one steady state; and if $B_2 > 0$, there are two steady states with $x = 0$. Thus, at the bifurcation point $B_2 = 0$, a functional relation exists between the parameters $k_1^+ p_{O_2}, k_2^+ p_{CO}$ and k_1^-. This is another critical simplification. A change of sign of B_2 from positive to negative corresponds to ignition.

11.1.6 Calculation of Reaction Rates

Steady-state reaction rates can be calculated on both branches of the slow curve taking into account the following relationships:

$$r_3 = r_3^+ = 2r_1^+ = 2r_1 \tag{11.21}$$

$$r_3 = r_3^+ = r_2^+ - r_2^- = r_2 \tag{11.22}$$

Then the steady-state rate of the overall reaction equals

$$r = r_3 = r_2 = 2r_1 \tag{11.23}$$

We use the following standard representation of the steady-state coordinates x_{ss} and y_{ss}, based on the regular perturbation theory: $(x_{ss}, y_{ss}) = (x_0 + \varepsilon x_1 + \cdots, y_0 + \varepsilon y_1 + \cdots)$. Thus,

$$r_3 = k_3^+ x_{ss} y_{ss} = \frac{1}{\varepsilon}(x_0 + \varepsilon x_1 + \cdots)(y_0 + \varepsilon y_1 + \cdots) \tag{11.24}$$

In the zero approximation, $\varepsilon = 0$, $x_0 = 0$ or $y_0 = 0$, so $r_3 = x_1 y_0$ or $r_3 = x_0 y_1$.

11.1.6.1 First branch of the slow curve

In this case, no carbon monoxide is adsorbed on the catalyst surface, so $y = 0$. At the first steady state, $(1,0)$, the reaction rate equals zero. The second steady state, $\left(1 - \dfrac{k_2^+ p_{CO}}{2k_1^+ p_{O_2}}, 0\right)$, exists for $B_1 > 0$ and in this steady state the rates of the elementary reactions are:

$$r_1^+ = \frac{\left(k_2^+\right)^2 p_{CO}^2}{4k_1^+ p_{O_2}} \tag{11.25}$$

$$r_2^+ = \frac{\left(k_2^+\right)^2 p_{CO}^2}{2k_1^+ p_{O_2}} \tag{11.26}$$

$$r_2^- = 0 \tag{11.27}$$

$$r_3^+ = 2r_1^+ = \frac{\left(k_2^+\right)^2 p_{CO}^2}{2k_1^+ p_{O_2}} \tag{11.28}$$

Clearly, the reaction rate of the overall reaction, Eq. (11.23) is indeed given by

$$r = r_3 = r_2 = 2r_1 = \frac{\left(k_2^+\right)^2 p_{CO}^2}{2k_1^+ p_{O_2}} \tag{11.29}$$

In this approximation—the first regular approximation $(x_0 + \varepsilon x_1)$—the rate of the reverse reaction of the second step, r_2^-, is zero. Therefore, on this branch, all steps of the detailed mechanism can be considered to be irreversible. A more accurate approximation would yield a positive but sufficiently small rate for this elementary reaction.

11.1.6.2 Second branch of the slow curve

In this case, no oxygen is adsorbed on the catalyst surface, so $x = 0$. For $B_2 > 0$, on this branch two steady states exist: $\left(0, 1 - \dfrac{k_2^+ p_{CO} + k_2^- \pm \sqrt{B_2}}{4k_1^+ p_{O_2}}\right)$. The corresponding rates of the elementary reactions are:

$$r_1^+ = k_1^+ p_{O_2}\left(\frac{k_2^+ p_{CO} + k_2^- \pm \sqrt{B_2}}{4k_1^+ p_{O_2}}\right)^2 \tag{11.30}$$

$$r_2^+ = k_2^+ p_{CO}\left(\frac{k_2^+ p_{CO} + k_2^- \pm \sqrt{B_2}}{4k_1^+ p_{O_2}}\right) \tag{11.31}$$

$$r_2^- = k_2^-\left(1 - \frac{k_2^+ p_{CO} + k_2^- \pm \sqrt{B_2}}{4k_1^+ p_{O_2}}\right) \tag{11.32}$$

$$r_3^+ = 2k_1^+ p_{O_2}\left(\frac{k_2^+ p_{CO} + k_2^- \pm \sqrt{B_2}}{4k_1^+ p_{O_2}}\right)^2 \tag{11.33}$$

Also for this branch, $r = r_3 = r_2 = 2r_1$, so Eq. (11.23) is valid again.

11.1.6.3 Reaction rates at bifurcation points

We define the bifurcation points as follows (see Fig. 11.2):

   A: extinction point with reaction rate r_A
   B: after-extinction point with reaction rate r_B
   C: ignition point with reaction rate r_C
   D: after-ignition point with reaction rate r_D

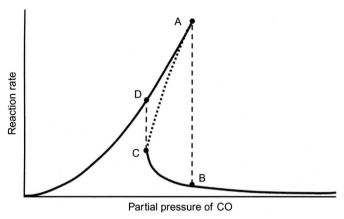

Fig. 11.2

Multiple steady states; reaction rate versus p_{CO}; irreversible adsorption of O_2 and reversible adsorption of CO; (A) extinction point; (B) after-extinction point; (C) ignition point; (D) after-ignition point.

At the first bifurcation point, extinction point A, $B_1 = 0$ and the rates of the elementary reactions for the branch $y = 0$ are

$$r_1^+ = k_1^+ p_{O_2} \tag{11.34}$$

$$r_2^+ = 2k_1^+ p_{O_2} \tag{11.35}$$

$$r_2^- = 0 \tag{11.36}$$

$$r_3^+ = 2k_1^+ p_{O_2} \tag{11.37}$$

and the overall reaction rate is

$$r_A = r_3 = r_2 = 2r_1 = 2k_1^+ p_{O_2} \tag{11.38}$$

After this bifurcation, the system jumps to the stable steady state on the branch $x = 0$. On this branch, the bifurcation condition $B_1 = 0$ is fulfilled at the reaction rate

$r = 2k_1^+ p_{O_2} \left(\dfrac{k_2^+ p_{CO} + k_2^- - \sqrt{B_2}}{4k_1^+ p_{O_2}} \right)^2$ after the extinction point. Thus, the reaction rate r_B after the

extinction point is given by

$$r_B = \frac{\left(k_2^-\right)^2}{2k_1^+ p_{O_2}} \tag{11.39}$$

The relation between the reaction rates at the extinction point and the after-extinction point is then obviously given by

$$r_A r_B = \left(k_2^-\right)^2 \tag{11.40}$$

At the second bifurcation point, ignition point C, $B_2 = 0$, so $\left(k_2^+ p_{CO} + k_2^-\right)^2 = 8k_2^- k_1^+ p_{O_2}$ and the rates of the elementary reactions for the branch $x = 0$ are

$$r_1^+ = \frac{k_2^-}{2} \tag{11.41}$$

$$r_2^+ = \frac{k_2^+ p_{CO}}{4k_1^+ p_{O_2}} \left(k_2^+ p_{CO} + k_2^-\right) = \frac{k_2^+ p_{CO}}{4k_1^+ p_{O_2}} \sqrt{8k_1^+ p_{O_2} k_2^-} \tag{11.42}$$

$$r_2^- = \frac{k_2^-}{4k_1^+ p_{O_2}} \left[4k_1^+ p_{O_2} - \left(k_2^+ p_{CO} + k_2^-\right)\right] = \frac{k_2^-}{4k_1^+ p_{O_2}} \left(4k_1^+ p_{O_2} - \sqrt{8k_1^+ p_{O_2} k_2^-}\right) \tag{11.43}$$

$$r_3^+ = 2r_1^+ = k_2^- \tag{11.44}$$

and the overall reaction rate is

$$r_C = r_3 = r_2 = 2r_1 = k_2^- \tag{11.45}$$

After this bifurcation, the system jumps to the stable steady state D, the after-ignition point, on the branch $y = 0$. The rate of the overall reaction at point D is

$$r_D = \frac{\left(k_2^+\right)^2 p_{CO}^2}{2k_1^+ p_{O_2}} \tag{11.46}$$

Then

$$r_C r_D = \frac{k_2^- \left(k_2^+\right)^2 p_{CO}^2}{2k_1^+ p_{O_2}} \tag{11.47}$$

At the bifurcation point of maximum complexity, $B_1 = B_2 = 0$, a further simplification for the reaction rate is

$$r = k_2^+ p_{CO} = k_2^- = 2k_1^+ p_{O_2} \tag{11.48}$$

From Eq. (11.48), it follows that

$$p_{CO} = \frac{k_2^-}{k_2^+} = \frac{1}{K_{eq,2}} \tag{11.49}$$

with $K_{eq,2}$ the equilibrium coefficient of the carbon monoxide adsorption step. This is a critical value of the carbon monoxide partial pressure, on the boundary of two critical phenomena, ignition and extinction. As the equilibrium coefficient is a function of temperature, this critical value of p_{CO} also depends on the temperature.

The partial pressure of carbon monoxide at the extinction and ignition points can be determined as follows. Extinction:

$$p_{CO,ext} = \frac{2k_1^+ p_{O_2}}{k_2^+} \tag{11.50}$$

Ignition:

$$p_{CO,ign} = \frac{\sqrt{8k_2^- k_1^+ p_{O_2}} - k_2^-}{k_2^+} = 2\sqrt{\frac{2k_1^+ p_{O_2}}{k_2^+} \frac{1}{K_{eq,2}} - \frac{1}{K_{eq,2}}}$$

$$= 2\sqrt{\frac{1}{K_{eq,2}}} \left(\sqrt{p_{CO,ext}} - \frac{1}{2\sqrt{K_{eq,2}}} \right) \tag{11.51}$$

11.1.7 Dynamics

11.1.7.1 Stability analysis

Fig. 11.3 shows the dynamics for different values of the bifurcation parameters. An analysis of the stability of the steady states of the carbon monoxide oxidation system shows that for $B_1 > 0$ and $B_2 > 0$ (Fig. 11.3A), the steady states $(1,0)$ and $\left(0, 1 - \frac{k_2^+ p_{CO} + k_2^- + \sqrt{B_2}}{4k_1^+ p_{O_2}}\right)$ are saddle points while the steady states $\left(1 - \frac{k_2^+ p_{CO}}{2k_1^+ p_{O_2}}, 0\right)$ and $\left(0, 1 - \frac{k_2^+ p_{CO} + k_2^- - \sqrt{B_2}}{4k_1^+ p_{O_2}}\right)$ are stable nodes. All trajectories are attracted to the latter node for initial conditions satisfying the inequality

$$x < y - 1 + \frac{k_2^+ p_{CO} + k_2^- - \sqrt{B_2}}{4k_1^+ p_{O_2}} \tag{11.52}$$

while for any other initial conditions all trajectories are attracted to the former node.

For $B_1 > 0$ and $B_2 < 0$ (Fig. 11.3B), there are two steady states on the branch $y = 0$ and none on the branch $x = 0$ Therefore, all trajectories are attracted to the node $\left(1 - \frac{k_2^+ p_{CO}}{2k_1^+ p_{O_2}}, 0\right)$.

For $B_1 < 0$ and $B_2 > 0$ (Fig. 11.3C), the only steady state is $\left(0, 1 - \frac{k_2^+ p_{CO} + k_2^- + \sqrt{B_2}}{4k_1^+ p_{O_2}}\right)$ and as this state does not belong to the invariant domain U, there are just two singular points. The first, $(1,0)$, is a saddle point, while the second, $\left(0, 1 - \frac{k_2^+ p_{CO} + k_2^- - \sqrt{B_2}}{4k_1^+ p_{O_2}}\right)$, is a stable node. All trajectories are attracted to this node.

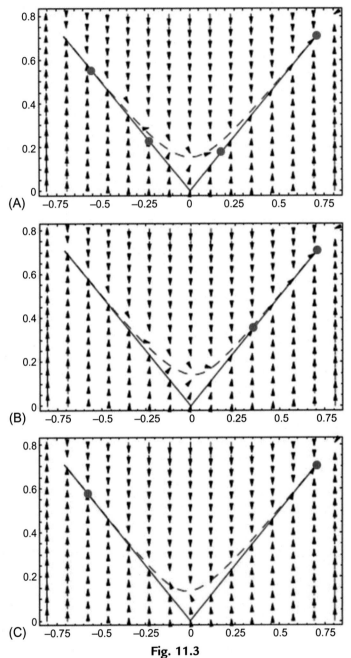

Fig. 11.3

Dynamics of the system for (A) $B_1>0$, $B_2>0$; (B) $B_1>0$, $B_2<0$; (C) $B_1<0$, $B_2>0$. *Solid line:* zero approximation ($\varepsilon=0$) of slow manifold; *dashed line:* first approximation of slow manifold; *gray dots:* steady states; *arrows:* system trajectories. *From Gol'dshtein, V., Krapivnik, N., Yablonsky, G., 2015. About bifurcational parametric simplification. Math. Model. Nat. Pheno. 10, 168–185. Copyright 2015, with permission from EDP Sciences; http://www.mmnp-journal.org/.*

In conclusion, if $B_2 > 0$ and the sign of B_1 changes from positive to negative, the rate of the overall reaction decreases dramatically (compare Fig. 11.3A and C), signifying that $B_1 = 0$ corresponds to the extinction point. Similarly, if $B_1 > 0$ and the sign B_2 changes from positive to negative, the overall reaction rate jumps from small to large values (compare Fig. 11.3A and B), indicating that $B_2 = 0$ is the ignition point. The point of maximum bifurcation complexity represents the coincidence of ignition and extinction points.

11.1.7.2 Physicochemical meaning of bifurcation parameters and bifurcation conditions

In explaining the physicochemical meaning of the bifurcation parameters and conditions, we will follow Chapter 9 of the book by Marin and Yablonsky (2011).

The condition $B_1 = 0$ is equivalent to

$$z_{ss} = \frac{k_2^+ p_{CO}}{2k_1^+ p_{O_2}} = 1 \tag{11.53}$$

for the corresponding steady state, that is, the entire catalyst surface is empty at steady state. Therefore, the bifurcation parameter B_1 defined in Eq. (11.9) determines the difference between the total normalized concentration of catalytic sites (unity) and the steady-state concentration of empty catalytic sites (z_{ss}), $1 - z_{ss}$.

The condition $B_2 = 0$ can be reformulated in terms of the parameter H:

$$H = \frac{\left(k_2^+ p_{CO} + k_2^-\right)^2}{8k_1^+ p_{O_2} k_2^-} \tag{11.54}$$

The condition $B_2 = 0$ is equivalent to $H = 1$.

As shown in the previous analysis, the condition $B_1 = 0$ determines the extinction point, while the condition $B_2 = 0$ (or $H = 1$) determines the ignition point.

Let us now discuss the conditions of the ignition phenomenon in more detail. At the ignition point, the rate of the overall reaction equals

$$r = k_2^- = 2k_1^+ p_{O_2} z_{ss}^2 \tag{11.55}$$

Thus, at ignition

$$z_{ss}^2 = \frac{k_2^-}{2k_1^+ p_{O_2}} \tag{11.56}$$

Now suppose that in our system only the reversible carbon monoxide adsorption step occurs, that is, the reversible interaction between gaseous carbon monoxide and empty catalyst sites. Then at equilibrium, which also is the steady state,

$$k_2^+ p_{CO} z_{eq} = k_2^- y_{eq} \tag{11.57}$$

With $y_{eq} + z_{eq} = 1$, Eq. (11.57) can be rewritten as

$$z_{eq} = \frac{k_2^-}{k_2^+ p_{CO} + k_2^-} = \frac{1}{K_{eq,2} p_{CO} + 1} \tag{11.58}$$

The condition $H = 1$ can be transformed as follows:

$$\frac{1}{H} = \frac{8 k_1^+ p_{O_2} k_2^-}{\left(k_2^+ p_{CO} + k_2^-\right)^2} = 1 \tag{11.59}$$

so that

$$\frac{4\left(k_2^-\right)^2}{\left(k_2^+ p_{CO} + k_2^-\right)^2} = \frac{k_2^-}{2 k_1^+ p_{O_2}} \tag{11.60}$$

Thus, with Eqs. (11.56), (11.58) it follows that

$$z_{ss} = 2 z_{eq} \tag{11.61}$$

This means that at the ignition point, the steady-state surface concentration of empty catalyst sites is double that of the equilibrium concentration. Before ignition, the steady-state concentration of adsorbed oxygen, x_{ss}, is very low. Therefore, the approximate surface concentration of adsorbed carbon monoxide in the vicinity of the ignition point is

$$y_{ss} = 1 - z_{ss} = 1 - 2 z_{eq} \tag{11.62}$$

At the extinction point, the surface composition is $x = 0$, $y = 0$, $z = 1$. At the point of maximum bifurcational complexity, where both $B_1 = 0$ and $H = 1$ are satisfied, two critical phenomena, ignition and extinction, merge. In this situation, all catalyst sites are empty, that is, $z = 1$. From the other side,

$$z_{ss} = 2 z_{eq} = \frac{2}{K_{eq,2} p_{CO} + 1} \tag{11.63}$$

Therefore,

$$\frac{2}{K_{eq,2} p_{CO} + 1} = 1 \tag{11.64}$$

and

$$K_{eq,2} p_{CO} = 1 \tag{11.65}$$

The same equation can be obtained by taking into account that at this point of maximum bifurcation complexity the rate of ignition equals the rate of extinction,

$$r_{ign} = r_{ext} \tag{11.66}$$

so

$$k_2^- = k_2^+ p_{CO} \tag{11.67}$$

which can be considered as a special condition of the parametric balance and after rearrangement is the same equation as Eq. (11.65), a rather interesting equation. It can be considered as a twisted classical adsorption equation. It also offers a unique opportunity to check the value of the equilibrium coefficient based on the experimentally observed partial pressure of carbon monoxide at the point of MBC:

$$K_{eq,2} = \frac{1}{p_{CO,\,MBC}} \tag{11.68}$$

It should be noted that the reaction rate at the ignition and extinction points are ill-observed values, whereas the reaction rates at the after-ignition and after-extinction points, r_D and r_B, are well-observed values. Assuming that the critical values of the partial carbon monoxide pressure ($p_{CO,ign}$ and $p_{CO,ext}$) are well-observed, the ratio of the rates, $\dfrac{r_{ext}}{r_{ign}} = \dfrac{r_A}{r_C}$, can be easily estimated from the known ratio $\dfrac{r_D}{r_B}$ as follows. The reaction rates at extinction point A and ignition point C are given by:

$$r_A = 2k_1^+ p_{O_2} = k_2^+ p_{CO,ext} \tag{11.69}$$

$$r_C = k_2^- \tag{11.70}$$

so

$$\frac{r_A}{r_C} = \frac{k_2^+ p_{CO,ext}}{k_2^-} = K_{eq,2} p_{CO,ext} \tag{11.71}$$

The reaction rates at after-extinction point B and after-ignition point D are given by:

$$r_B = \frac{\left(k_2^-\right)^2}{2k_1^+ p_{O_2}} = \frac{\left(k_2^-\right)^2}{k_2^+ p_{CO,ext}} \tag{11.72}$$

and

$$r_D = \frac{\left(k_2^+\right)^2 p_{CO,ign}^2}{2k_1^+ p_{O_2}} = \frac{\left(k_2^+\right)^2 p_{CO,ign}^2}{k_2^+ p_{CO,ext}} \tag{11.73}$$

so

$$\frac{r_D}{r_B} = \frac{\left(k_2^+\right)^2}{\left(k_2^-\right)^2} p_{CO,ign}^2 \quad \Leftrightarrow \quad K_{eq,2} = \sqrt{\frac{r_D}{r_B}} \frac{1}{p_{CO,ign}} \tag{11.74}$$

From Eq. (11.71), (11.74), it follows that

$$\frac{r_A}{r_C} = \sqrt{\frac{r_D \, p_{CO,ext}}{r_B \, p_{CO,ign}}} \tag{11.75}$$

$$\frac{r_{ext}}{r_{ign}} = \frac{r_A}{r_C} = K_{eq,2} \, p_{CO,ext} = \sqrt{\frac{r_D \, p_{CO,ext}}{r_B \, p_{CO,ign}}} \tag{11.76}$$

11.1.8 An Attempt at Generalization

The example of critical simplification analyzed in this chapter may serve as a good subject for some generalization. Let us assume that in our system there are three independent variables (chemical concentrations), one mass conservation balance, and one infinitely fast reaction between two surface species. As usual, our system is governed by the mass-action law and there is no question of autocatalysis. In this case, the steady-state value of one variable will be zero (or rather, negligible) and the dynamics of the system will be two-dimensional. Consequently, critical simplification will be observed for this system as well.

11.2 "Kinetic Dance": One Step Forward—One Step Back

11.2.1 History

Relaxation methods in chemical kinetics were proposed by Manfred Eigen in the 1950s (Eigen, 1954). The main methodology and mathematical idea of such methods is an analysis of transient regimes, so-called relaxation, that is, the behavior of chemical systems perturbed near equilibrium. In 1967, Eigen, together with Norrish and Porter, received the Nobel Prize in Chemistry for their theoretical development of this method and its application, in particular to acid-base neutralization reactions. Before that, these reactions were considered to be "immeasurably fast," and Eigen's Nobel lecture was aptly titled "Immeasurably Fast Reactions."

Eigen's ideas also sparked an interest in relaxation studies in heterogeneous catalysis. Bennett (1967) proposed relaxation experiments for continuous stirred-tank reactors (CSTRs). An experimental implementation was performed some years later by Kobayashi and Kobayashi (1972a–c) and by Yang et al. (1973).

Vast information on nonsteady-state (dynamic) methods of studying heterogeneous catalytic reactions was presented in reviews by Kobayashi and Kobayashi (1974), Bennett (1976, 2000), Bennett et al. (1972), and Berger et al. (2008).

For almost 50 years of nonsteady-state relaxation in the field of heterogeneous catalysis, different experimental and theoretical procedures have been tested, and a huge amount of results have been accumulated. A variety of experimental set-ups, such as batch and semi-batch reactors, CSTRs and plug-flow reactors (PFRs) have been used with different modes of operation, for example, step- or pulse-response techniques, cyclic feeding, steady-state isotopic transient kinetic analysis, the tapered element oscillating microbalance method for measuring the mass change of a sample bed, and so on.

In this battery of nonsteady-state devices and approaches, the temporal analysis of products (TAP) approach occupies a special place. Originally created by John Gleaves in 1988 (Gleaves et al., 1988) and modified in 1997 (Gleaves et al., 1997), TAP has found applications in many areas of chemical kinetics and chemical engineering. The TAP method provides unique information on the nonsteady-state catalyst state under conditions of insignificant change of the catalyst composition. Using the thin-zone TAP reactor configuration (Gleaves et al., 1997), this approach allows extracting nonsteady-state kinetic information (reaction rate data) without any a priori assumption on the kinetic model, which is similar to the use of a CSTR model for extracting steady-state kinetic information.

All of the methods mentioned here have proven their effectiveness and exhibited their limitations. Still, this area invites novel ideas to be tested.

11.2.2 Experimental Procedures

Before studying nonideal reactors and other generalizations, we will first analyze two types of ideal chemical reactors, the CSTR and the PFR. We will assume that:

- the reactions occur under isothermal conditions;
- the kinetic model is linear, so all reactions are first order or pseudo-first order;
- the flow rate is constant.

In the analysis of both reactor models, special attention will be paid to the following characteristics: the initial, inlet, and steady-state composition of the reacting mixture (or just the initial, inlet, and steady-state concentrations of one reactant).

For both reactor types, we describe experimental procedures that are sequences of different experiments. The first procedure is a *parallel* one and has to be performed in two separate reactors, R_I and R_{II}. Both reactors, which have an identical construction and are filled with the same amount of the same catalyst, are fed with a gas mixture of the same initial concentration, c_{in}, and are then allowed to reach the steady state. Ideally, in both reactors the same steady state concentration, c_{ss}, will be achieved. Next, the inlet concentration of the first reactor is instantaneously changed to $c_{in} + \Delta$ and the response, $c^+(t)$, is measured. Similarly, the inlet concentration of the second reactor is instantaneously changed from the same c_{in} to $c_{in} - \Delta$ and the response, $c^-(t)$, is also measured. The parallel procedure is presented qualitatively in Fig. 11.4.

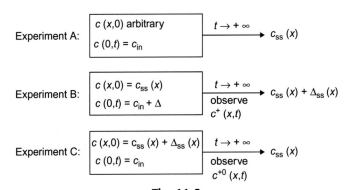

Fig. 11.4

Parallel experimental procedure.

The second procedure is a *sequential* one (see Fig. 11.5) and consists of three experiments (A, B, and C). During experiment A, a mixture at steady state is prepared. In this experiment, we can use a chosen inlet and arbitrary initial concentration, $c_{A,in}$ and c_{A0}. After some time, the steady-state concentration, $c_{A,ss}$, is achieved as a result of experiment A. Then this steady-state concentration is used as initial concentration for experiment B: $c_{B0} = c_{A,ss}$. Then the inlet concentration for this experiment is chosen as follows: $c_{B,in} = c_{A,in} + \Delta$. Then during this transient experiment, $c^+(t)$ is monitored until a new steady state is reached with concentration $c_{B,ss}$. Similarly, for experiment C, the initial concentration is the steady-state concentration of experiment B, $c_{C0} = c_{B,ss}$, while now the inlet concentration is set equal of that of experiment A, $c_{C,in} = c_{A,in}$. Then in a new transient experiment $c^{+,0}(t)$ is measured until a new steady-state concentration, $c_{C,ss}$, is achieved, which will equal $c_{A,ss}$.

Fig. 11.5

Sequential experimental procedure.

We will show that for the parallel procedure in the linear case

$$c^+(t) + c^-(t) = 2c_{ss} \tag{11.77}$$

which is constant in time and does not depend on the perturbation Δ.

For the sequential procedure, we will show that in the linear case

$$c^+(t) + c^{+,0}(t) = \text{constant} \tag{11.78}$$

but the constant does depend on the perturbation Δ.

11.2.2.1 Ideal CSTR

The differential mass balance equation for the CSTR is

$$\frac{dc(t)}{dt} = \frac{c_{\text{in}} - c(t)}{\tau} - kc(t) \tag{11.79}$$

with initial condition $c(t=0) = c_0$.

The unique solution of Eq. (11.79) is

$$c(t) = \frac{c_{\text{in}}}{1+k\tau} + e^{-\left(k+\frac{1}{\tau}\right)t}\left(c_0 - \frac{c_{\text{in}}}{1+k\tau}\right) \tag{11.80}$$

As $t \to +\infty$, this solution tends to

$$c = \frac{c_{\text{in}}}{1+k\tau} = c_{\text{ss}} \tag{11.81}$$

which is independent on c_0. The general equation, Eq. (11.80), can be applied to all experiments performed in an ideal CSTR.

11.2.2.1.1 Parallel procedure

Applying Eq. (11.80) to the first reactor (R_I) in the parallel procedure, we find that

$$c^+(t) = \frac{c_{\text{in}} + \Delta}{1+k\tau} + e^{-\left(k+\frac{1}{\tau}\right)t}\left(\frac{c_{\text{in}}}{1+k\tau} - \frac{c_{\text{in}}+\Delta}{1+k\tau}\right) = \frac{c_{\text{in}} + \Delta - \Delta e^{-\left(k+\frac{1}{\tau}\right)t}}{1+k\tau} \tag{11.82}$$

Similarly, for the second reactor (R_{II}),

$$c^-(t) = \frac{c_{\text{in}} - \Delta + \Delta e^{-\left(k+\frac{1}{\tau}\right)t}}{1+k\tau} \tag{11.83}$$

Consequently,

$$c^+(t) + c^-(t) = \frac{2c_{\text{in}}}{1+k\tau} = 2c_{\text{ss}} \tag{11.84}$$

11.2.2.1.2 Sequential procedure

Since for experiment B $c_{B,in} = c_{A,in} + \Delta = c_{in} + \Delta$, applying Eq. (11.80) leads to

$$c^+(t) = \frac{c_{in}}{1+k\tau} + \left(1 - e^{-\left(k+\frac{1}{\tau}\right)t}\right)\frac{\Delta}{1+k\tau} = \frac{c_{in} + \Delta - \Delta e^{-\left(k+\frac{1}{\tau}\right)t}}{1+k\tau} \tag{11.85}$$

The steady-state concentration for this experiment is

$$c_{B,ss} = \frac{c_{in} + \Delta}{1+k\tau} \tag{11.86}$$

This concentration is taken as the initial concentration for experiment C and the inlet concentration is the inlet concentration of experiment A, so the corresponding solution is

$$c^{+0}(t) = \frac{c_{in}}{1+k\tau} + e^{-\left(k+\frac{1}{\tau}\right)t}\frac{\Delta}{1+k\tau} = \frac{c_{in} + \Delta e^{-\left(k+\frac{1}{\tau}\right)t}}{1+k\tau} \tag{11.87}$$

Adding expressions (11.85) and (11.87) yields

$$c^+(t) + c^{+0}(t) = \frac{2c_{in} + \Delta}{1+k\tau} \tag{11.88}$$

In the sequential procedure, the combination of experiments A and B is the same as the experiment in the first reactor in the parallel procedure. The $c^+(t)$ dependence is given by the same equation, Eq. (11.82) (or Eq. 11.85) and the steady-state concentration for both experimental procedures is given by Eq. (11.86).

The difference between the results obtained by the parallel and the sequential procedure is that for the latter the sum of the trajectories is constant but depends on the perturbation value Δ (Eq. 11.88). For the parallel procedure, such a dependence is absent (Eq. 11.84). Furthermore, the dependence of the sum of the concentration trajectories on experiment A can be eliminated by running two sequential procedures with opposite values of Δ ("back-to-back"). Writing $c^-(t)$ and $c^{-0}(t)$ for the concentrations during the second sequential procedure we obtain

$$c^-(t) + c^{-0}(t) = \frac{2c_{in} - \Delta}{1+k\tau} \tag{11.89}$$

and adding Eqs. (11.88), (11.89) then yields

$$c^+(t) + c^{+0}(t) + c^-(t) + c^{-0}(t) = \frac{4c_{in}}{1+k\tau} = 4c_{ss}, \tag{11.90}$$

which is independent on Δ.

11.2.2.2 Ideal PFR

We are now going to prove the same result for the PFR. The differential mass balance equation for the PFR is

$$\frac{\partial c}{\partial t} + v\frac{\partial c}{\partial x} = -kc \tag{11.91}$$

where $c(0, t) = c_{in}$ is the boundary condition, v is the linear velocity, and x is the longitudinal position in the reactor. The steady-state concentration profile corresponding to the given boundary condition does not depend on initial concentration values and can be calculated by setting $\frac{\partial c}{\partial t} = 0$ in Eq. (11.91):

$$v\frac{\partial c_{ss}}{\partial x} = -kc_{ss} \tag{11.92}$$

with boundary condition $c_{ss}(x=0) = c_{in}$.

The unique solution of Eq. (11.92) is given by

$$c_{ss}(x) = c_{in}e^{-\frac{kx}{v}} \tag{11.93}$$

This concentration profile is the initial condition for performing an experiment in which $c^+(x, t)$ is observed. Increasing c_{in} to $c_{in} + \Delta$ leads to the following response:

$$c^+(x, t) = e^{-\frac{kx}{v}}\left(c_{in} + \Delta H\left(t - \frac{x}{v}\right)\right) \tag{11.94}$$

where H is the Heaviside function. The steady-state concentration achieved in this experiment is

$$c_{ss}^+(x) = e^{-\frac{kx}{v}}(c_{in} + \Delta) \tag{11.95}$$

In the sequential procedure, this concentration profile is used as the initial condition for another experiment (experiment C), in which the inlet concentration is restored to its initial value, c_{in}, and the dependence $c^{+0}(x, t)$ is observed:

$$c^{+0}(x, t) = e^{-\frac{kx}{v}}\left(c_{in} + \Delta H\left(\frac{x}{v} - t\right)\right) \tag{11.96}$$

Adding Eqs. (11.94), (11.96) yields

$$c^+(x, t) + c^{+0}(x, t) = e^{-\frac{kx}{v}}(2c_{in} + \Delta) \tag{11.97}$$

which is independent on time, but does depend on Δ.

Experiments in which $c^-(x, t)$ and $c^{-0}(x, t)$ are observed, can be described by similar equations as Eqs. (11.94)–(11.97) but with Δ replaced with $(-\Delta)$. The equivalent of Eq. (11.97) then is

$$c^-(x, t) + c^{-0}(x, t) = e^{-\frac{kx}{v}}(2c_{in} - \Delta) \tag{11.98}$$

and the sum of Eqs. (11.97), (11.98) equals

$$c^+(x, t) + c^{+0}(x, t) + c^-(x, t) + c^{-0}(x, t) = 4c_{in}e^{-\frac{kx}{v}} \tag{11.99}$$

This is the result of a four-step procedure consisting of two two-step sequential procedures. It is remarkable that the sum of four terms in Eq. (11.99) neither depends on time nor on Δ. This latter fact is also valid for the parallel procedure in the PFR since in this case

$$c^+(x, t) + c^-(x, t) = 2c_{in}e^{-\frac{kx}{v}} \tag{11.100}$$

11.2.2.3 Nonideal PFR

In the case of a nonideal PFR, Eq. (11.91) becomes

$$\frac{\partial c}{\partial t} = D\frac{\partial^2 c}{\partial x^2} - v\frac{\partial c}{\partial x} - kc = \tilde{L}(c) \tag{11.101}$$

where D is the diffusion coefficient and $\tilde{L}(c)$ is a linear operator of c acting on the spatial coordinate only. Eq. (11.101) is the generalized form for which we will prove similar results as for the ideal PFR.

Again, the first step is to determine the steady-state solution $c_{ss}(x)$ of Eq. (11.101) with a prescribed c_{in}. This means solving

$$\tilde{L}(c_{ss}) = 0 \tag{11.102}$$

with $c_{ss}(x = 0) = c_{in}$.

We also need to define $\Delta_{ss}(x)$ as the solution of the equations

$$\tilde{L}(\Delta_{ss}) = 0 \tag{11.103}$$

with $\Delta_{ss}(x = 0) = \Delta$.

In this case, $c^+(x, t)$ will satisfy three equations:

$$\begin{cases} \dfrac{\partial c^+(x, t)}{\partial t} = \tilde{L}(c^+(x, t)) \\ c^+(0, t) = c_{in} + \Delta \\ c^+(x, 0) = c_{ss}(x) \end{cases} \tag{11.104}$$

The steady-state concentration reached by $c^+(x, t)$ will be $c_{ss}(x) + \Delta_{ss}(x)$. Consequently, $c^{+0}(x, t)$ will satisfy

$$
\begin{cases}
\dfrac{\partial c^{+0}(x, t)}{\partial t} = \tilde{L}\big(c^{+0}(x, t)\big) \\[2mm]
c^{+0}(0, t) = c_{in} \\[2mm]
c^{+0}(x, 0) = c_{ss}(x) + \Delta_{ss}(x)
\end{cases}
\tag{11.105}
$$

Adding Eqs. (11.104), (11.105) with $c^+(x, t) + c^{+0}(x, t) = w(x, t)$ yields

$$
\begin{cases}
\dfrac{\partial w}{\partial t} = \tilde{L}(w) \\[2mm]
w(0, t) = 2c_{in} + \Delta \\[2mm]
w(x, 0) = 2c_{ss}(x) + \Delta_{ss}(x)
\end{cases}
\tag{11.106}
$$

From Eqs. (11.102), (11.103) it is clear that $w(x, t) = 2c_{ss}(x) + \Delta_{ss}(x)$ is the solution of Eq. (11.106). This proves that for the nonideal PFR and any generalization, $c^+(x, t) + c^{+0}(x, t)$ is independent of time.

11.2.3 Concluding Remarks

The new nonsteady-state procedures described here, namely parallel and sequential procedures, may find wide application for many purposes:

- They open a perspective for performing a new class of controlled nonsteady-state kinetic experiments, which can be termed "joint kinetic experiments."
- They enable distinguishing linear processes from nonlinear ones experimentally; presently this differentiation is made only statistically using model discrimination.
- Under the assumption of linearity it becomes possible to predict or estimate a transient regime, say $c^-(t)$ based on a known steady-state concentration and known symmetrical transient regime, $c^+(t)$.

11.3 Intersections and Coincidences

11.3.1 Introduction

A rigorous mathematical analysis, in particular an analysis using computer algebra methods, offers opportunities not only to describe different observed phenomena, but also to pose new problems. These problems are related to new questions that do not necessarily lend themselves directly to simple solutions, but we hope can prove valuable in practice and interesting from a theoretical viewpoint.

In Chapter 6, we explained the concept of time invariances obtained in dual (or reciprocal) kinetic experiments. Such experiments can be performed in batch reactors or TAP reactors and start from special, reciprocal, initial conditions. This results in certain mixed quotientlike functions of selected nonsteady-state normalized concentrations from both experiments always being equal to the equilibrium coefficient of the reaction, during the whole course of the temporal experiment, that is, not only at the end, when equilibrium conditions are reached. Examples of such quotientlike functions are presented in Chapter 6. In this section, two new patterns will be presented, namely intersections and coincidences.

An *intersection* of two temporal concentrations, for instance $c_A(t)$ and $c_B(t)$, means that these concentrations can be considered equal at some point in time, that is, $c_A(t) = c_B(t)$. It is a well-known mathematical fact that phase trajectories do not intersect or merge; nevertheless, the temporal trajectories may well intersect. A special case is osculation, in which not only the concentrations but also the temporal slopes coincide: $c_A(t) = c_B(t)$ and $dc_A(t)/dt = dc_B(t)/dt$.

A *coincidence in time* means that at least two special events occur at the same point in time, while the occurrence of two special events at the same value of concentrations represents a *coincidence in value*. For example, the maximum of a concentration dependence may coincide with the point at which this dependence intersects with another concentration dependence; or the intersection between three concentration dependences may occur at the same moment in time providing a triple intersection.

11.3.2 Two-Step Consecutive Mechanism With Two Irreversible Steps

11.3.2.1 Kinetic model and maximum concentration of B

Consecutive reaction mechanisms are well-known basic mechanisms in chemical kinetics. The simplest example of such a reaction scheme is $A \xrightarrow{k_1} B \xrightarrow{k_2} C$. Its kinetic model is represented as follows

$$\frac{dc_A}{dt} = -k_1 c_A \tag{11.107}$$

$$\frac{dc_B}{dt} = k_1 c_A - k_2 c_B \tag{11.108}$$

$$\frac{dc_C}{dt} = k_2 c_B \tag{11.109}$$

with c_A, c_B, and c_C the concentrations of A, B, and C and k_1 and k_2 the rate coefficients of the first and second reaction, which are of the Arrhenius type:

$$k_i = k_{i,0} \exp\left(-\frac{E_{ai}}{R_g T}\right) \tag{11.110}$$

The solution of this set of equations in the case that c_{C0} and c_{B0} equal zero can be found in many popular textbooks. If the reaction takes place in a batch reactor or PFR τ and $k_1 \neq k_2$ the solution in terms of the normalized concentrations is given by

$$A = \exp(-k_1 t) \tag{11.111}$$

$$B = k_1 \left[\frac{\exp(-k_1 t) - \exp(-k_2 t)}{k_2 - k_1} \right] \tag{11.112}$$

$$C = 1 - \frac{k_2 \exp(-k_1 t) - k_1 \exp(-k_2 t)}{k_2 - k_1} \tag{11.113}$$

and

$$A + B + C = \frac{c_A}{c_{A0}} + \frac{c_B}{c_{A0}} + \frac{c_C}{c_{A0}} = 1; \quad t \geq 0 \tag{11.114}$$

In the case that $k_1 = k_2$, the solution is less straightforward but it can be obtained by direct solution or using the Laplace domain, see also Yablonsky et al. (2010):

$$A = \exp(-k_1 t) \tag{11.115}$$

$$B = k_1 t \exp(-k_1 t) \tag{11.116}$$

$$C = 1 - (1 + k_1 t) \exp(-k_1 t) \tag{11.117}$$

The consecutive reaction mechanism is characterized by a maximum in the concentration of intermediate B (in contrast with the parallel mechanism $A \rightarrow B$, $A \rightarrow C$), which can be found by equating the time derivative of B to zero. For $k_1 \neq k_2$, the maximum normalized concentration of B is given by

$$B_{\max} = \left(\frac{k_2}{k_1} \right)^{\frac{k_2}{k_1 - k_2}} = \rho^{\frac{\rho}{1 - \rho}} \tag{11.118}$$

Fig. 11.6 shows this normalized maximum concentration as a function of the ratio of the rate coefficients, ρ.

The time at which the maximum occurs for $k_1 \neq k_2$ is given by

$$t_{B,\max} = \frac{\ln\left(\dfrac{k_2}{k_1}\right)}{k_2 - k_1} = \frac{\ln k_2 - \ln k_1}{k_2 - k_1} \tag{11.119}$$

The physical requirement that $t_{B,\max} > 0$ is always satisfied because the natural logarithm is a monotone increasing function, so the sign of the numerator is always the same as that of the denominator, resulting in a positive quotient.

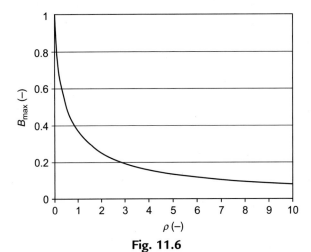

Fig. 11.6

Plot of B_{max} as a function of the ratio of rate coefficients, ρ.

In the case that $\rho = 1$ (so $k_1 = k_2$), a rather remarkable but not widely known expression for the maximum concentration of B holds (Kubasov, 2004; Yablonsky et al., 2010), which Yablonsky et al. called the *Euler point E*:

$$B_{max} = e^{-1}; \quad t_{B,max} = \frac{1}{k_1} = \frac{1}{k_2} \qquad (11.120)$$

11.3.2.2 Concentration patterns and comparison of rate coefficients

The normalized concentration of A at $t_{B,max}$ can be calculated from Eq. (11.111) and equals $\rho^{1/(\rho-1)}$ for $k_1 \neq k_2$ and $e^{-1} (= B_{max})$ for $k_1 = k_2$, and thus

$$B_{max} = (A(t_{B,max}))^{\rho} \qquad (11.121)$$

Consequently, if $k_1 = k_2$, the concentrations of A and B at $t_{B,max}$ are equal; if $k_1 > k_2$, the concentration of B exceeds that of A; and if $k_1 < k_2$, the concentration of B is smaller than that of A. Looking at experimental dependencies with these patterns, and assuming the consecutive reaction mechanism, we immediately obtain an indication of the relative values of k_1 and k_2.

The question now arises whether a temperature can be chosen such that some predetermined value of ρ is obtained, and more specifically a value of ρ corresponding to the Euler point. Solving the Arrhenius expressions (11.110) for k_1 and k_2 in terms of this target value, $\tilde{\rho}$, yields the following formal expression for the temperature:

$$T_{\tilde{\rho}} = \frac{1}{R_g} \frac{E_{a1} - E_{a2}}{\ln(\tilde{\rho} k_{1,0}/k_{2,0})} = \frac{1}{R_g} \frac{E_{a1} - E_{a2}}{\ln \tilde{\rho} k_{1,0} - \ln k_{2,0}} \qquad (11.122)$$

A necessary condition for this expression to have physicochemical meaning is that it must be positive, so either $E_{a1} > E_{a2}$ and $\tilde{\rho}k_{1,0} > k_{2,0}$ or $E_{a1} < E_{a2}$ and $\tilde{\rho}k_{1,0} < k_{2,0}$. In addition, for the Euler point to be observable at realistic temperatures ($\sim$100–500 K) there are limits on the values of the Arrhenius parameters, typically $0.01 < k_{1,0}/k_{2,0} < 100$ and $|E_{a1} - E_{a2}| < 20\,\text{kJ/mol}$ (Yablonsky et al., 2010). For an order of magnitude of the ratio $k_{1,0}/k_{2,0}$ of 10, the difference between the activation energies has to be of the order of 10 kJ/mol, which is a realistic value.

11.3.2.3 Intersections of temporal concentration curves

For the consecutive reaction $A \rightarrow B \rightarrow C$ two types of trajectories are studied: (i) phase trajectories in the complete variable (concentration) space and (ii) temporal trajectories of each variable separately. Phase trajectories neither intersect nor merge, but temporal trajectories may or may not intersect. At a point of intersection, the concentrations of two or three reaction components are equal. Yablonsky et al. (2010) performed an analysis of the theoretically possible types of temporal concentration intersections and the conditions of their occurrence.

11.3.2.3.1 Intersection of A and B curves, of A and C curves and of B and C curves

The time of intersection of the normalized concentrations of A and B can be found by equating Eqs. (11.111), (11.112) and solving for t

$$t_{A=B} = \frac{\ln\left(\dfrac{k_1}{2k_1 - k_2}\right)}{k_2 - k_1} = \frac{\ln k_1 - \ln(2k_1 - k_2)}{k_1 - (2k_1 - k_2)} \tag{11.123}$$

We can now distinguish the following cases for $k_1 \neq k_2$:

(1) Single intersection at finite time

For $k_2 < 2k_1$ and $k_1 \neq k_2$ there is a single intersection at finite time (eg, Fig. 11.7A). The value of the normalized concentrations at the intersection is given by

$$A = B = (2 - \rho)^{\frac{1}{\rho - 1}} \quad \rho > 0,\ \rho \neq 1 \tag{11.124}$$

If $k_2 \rightarrow 0$, $t_{A=B} \rightarrow \dfrac{\ln 2}{k_1}$ and the corresponding concentrations $A = B \rightarrow \dfrac{1}{2}$.

(2) For $k_1 = k_2$, Eqs. (11.115), (11.116) must be used and in this case also a single intersection occurs, at the Euler point, where $t_{A=B} = t_{B,\max}$ and the concentrations of A and B are given by Eq. (11.120) (Fig. 11.7B).

(3) Intersection achieved at time$+\infty$

If $k_2 \rightarrow 2k_1$, $t_{A=B} \rightarrow +\infty$ and the corresponding concentrations $A = B \rightarrow 0$ (Fig. 11.7C).

(4) No intersection

For $k_2 > 2k_1$ there is no intersection of the concentrations of A and B (Fig. 11.7D).

(5) Multiple nonosculating intersections

This situation is not physically possible because the equality $A = B$ has only one solution, Eq. (11.123).

(6) Osculating intersection

It is physically impossible for an osculating intersection to occur as the system $A = B, dA/dt = dB/dt$ has no finite solutions.

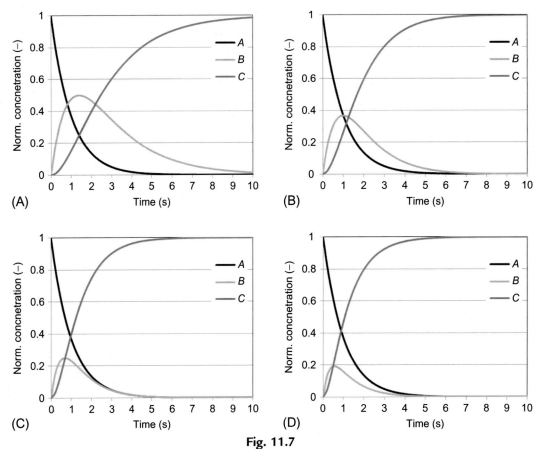

Fig. 11.7

Possible types of intersections between A and B as a function of time for different ratios of the rate coefficient. (A) Acme point, $\rho = 0.5$; (B) Euler point, $\rho = 1$; (C) osculation point, $\rho = 2$; (D) triad point, $\rho = 3$.

This analysis can be used to test a hypothesis about the type of mechanism: if more than one intersection of the concentration curves A and B is observed, this implies that the consecutive reaction scheme $A \rightarrow B \rightarrow C$ cannot be valid, and a more complex mechanism is needed. If, on the other hand, there is only one intersection, the consecutive scheme is likely and the fact that an intersection is indeed observed directly implies that $k_2 < 2k_1$.

As mentioned in the previous section, at certain values of k_1 and k_2, namely for $k_2 > 2k_1$, the concentration curves of A and B as a function of time do not intersect. However, there always is a point of intersection of curves A and C and of curves B and C, called an unavoidable intersection.

For the intersection of A and C, the normalized concentration A with $0 < A < 1$ solves one of the following equations:

$$A^\rho + A(1 - 2\rho) + (\rho - 1) = 0 \quad \rho \neq 1 \tag{11.125}$$

or

$$A \ln A - 2A + 1 = 0 \quad \rho = 1 \tag{11.126}$$

For determining the intersection of B and C, the normalized concentration A solves

$$2A^\rho - A(1 + \rho) + (\rho - 1) = 0 \quad \rho \neq 1 \tag{11.127}$$

or

$$2A \ln A - A + 1 = 0 \quad \rho = 1 \tag{11.128}$$

Then it follows that $B = C = (1 - A)/2$.

When $\rho \neq 1$ is a sufficiently simple integer or fraction, Eqs. (11.125), (11.127) can be solved exactly. The solutions of Eqs. (11.126), (11.128) can be obtained by using the Lambert function, see Yablonsky et al. (2010).

In all cases, the time of intersection is readily obtained as

$$t_{\text{int}} = \frac{1}{k_1} \ln \frac{1}{A} \tag{11.129}$$

11.3.2.3.2 Intersection of dA/dt and dB/dt and of dB/dt and dC/dt curves

Equating the derivatives of Eqs. (11.111), (11.112) and solving for t yields

$$t_{\text{A}'=\text{B}'} = \frac{1}{k_2 - k_1} \ln \frac{k_2}{2k_1 - k_2} = 2 \frac{(\ln k_2 - \ln(2k_1 - k_2))}{k_2 - (2k_1 - k_2)} \tag{11.130}$$

which is well defined for $k_2 < 2k_1$, or $\rho < 2$, and positive, and has the remarkable property that

$$t_{\text{A}'=\text{B}'} = t_{\text{B,max}} + t_{\text{A}=\text{B}} \tag{11.131}$$

Similarly, we obtain

$$t_{\text{B}'=\text{C}'} = \frac{1}{k_2 - k_1} \ln \frac{2k_2}{k_1 + k_2} = \frac{\ln 2k_2 - \ln(k_1 + k_2)}{2k_2 - (k_1 + k_2)} \tag{11.132}$$

At the Euler point, the corresponding values are $t_{\text{A}'=\text{B}'} = 2/k_1$ and $t_{\text{B}'\_\text{C}'} = 1/2k_1$.

11.3.2.4 Special points

During their analysis, Yablonsky et al. (2010) found a number of special points, which they named Acme, Golden, Euler, Lambert, Osculation, and Triad (AGELOT) points. Table 11.1 summarizes some characteristics of these points. For more details see Yablonsky et al. (2010).

Table 11.1 Special points and their characteristics

Point	$\rho = k_2/k_1$	$A = B$ $t_{A=B}$	$A = C$ $t_{A=C}$	$B = C$ $t_{B=C}$	B_{max} $t_{B,max}$
Acme	1/2	4/9	1/4	4/9	½
		$\frac{1}{k_1}\ln\frac{9}{4}$	$\frac{1}{k_1}\ln 4$	$\frac{1}{k_1}\ln 9$	$\frac{1}{k_1}\ln 4$
Golden[a]	$1/\phi = 0.618034$	$(3-\phi)^{-\phi^2}$	0.270633	0.422003	$\phi^{-\phi}$
		$\frac{\phi^2\ln(3-\phi)}{k_1}$	$\frac{1.306989}{k_1}$	$\frac{1.857946}{k_1}$	$\frac{\phi\ln\phi}{k_1}$
Euler	1	$\frac{1}{e}$	0.317844	0.357665	$\frac{1}{e}$
		$\frac{1}{k_1}=\frac{1}{k_2}$	$\frac{1.146193}{k_1}$	$\frac{1.256431}{k_1}$	$\frac{1}{k_1}=\frac{1}{k_2}$
Lambert	1.1739824...	1/3	1/3	1/3	0.38800
		$\frac{\ln 3}{k_1}$	$\frac{\ln 3}{k_1}$	$\frac{\ln 3}{k_1}$	$\frac{0.9219423}{k_1}$
Osculation[a]	2	0	$\frac{1}{\phi^2}=2-\phi$	1/4	1/4
		$+\infty$	$\frac{2\ln\phi}{k_1}$	$\frac{\ln 2}{k_1}$	$\frac{\ln 2}{k_1}$
Triad[a]	3	—	$\sqrt{2}-1$	$1-\phi/2$	$\sqrt{3}/9$
			$\frac{\ln(\sqrt{2}+1)}{k_1}$	$\frac{\ln\phi}{k_1}$	$\frac{\ln 3}{2k_1}$

[a]ϕ denotes the golden ratio, $(\sqrt{5}+1)/2=1.618034$.

The *Acme* point occurs at $\rho = 1/2$, see Fig. 11.7A. Some remarkable properties of this point are the following:

- B reaches its maximum value (of 1/2) at the point where A and C intersect;
- at the point of intersection of B and C, $dA/dt = dB/dt$, so the upward slope of C at this point equals twice the downward slope of B;
- a remarkable additivity property for the times of intersection exists: $t_{A=B}+t_{A=C}=t_{B=C}$;
- the values of A at $t_{A=B}$ and B at $t_{B=C}$ are the same.

At the Acme point, Eqs. (11.115)–(11.117) can be rewritten for all times t as

$A = \exp(-k_1 t) = \left(\sqrt{A}\right)^2, B = 2\sqrt{A}\left(1 - \sqrt{A}\right), C = \left(1 - \sqrt{A}\right)^2$. Thus, for every set of concentrations (A, B, C), there also is a point on the trajectory with the concentration values of A and B interchanged, that is, the phase portrait of the trajectory is symmetrical. Consequently, if we were to base time on B (instead of on A or, equivalently, C, as is more usual), each value B would correspond to two real times t_1 and t_2, with $(A(t_1), B(t_1), C(t_1))$ equal to $(C(t_2), B(t_2), A(t_2))$.

The most interesting feature of the *Euler* point (Fig. 11.7B), at which $\rho = 1$, is that the maximum value of B $(1/e)$ is reached at its point of intersection with A.

It is possible for all three concentrations to be equal at the same time, that is $A = B = C = 1/3$. The condition required for achieving this is what Yablonsky et al. (2010) call the *Lambert* point ratio ρ_L, the value of which can be determined to be 1.1739824....

At the *osculation* point, $\rho = 2$. Experimentally, we are better able to observe peak and intersection values than osculation at infinity, because tail analysis can be marred by experimental error.

We can also consider the angle of intersection between concentration curves, but this only makes sense when both axes are commensurate, as can be obtained, for instance, by replacing time by C. If t is the time of intersection, then the angle of intersection, γ, is given by the difference between angles α and β

$$\tan\gamma = \tan(\alpha - \beta) = \frac{\tan\alpha - \tan\beta}{1 + \tan\alpha\tan\beta} = \left(\frac{A'}{C'} - \frac{B'}{C'}\right)\frac{1}{1 + \frac{A'B'}{C'^2}} \tag{11.133}$$

with α and β the angles between the horizontal axis and the tangents to respectively the A versus C curve and the B versus C curve.

Simplification and adding absolute value signs to obtain a geometrically meaningful angle between 0 and 90 degrees then leads to

$$\tan\gamma = \left|\frac{\rho(\rho - 2)}{\rho^2 + \rho - 1}\right| \tag{11.134}$$

So-called *orthogonal intersection* occurs for an infinite value of this tangent, that is, for $\rho = \left(\sqrt{5} - 1\right)/2 = 1/\phi$, the inverse of the golden ratio, which Yablonsky et al. (2010) call the *golden point*.

The *triad* point occurs at $\rho = 3$ (see Fig. 11.7D). At this point, A and B do not intersect at all.

11.3.2.5 Sequence of times and of values of B$_{max}$ and at intersections

For different parametric domains (D1–D5), that is, for different values of ρ, the sequences (T1–T5) of the times at which the maximum in B and the intersections occur is different, while the sequences of concentration values (V1–V5) are also different, as summarized qualitatively in Tables 11.2 and 11.3. Confronting the experimental data with these lists of possibilities, we may draw conclusions on the validity of the kinetic model relating to the proposed reaction scheme.

Table 11.2 Qualitative list of possible sequences in time

Domain	ρ	Time	Order (Earliest to Latest Time)
D1:	$0<\rho<1/2$	T1:	$A=B, A=C, B_{max}, B=C$
D2:	$1/2<\rho<1$	T2:	$A=B, B_{max}, A=C, B=C$
D3:	$1<\rho<1.17...$	T3:	$B_{max}, A=B, A=C, B=C$
D4:	$1.17...<\rho<2$	T4:	$B_{max}, B=C, A=C, A=B$
D5:	$2<\rho$	T5:	$B=C, B_{max}, A=C,$ no $A=B$

Table 11.3 Qualitative list of possible sequences of concentration values

Domain	ρ	Time	Order (Smallest to Largest Value)
D1:	$0<\rho<1/2$	V1:	$A=C, A=B, B=C, B_{max}$
D2:	$1/2<\rho<1$	V2:	$A=C, B=C, A=B, B_{max}$
D3:	$1<\rho<1.17...$	V3:	$A=C, B=C, A=B, B_{max}$
D4a:	$1.17...<\rho<1.19...$	V4:	$A=B, B=C, A=C, B_{max}$
D4b:	$1.19...<\rho<2$	V4:	$A=B, B=C, B_{max}, A=C$
D5:	$2<\rho$	V5:	$B=C, B_{max}, A=C,$ no $A=B$

A change of sequence indicates a point of intersection between curves. For example, in Table 11.2, between D1 and D2, the values $A=C$ and B_{max} change position, so these curves intersect at the Acme point with $\rho=0.5$ the final value of domain D1 and the starting value of D2. Note that the same value sequence is obtained for the intervals $1/2<\rho<1$ and $1<\rho<1.17...$. This is due to the osculation at the Euler point of the $A=B$ and B_{max} curves. A special point is that where the $A=C$ and B_{max} value curves intersect, namely at the unique value of ρ that satisfies the transcendental equation

$$\rho^{\rho/(\rho-1)}+\frac{1-\rho^{-\rho}}{1-\rho}=2 \tag{11.135}$$

This value is $\rho=1.197669$.

Tables 11.2 and 11.3 present all possible scenarios of observed characteristics of transient behavior in terms of the times of intersection and their values. The relation between time sequences T1–T5 and the value sequences V1–V5 is not always one to one.

T1 is related to V1, T2 to V2, and T3 to V2, as well. T3 is also related to V3 or V4, domain D4 being split in two subdomains D4a and D4b corresponding to V3 and V4, respectively. Finally, T5 is related to V5. In fact, D4a is only a tiny domain (2% relative size), which may be hard to observe. The total amount of possible permutations of times and of values is $24+6=30$ each (24 if $A=B$ occurs, 6 otherwise), there are $24 \times 24 + 6 \times 6 = 612$ joint possibilities, of which surprisingly only six actually occur.

The equations for consecutive reactions presented in the literature are correct. In some textbooks, expressions for $t_{B,max}$ are given. However, no classification and understanding is given of special points and the regimes and scenarios of behavior governed by these points. Consequently, the illustrations are sometimes wrong, showing impossible sequences of times at which intersections occur, for example, $B=C$, B_{max}, $A=C$, $A=B$ or B_{max}, $B=C$, $A=C$.

Furthermore, in some cases, the illustrations unintentionally correspond to special points, especially the Euler and Acme points, without being acknowledged as such.

The mistake most often found, however, is the identification of the second exponent being dominant as a quasi-steady state. This is the result of failing to understand that there are only two extreme cases of domination in this reaction, that is, domination of the first exponent or domination of the second exponent, see Eq. (11.112), and the other exponent has nothing to do with a quasi-steady-state regime. The first regime, in which k_1 is very large compared to k_2, can be considered as the $B \rightarrow C$ regime, and the second regime, in which k_2 is very large, as $A \rightarrow C$.

11.3.2.6 Applications

This analysis, first presented by Yablonsky et al. (2010), can significantly improve our ability to distinguish between different reaction mechanisms. Previously, the observation of a maximum concentration of B in a reaction starting from A and involving B and C was considered to be a sufficient indication for distinguishing the consecutive scheme $A \rightarrow B \rightarrow C$ from the parallel one, $A \rightarrow B$, $A \rightarrow C$. Now, based on a number of possible types of intersection, we can falsify the assumption of the consecutive scheme (and propose a more complicated alternative) or justify it and deduce certain additional information regarding the rate coefficients.

This analysis can, for example, be applied to multistep radioactive decay reactions and to isomerization reactions. In such multistep processes, every step is by definition a first-order process. An example of multistep radioactive decay is the Actinium series (see Lederer et al., 1968), in which ^{211}Bi alpha-decays to ^{207}Tl, which beta-decays to ^{207}Pb with respective half-lives of 2.14 and 4.77 min. Therefore, in this two-step consecutive process, $k_2/k_1 = \rho = 2.14/4.77 = 0.449$, very close to the Acme point. Similarly, in the Radium series, ^{214}Pb beta-decays to ^{214}Bi, which beta-decays to ^{214}Po, which then alpha-decays very rapidly (with a half-life of only 0.16 ms) to ^{201}Pb. This multistep decay can be closely approximated by two steps, the first with a half-life of 27 min, the second with a half-life of 20 min. The corresponding ratio of rate coefficients is $\rho = 27/20 = 1.35$, which is between the Lambert and the osculation points.

11.3.2.7 Experimental error

When investigating a reactor set-up in practice, measurements will be affected by experimental error. For instance, determining the intersection between two curves may be done through linear interpolation by first identifying two data points (t_1, c_1) and (t_2, c_2) on one curve, and two (t_1', c_1') and (t_2', c_2') on another, such that the two line segments defined by these data points will intersect. This intersection is then defined by $t = D_t/D_1$ and $c = -D_c/D_1$, where

$$D_1 = \begin{vmatrix} t_1 & t_2 & 0 & 0 & 1 & 0 \\ c_1 & c_2 & 0 & 0 & 0 & 1 \\ 1 & 1 & 0 & 0 & 0 & 0 \\ 0 & 0 & t_1' & t_2' & 1 & 0 \\ 0 & 0 & c_1' & c_2' & 0 & 1 \\ 0 & 0 & 1 & 1 & 0 & 0 \end{vmatrix} \tag{11.136}$$

and

$$D_t = \begin{vmatrix} t_1 & t_2 & 0 & 0 & 0 & 0 \\ c_1 & c_2 & 0 & 0 & 1 & 0 \\ 1 & 1 & 0 & 0 & 0 & 1 \\ 0 & 0 & t_1' & t_2' & 0 & 0 \\ 0 & 0 & c_1' & c_2' & 1 & 0 \\ 0 & 0 & 1 & 1 & 0 & 1 \end{vmatrix}, \quad D_c = \begin{vmatrix} t_1 & t_2 & 0 & 0 & 1 & 0 \\ c_1 & c_2 & 0 & 0 & 0 & 0 \\ 1 & 1 & 0 & 0 & 0 & 1 \\ 0 & 0 & t_1' & t_2' & 1 & 0 \\ 0 & 0 & c_1' & c_2' & 0 & 0 \\ 0 & 0 & 1 & 1 & 0 & 1 \end{vmatrix} \tag{11.137}$$

Since determinants are multilinear, a straightforward application of error analysis is to propagate estimates on the time and concentration errors through these equations, whether using a statistical or an interval approach.

Experimental errors can lead to a false value or obscure the true value of a maximum. Typically, one would then fit a least-squares parabola through sufficient data points and estimate the maximum from it.

More generally, one may wish to make maximum use of the information provided by all the data points. This can readily be done by fitting the exact solution to the data via the parameters k_1 and k_2, ideally taking into account a full statistical description including repeat experiments, estimated (co)variances, and so on. If the resulting residues show no persistence of systematic errors (which may require substantial correction efforts), the nonlinear least-squares method will yield aggregate estimates of the standard deviations $\sigma_{1,2}$ and correlation r on the values of the rate coefficients, which can then be propagated through all equations to determine, for example, confidence intervals for the identification of special points. In particular, the corresponding variance of ρ is given by

$$\sigma_\rho^2 = \frac{k_1^2\sigma_2^2 + k_2^2\sigma_1^2 - 2rk_1k_2\sigma_1\sigma_2}{k_1^4} \tag{11.138}$$

11.3.3 Two-Step Consecutive Mechanism With One Reversible and One Irreversible Step

11.3.3.1 Kinetic model

The two-step consecutive mechanism represented by $A \underset{k_1^-}{\overset{k_1^+}{\rightleftharpoons}} B \overset{k_2}{\rightarrow} C$ is widely used in chemical kinetics, for example, as a model for a complex reaction with a rate-limiting step or quasi equilibrium. The kinetic model for this reaction scheme is represented as follows:

$$\frac{dc_A}{dt} = -k_1^+ c_A + k_1^- c_{|B} \tag{11.139}$$

$$\frac{dc_B}{dt} = k_1^+ c_A - \left(k_1^- + k_2\right)c_B \tag{11.140}$$

$$\frac{dc_C}{dt} = k_2 c_B \tag{11.141}$$

with k_1^+ and k_1^- the rate coefficients of the forward and reverse reaction of the first step and k_2 the rate coefficient of the second step.

Constales et al. (2013) have studied the intersections, coincidences, and reciprocal symmetries of this reaction scheme. They considered two trajectories, defined by extreme initial conditions to ensure reciprocity: a trajectory starting from A with $(A_0, B_0, C_0) = (1, 0, 0)$ and a trajectory starting from B with $(A_0, B_0, C_0) = (0, 1, 0)$.

The kinetic model has three eigenvalues, one of which equals zero while the other two are given by

$$\lambda_{p,m} = \frac{k_1^+ + k_1^- + k_2 \pm \sqrt{\left(k_1^+ + k_1^- + k_2\right)^2 - 4k_1^+ k_2}}{2} \tag{11.142}$$

where subscripts p and m denote the roots with respectively the plus and minus signs. The values of these roots can be shown to satisfy $\lambda_p > k_2 > \lambda_m > 0$ (see Appendix A in Chapter 6). The exact solutions for the trajectories corresponding to the dual experiments are given by

$$A_A(t) = \frac{\lambda_p(k_2 - \lambda_m)\exp(-\lambda_m t) + \lambda_m(\lambda_p - k_2)\exp(-\lambda_p t)}{k_2(\lambda_p - \lambda_m)} \tag{11.143}$$

$$B_A(t) = \frac{\lambda_p \lambda_m(\exp(-\lambda_m t) - \exp(-\lambda_p t))}{k_2(\lambda_p - \lambda_m)} \tag{11.144}$$

$$C_A(t) = 1 - \frac{\lambda_p(\exp(-\lambda_m t) - \lambda_m \exp(-\lambda_p t))}{\lambda_p - \lambda_m} \tag{11.145}$$

$$A_B(t) = \frac{(\lambda_p - k_2)(k_2 - \lambda_m)(\exp(-\lambda_m t) - \exp(-\lambda_p t))}{k_2(\lambda_p - \lambda_m)} \tag{11.146}$$

$$B_B(t) = \frac{\lambda_m(\lambda_p - \lambda_2)\exp(-\lambda_m t) + \lambda_p(k_2 - \lambda_m)\exp(-\lambda_p t)}{k_2(\lambda_p - \lambda_m)} \tag{11.147}$$

$$C_B(t) = 1 - \frac{(\lambda_p - k_2)(\exp(-\lambda_m t) + (k_2 - \lambda_m)\exp(-\lambda_p t))}{\lambda_p - \lambda_m} \tag{11.148}$$

11.3.3.2 Maxima and intersections

We consider the curves $A_A(t)$, $B_A(t)$, $C_A(t)$, $A_B(t)$, $B_B(t)$, and $C_B(t)$, their maxima and their intersections. It is intuitively clear, and easy to show from the explicit solution, that $A_A(t)$ and $B_B(t)$ are monotonically decreasing functions with a maximum at $t=0$, and that $C_A(t)$ and $C_B(t)$ are monotonically increasing functions for which the maximum is reached at $t=+\infty$. A nontrivial maximum of the concentration dependence A_B always exists, and this maximum always occurs at the same time as that of B_A, which is one of the fingerprints of this consecutive mechanism. Furthermore, since the proportion between B_A and A_B is always fixed (at $K_{eq,1} = k_1^+/k_1^-$) for $t>0$ (see Section 6.4), these dependences also reach their maximum at the same time, and they either have the same value at $t=0$ only (for $k_1^+ \neq k_1^-$) or have identical values for all $t \geq 0$ (for $k_1^+ = k_1^-$).

In order to determine the conditions for the existence of intersections between concentration curves, the explicit equations, Eqs. (11.143)–(11.148) can be analyzed in detail. As an example, the time at which $A_A = B_A$ follows from equating Eqs. (11.143), (11.144), which yields

$$t = \frac{1}{\lambda_p - \lambda_m} \ln \frac{2 - \dfrac{k_2}{\lambda_p}}{2 - \dfrac{k_2}{\lambda_m}} \tag{11.149}$$

As $\lambda_p > k_2 > \lambda_m > 0$ the numerator in the natural logarithm is larger than one for all values of k_2 and is larger than the denominator. Thus, in order for intersection to occur, the only requirement is that the denominator be positive, so $k_2/\lambda_m < 2$, which leads to the condition that $k_1^+ - k_1^- - k_2/2 > 0$ (see Constales et al., 2013).

The 15 intersections that are possible, in principle, for the reaction scheme $A \underset{k_1^-}{\overset{k_1^+}{\rightleftarrows}} B \overset{k_2}{\rightarrow} C$ are listed in Table 11.4. Of these, seven always occur, four never occur, and four occur conditionally, depending on simple inequalities involving the rate coefficients. Based on these inequalities, seven domains can be distinguished, as listed in Table 11.5, where the first three rows define which inequalities hold for these domains and the next four rows list whether or not the given equality occurs in a certain domain.

Table 11.4 Possible intersections of concentration curves and conditions for their occurrence

Intersection	Condition
$A_A = A_B$	Never, only at $t = +\infty$
$A_A = B_A$	$k_1^+ - k_1^- - k_2/2 > 0$
$A_A = B_B$	Never, only at $t = 0$ and $t = +\infty$
$A_A = C_A$	Always
$A_A = C_B$	Always
$A_B = B_A$	Never, only at $t = 0$ and $t = +\infty$ for $k_1^+ \neq k_1^-$; at all t for $k_1^+ = k_1^-$
$A_B = B_B$	$k_1^+ - k_1^- - k_2/2 < 0$
$A_B = C_B$	$k_1^- - k_2 > 0$
$B_A = B_B$	Always
$B_A = C_A$	Always
$B_A = C_B$	$k_1^+ - k_2 > 0$
$B_B = C_B$	Always
$C_A = A_B$	Always
$C_A = B_B$	Always
$C_A = C_B$	Never, only at $t = 0$ and $t = +\infty$

Table 11.5 Domains for the occurrence of intersections of concentration curves

	D1	D2	D3	D4	D5	D6	D7
$k_1^+ - k_1^- - k_2/2 > 0$	+	+	−	−	+	−	−
$k_1^+ - k_2 > 0$	+	−	−	+	+	+	−
$k_1^- - k_2 > 0$	−	−	−	−	+	+	+
$A_A = A_B$	+	+	−	−	+	−	−
$A_B = B_B$	−	−	+	+	−	+	+
$A_B = C_B$	−	−	−	−	+	+	+
$B_A = C_B$	−	−	−	+	+	+	−

The existence of maxima and intersections can be presented as a set of obligations (always), prohibitions (never), and prescriptions (sometimes, according to rules). This set can be useful for validating mechanisms and determining parametric domains.

11.3.3.3 Ordering domains and coincidences

Each maximum concentration and each intersection of concentration curves is characterized by a time and a concentration value. When two of these times, or two of these concentration values, coincide for different events, there is a coincidence. Constales et al. (2013) have calculated all different ordering domains of intersections and maxima, both in time and in value, and ordered all possible pairs (132) of these coincidences in tables and graphically represented their relations. This revealed a surprising richness of the coincidence structure, and gave rise to a very intricate puzzle of subdomains defined by a unique combination of existence and ordering of intersections and maxima.

Fig. 11.8 shows a few examples, which are discussed in more detail below.

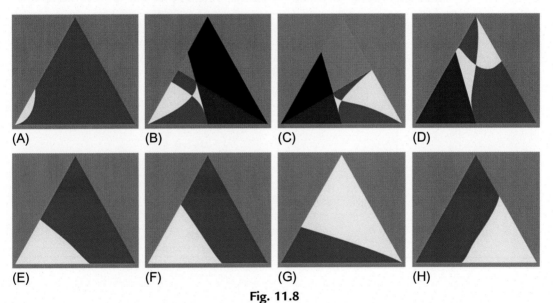

(A)   (B)   (C)   (D)

(E)   (F)   (G)   (H)

Fig. 11.8

Curvilinear cases (A) $t_{A_A=C_A}, t_{A_B,\max}$; (B) $A_A=B_A$, $B_A=C_B$; (C) $A_B=B_B$, $A_B=C_B$; (D) $A_B=B_B$, $A_B=C_A$; (E) $A_A=C_B$, $B_A=B_B$; (F) $A_A=C_A$, $B_A=B_B$; (G) $t_{B_B=C_B}, t_{A_B,\max}$; (H) $B_B=C_B$, $A_{B,\max}$.

The equilateral triangles represent triples (k_1^+, k_1^-, k_2) in barycentric coordinates; the left vertex represents pure k_1^+, that is $(1, 0, 0)$, the right vertex pure k_1^-, that is $(0, 1, 0)$, and the top vertex represents pure k_2, that is $(0, 0, 1)$. The color coding used in the figures is as follows: if both events exist and the left concentration or time mentioned in the caption is smaller, the point is blue, but if the right is smaller, it is yellow. Dark blue means that the left-side event alone exists, and dark yellow that the right-side event alone exists. Black means that neither side exists. Although some cases look similar, they are all different when inspected at sufficient magnification. The case $A \to B \to C$ studied by Yablonsky et al. (2010) corresponds to $k_1^- = 0$, that is, the top-left side of the equilateral triangle. Similarly, quasi-equilibrium of the first reaction corresponds to very small k_2, that is, the bottom side of the equilateral triangle.

Fig. 11.8A shows the curvilinear case of the coincidence between the time of intersection $t_{A_A=C_A}$ and the time of maximum $t_{AB,\max}$. The coloring of this figure is mostly blue (grey), with a small yellow (white) domain near the left-bottom vertex. The top-left side of the triangle corresponds to the irreversible case studied by Yablonsky et al. (2010). Visually we can identify the boundary between yellow and blue on that side as $k_2 = k_1^+/2$. When the size of subdomains is very different, as in this case, the ordering of the intersections is a strong tool for identifying the parameters when they are part of the small subdomain.

Fig. 11.8B shows the curvilinear case of the coincidence of the concentration value at intersection $A_A=B_A$ with that at the other intersection, $B_A=C_B$. Five different domains occurs, because all possible cases arise, depending on the relative values of the parameters: (1) in the top right domain, neither intersection exists; (2) in the bottom right domain, only intersection $A_A=B_A$, and (3) in the appr. middle left, only intersection $B_A=C_B$; (4) finally in the bottom left and appr. middle left domain, the concentration $A_A=B_A$ has the smaller value, while (5) in the remaining domain the concentration $B_A=C_B$ is smallest.

There is an interesting checkered domain shape, showing that two different inequalities are at the origin of the ordering cases. The intersection point between the cross-shaped blue and yellow subdomains is a special point of the reaction mechanism, an intersection of coincidence curves. Such points are of great importance in characterizing all subdomains defined by full sets of orderings. Qualitatively similar results are shown in Fig. 11.8C and D but for quite different intersections.

Fig. 11.8E–H exhibit domains that are separated by nearly straight curves, which indicates that the solutions to the corresponding equations are close to a simple linear relationship, but do not quite coincide with it.

The kinetic behavior of complex chemical reactions may, in some temporal or parametric domains, be approximated by $A \rightleftarrows B \rightarrow C$, or by an evolution of such simple mechanisms: first $A \rightleftarrows B \rightarrow C$, then $C \rightleftarrows D \rightarrow E$, and so on. The discovered patterns can be used for recognizing submechanisms, and for estimating their parameter values and the evolution of the submechanisms. Generally, such properties of simple linear and nonlinear systems in fact reflect their unexpected complexity. Constales et al. (2013) have proposed to use a special term for defining this phenomenon: simplexity.

11.3.4 Concluding Remarks

A similar analysis can be applied to other reaction mechanism, such as parallel irreversible first-order reactions $A \rightarrow B$ and $A \rightarrow C$, a triangle of reversible first-order reactions and simple models with nonlinear dependencies. The results obtained for the temporal changes of concentrations can be directly applied to steady-state PFR models, with the astronomic time replaced by the space time or residence time τ. Similar analyses can also be done for CSTRs.

An interesting opportunity for the application of these models is provided by thermodesorption and thermogravimetry problems, which both consist of a sequence of linear processes, respectively a change of temperature and a loss of weight with time.

As the systems considered grow more complex, the number of intersection and maximum points and the number o sequences of such points will grow rapidly: polynomially for the former, combinatorially for the latter. This increasing complexity can to some extent be matched by automated methods. Ultimately, we expect unifying properties to emerge. Indeed, we believe that the analysis of intersections and coincidences presents a source of vast information on new families of patterns for distinguishing mechanisms and parametric domains, and might even find applications outside the domain of chemical kinetics.

Previously, from the point of the intensive studying of nonlinear behavior (see Chapter 7), by default it was considered that the properties of simple kinetic models, in particular linear kinetic models of many reactions and models of single nonlinear first- second-, and third-order reactions, are completely known from the literature and therefore there is nothing interesting in studying these models. This was our opinion as well, until recent results regarding simple linear models and some nonlinear models of single reactions were revealed (Yablonsky et al., 2010, 2011a,b).Three classes of models have been discovered:

*   time invariances observed in dual kinetic experiments
*   intersections of kinetic dependences
*   coincidences of special events like two intersections, an intersection, and a concentration maximum, and so on

All these patterns, which have been discussed in this chapter and in Chapter 6 in detail, can be used for many purposes, such as testing the validity of an assumed reaction scheme and its corresponding model and estimating parameters of a kinetic model based on the occurrence of patterns and predicting concentration dependences. This approach can be termed *pattern kinetics* or *event-based kinetics*. It is interesting that there is a remarkable resemblance between our patterns of coincidences, and the abstract and conceptual art of, for instance, Felix De Boeck and Sol LeWitt.

Nomenclature

A, B, C	normalized concentrations of A, B, C $(-)$
A_A, A_B	normalized concentration of A starting from resp. A, B $(-)$
B_1, B_2	bifurcation parameters
c_i	concentration of component i (mol/m^3)
D	diffusion coefficient (m^2/s)

$D_{1,c,t}$	determinants
E_a	activation energy (J/mol)
H	parameter
H	heaviside function
K_{eq}	equilibrium coefficient ($-$)
k_i	rate coefficient of reaction i (s^{-1})
$k_{i,0}$	preexponential factor for reaction i (s^{-1})
$\tilde{L}$	linear operator
p_i	partial pressure of component i (Pa)
R_g	universal gas constant (J/mol/K)
R	reaction rate (s^{-1})
r_s	rate of elementary step s (s^{-1})
T	temperature (K)
T	time (s)
u, v	new coordinates, defined by Eq. (11.11) ($-$)
v	velocity (m/s)
x	distance from reactor inlet (m)
x, y, z	normalized/fractional surface concentrations ($-$)

Greek Symbols

α	angle between the horizontal axis and the tangent to the A versus C curve
β	angle between the horizontal axis and the tangent to the B versus C curve
γ	angle of intersection, given by the difference between angles α and β
Δ	discriminant
Δ	concentration perturbation (mol/m^3)
E	small parameter (s)
θ_j	fractional surface coverage of intermediate j ($-$)
λ	eigenvalue
ρ	ratio of rate coefficients k_2/k_1 ($-$)
σ	variance
τ	space time (s)
ϕ	golden ratio, $\frac{1}{2}(\sqrt{5}+1)$ ($-$)

Subscripts

0	initial
eq	equilibrium
ext	extinction

ign	ignition
in	inlet
m	root with minus sign
max	maximum
MBC	point of maximum bifurcational complexity
p	root with plus sign
ss	steady state

Superscripts

+	of forward reaction
+	change to higher concentration
−	of reverse reaction
−	change to lower concentration

References

Bennett, C.O., 1967. A dynamic method for the study of heterogeneous catalytic kinetics. AICHE J. 13, 890–895.

Bennett, C.O., 1976. The transient method and elementary steps in heterogeneous catalysis. Catal. Rev. 13, 121–148.

Bennett, C.O., 2000. Experiments and processes in the transient regime for heterogeneous catalysis. In: Haag, W.O., Gates, B., Knözinger, H. (Eds.), Advances in Catalysis. Elsevier Sciences, Dordrecht, pp. 329–416.

Bennett, C.O., Cutlip, M.B., Yang, C.C., 1972. Grandientless reactors and transient methods in heterogeneous catalysis. Chem. Eng. Sci. 27, 2255–2264.

Berger, R.J., Kapteijn, F., Moulijn, J.A., et al., 2008. Dynamic methods for catalytic kinetics. Appl. Catal. A 342, 3–28.

Bykov, V.I., Goldfarb, I., Gol'dshtein, V.M., 2006. Singularly perturbed vector fields. J. Phys. Conf. Ser. 55, 28.

Bykov, V., Gol'dshtein, V., Maas, U., 2008. Simple global reduction technique based on decomposition approach. Combust. Theor. Model. 12, 389–405.

Constales, D., Yablonsky, G.S., Marin, G.B., 2013. Intersections and coincidences in chemical kinetics: linear two-step reversible-irreversible reaction mechanism. Comput. Math. Appl. 65, 1614–1624.

Eigen, M., 1954. Über die Kinetik sehr schnell verlaufender Ionenreaktionen in wässeriger Lösung. Z. Phys. Chem. N.F. 1, 176–200.

Gleaves, J.T., Ebner, J.R., Kuechler, T.C., 1988. Temporal analysis of products (TAP)—a unique catalyst evaluation system with submillisecond time resolution. Catal. Rev. Sci. Eng. 30, 49–116.

Gleaves, J.T., Yablonskii, G.S., Phanawadee, P., Schuurman, Y., 1997. TAP-2: an interrogative kinetics approach. Appl. Catal. A 160, 55–88.

Gol'dshtein, V., Krapivnik, N., Yablonsky, G., 2015. About bifurcational parametric simplification. Math. Model. Nat. Pheno. 10, 168–185.

Kobayashi, M., Kobayashi, H., 1972a. Application of transient response method to the study of heterogeneous catalysis: I. Nature of catalytically active oxygen on manganese dioxide for the oxidation of carbon monoxide at low temperatures. J. Catal. 27, 100–107.

Kobayashi, M., Kobayashi, H., 1972b. Application of transient response method to the study of heterogeneous catalysis: II. Mechanism of catalytic oxidation of carbon monoxide on manganese dioxide. J. Catal. 27, 108–113.

Kobayashi, M., Kobayashi, H., 1972c. Application of transient response method to the study of heterogeneous catalysis: III. Simulation of carbon monoxide oxidation under an unsteady state. J. Catal. 27, 114–119.

Kobayashi, H., Kobayashi, M., 1974. Transient response method in heterogeneous catalysis. Catal. Rev. Sci. Eng. 10, 139–176.

Kubasov, A.A., 2004. Chemical Kinetics and Catalysis. Part I. Phenomenological Kinetics. Moscow University Publishing, Moscow.

Lederer, C.M., Hollander, J.M., Perlman, I., 1968. Table of Isotopes, sixth ed. Wiley, New York.

Marin, G.B., Yablonsky, G.S., 2011. Kinetics of Chemical Reactions—Decoding Complexity. Wiley-VCH, Weinheim.

Yablonskii, G.S., Lazman, M.Z., 1996. New correlations to analyze isothermal critical phenomena in heterogeneous catalysis reactions ("Critical simplification", "hysteresis thermodynamics"). React. Kinet. Catal. Lett. 59, 145–150.

Yablonsky, G.S., Mareels, I.M.Y., Lazman, M., 2003. The principle of critical simplification in chemical kinetics. Chem. Eng. Sci. 58, 4833–4842.

Yablonsky, G.S., Constales, D., Marin, G.B., 2010. Coincidences in chemical kinetics: surprising news about single reactions. Chem. Eng. Sci. 65, 6065–6076.

Yablonsky, G.S., Constales, D., Marin, G.B., 2011a. Equilibrium relationships for nonequilibrium chemical dependencies. Chem. Eng. Sci. 66, 111–114.

Yablonsky, G.S., Gorban, A.N., Constales, D., Galvita, V.V., Marin, G.B., 2011b. Reciprocal relations between kinetic curves. Europhys. Lett. 93, 2004–2007.

Yang, C.C., Cutlip, M.B., Bennett, C.O., 1973. A study of nitrous oxide decomposition on nickel oxide by a dynamic method. In: Proceedings of 5th Congress on Catalysis, Palm Beach.

Index

Note: Page numbers followed by *f* indicate figures, *t* indicate tables, and *s* indicate schemes.

A

Acme, Golden, Euler, Lambert, Osculation, and Triad (AGELOT) points, 380, 380*t*
Adsorption
  irreversible, 123–124
  mechanism, 244, 248*f*, 249–251, 250–251*t*, 351–353
  (*see also* Langmuir-Hinshelwood adsorption mechanism)
  reversible, 124–126
Arrhenius equation, 294–295, 295*f*
Augmented molecular matrix, 20–21, 30*t*
Augmented stoichiometric matrix, 24–26
  enthalpy change and equilibrium coefficients, 26
  key and nonkey reactions, 25–26
  reduced row echelon form (RREF) of, 25, 31*t*
Average chain length characteristics, 338–339

B

Batch reactor (BR), 35–36, 37*f*, 42–43
Belousov-Zhabotinsky reaction, 163, 251–252
Bifurcation parameters, 356–357, 363–366
Bifurcation theory, 4–5
Bipartite graph
  of complex reaction, 73–78
  irreversible reaction, 74*f*
  oxidation of CO on Pt, 76*f*

reaction mechanism, 76*f*
  reversible reaction, 75*f*
BR. *See* Batch reactor (BR)

C

Catalytic chemical oscillation model, 252–257
Catalytic reaction
  mechanism, 210
  single-route, 67–68
Chemical equilibrium and optimum, 166–174
Chemical kinetics
  law, 2–3
  relaxation methods, 366
Chemical kinetics, simplification in, 88–101
  based on abundance, 89–90
  defined, 83–84
  and experimental observations, 98–100
  lumping analysis, 100–101
  quasi-equilibrium approximation, 92–94
  quasi-steady-state approximation, 94–98
  rate-limiting step approximation, 90–92
Chemical oscillation
  in isothermal system, 251–257
  model, catalytic, 252–257
Chemical reaction, 221
  complex, 181–183
  dynamic behavior of, 222
  kinetic experiment, 39
Chemical thermodynamics, 159–166

equilibrium and principle of detailed balance, 165–166
  reversibility and irreversibility, 164
  steady state and equilibrium, 163
Chemistry and mathematics, 1–2
  historical aspects, 2–3
  new trends, 3–5
Coincidences, 373–390
Completeness test, 200–201, 201*f*
Complex reaction
  bipartite graph of, 73–78, 74–77*f*
  chemical, 181–183
  vs. elementary reaction, 55–56
  steady-state equation for, 62–68
Computational fluid dynamics (CFD), 330
Conical reactors, 147
Consecutive reaction mechanisms, 374–375
Continuous-flow reactor, 35–36
Continuous stirred-tank reactor (CSTR), 4–5, 40–41, 43, 173–174, 236, 237*f*, 269–270, 367, 369–370
  mathematical model of nonisothermal, 237–239
Cramer's rule, 64, 66
Critical simplification
  adsorption mechanism, 351–353
  bifurcation parameters, 356–357
  dynamics, 361–366
  generalization, 366
  principle, 351
  reaction rates, 357–361
  singular perturbation, zero approximation for, 354–355
  slow-fast system, 354

Cross rule, 1, 1*s*
CSTR. *See* Continuous stirred-tank
    reactor (CSTR)

D

Damköhler number, 38, 41,
    142–143, 237–238, 271–272,
    280–282
Detailed balance principle,
    165–166, 189–194
Deterministic modeling techniques,
    310–322
Diffusion
    multicomponent, 107–108
    in nonporous solids, 107
    one-zone reactor, 121–123
    radial, 139–141
    surface, 147–148
Diffusion model
    nonlinear, 109–110
    reaction, 83–84
Diffusion system equation,
    106–110, 129
Diffusion theory, 105
Discrete weighted Galerkin
    formulation. *See* Galerkin
    method
Distributed models, 38–39
Dynamic behavior, of chemical
    reaction system, 222

E

Einstein's mobility, 106–107
Elementary reaction, 174*t*
    vs. complex reaction, 55–56
    definition of, 45–46
Elements
    balance, molecular matrix to, 12–21
    combination of two groups of,
    20–21
    fixed ratio of, 17–18
Eley-Rideal mechanism, 76*f*, 77*f*,
    74–77
Emulsion polymerization, 334–339
Enthalpy change, 26
Enzyme-catalyzed reactions, 61
Equilibrium, 165–181
    conversion, 168–170, 169*t*, 171*f*,
    172, 173*f*
    vs. optimum regime, 168–174

Equilibrium coefficient, 26, 167
Estimating vector, 112
Event-based kinetics, 390
Experimental errors, 384–385
Extended method of moments,
    315–319

F

Fick's first law, 38, 106
Fick's second law, 106
Fixed pivot method, 319–321
Full CLD methods, 315–322

G

Galerkin method, 321–322
Gas-phase chain reaction, 86
Gas-solid catalytic system, 86,
    105, 107
Genetic algorithms, 302–304,
    303–304*f*
Gibbs free energy, 162, 176
Gillespie-based kinetic Monte Carlo
    (kMC) algorithm, 322–327,
    326*f*
Gradient method, 290, 291*f*
Graph theory, 61–79
    applications, 62–68
    trees in, 65–68, 66*f*

H

Heisenberg's matrix mechanics, 3
Heterogeneous catalysis, 62–63
    isothermal, 242–251
Heterogeneously catalyzed
    reactions
    rate laws for, 296–299
Heterogeneous polymerization
    dispersed media, 330–339
    solid catalysts, 339–342
Heterogeneous reaction system,
    47–48, 105
Homogeneous reaction, 46–47, 48*t*
Horiuti numbers, 26–30
Hysteresis, 243*f*, 246–247

I

Ideal reactor, 39–42
    net rate of production in, 45*t*
    with solid catalyst, 42–44

Ideal thermodynamic system, 160
Impact mechanism. *See* Eley-Rideal
    mechanism
Instability and stability, 221*f*
Intersections, 373–390
Isomerization reaction, 23, 58,
    64–65, 65–66*f*, 79, 176–181,
    178*f*, 203, 206, 383–384
    characteristic equation, 79
    dynamics of, 58–61
    King-Altman graph of, 65*f*
    mechanism of, 65*f*

K

Kinetic coupling, 68–73
Kinetic experiments, 35
    chemical reaction, 39
    ideal reactors, 39–42
    kinetic data analysis, 35–36
    material balances, 36–45
    reactors for, 35–36, 37*f*
    requirements of experimental
    information, 36
    transport in reactors, 37–39
Kinetic parameters, temperature
    dependence of, 301–302
Knudsen diffusion, 107, 114, 267
    coefficient, 111

L

Langmuir-Hinshelwood
    mechanism, 74–77, 74–77*f*,
    351
Langmuir-Hinshelwood-Hougen-
    Watson (LHHW) kinetics, 53*f*,
    53–55, 244, 296
Laplace transform, 115–116
Least-squares criterion, 285–286
Le Chatelier's principle, 166–168,
    168*t*, 169*f*
Levenberg-Marquardt compromise
    method, 291–294
LHHW kinetics. *See* Langmuir-
    Hinshelwood-Hougen-Watson
    (LHHW) kinetics
Linear algebra, 2
Local stability analysis, 224–225
    chemical oscillations, 235–236
    global dynamics, 230–231
    imaginary roots, 230

procedure, 225–226
real roots, 229
in system with two variables, 226–230
Lumped model, 38–39
Lyapunov function, 162
thermodynamic, 231–236

M

Macroscale modeling techniques, 328–330
Marcelin-de Donder kinetics, 163
Mass-action law, 46–47, 175–176
for surface diffusion, 109
Mass balance, 199, 200*f*
Matrix
definition, 9
molecular, 11–21
stoichiometric, 21–26
Matrix modeling, TAP reactor, 152–154
Matrix, molecular, 11–12
application of, 12–21
augmented, 20–21
Mendeleev's periodic table, 3
Method of moments, 310–315
Michaelis-Menten model, 87–88
Microscale modeling techniques, 310–328
Molecular matrix, 11–12
application of, 12–21
augmented, 20–21
Moving bed reactor, 159
Multigrain models, 340, 341*f*
Multiple reaction
irreversible and reversible, 57*f*
vs. single reaction, 55, 56*f*
Multiplicity of steady states, 236–251, 243*f*, 251*t*
Multiresponse TAP theory, 129–132
Multiroute mechanism
steady-state kinetic equation for, 68–73, 69*f*, 71–72*f*
Multizone configurations, optimization of, 267–283

N

Newton-Gauss algorithm, 286
Nonisothermal system

multiplicity of steady states in, 236–242
Nonlinear diffusion model, 109–110
Nonlinear reactions
symmetry relations for, 202–215
time invariances for, 214*t*
Non-steady-state kinetic behavior, 38–39, 58, 78, 83, 222–224

O

ODE. *See* Ordinary differential equation (ODE)
One active thin zone, 270
One-zone reactor
diffusion, 121–123
irreversible adsorption, 123–124
reversible adsorption, 124–126
Onsager's approach, 107–108
Optimal active zone configurations
equidistant configurations, 274–283
numerical experiments, 272–274
Optimum, 166–174
conversion, 170–172, 171*f*, 173*f*
Optimum regime *vs.* equilibrium, 168–174
Ordinary differential equation (ODE), 58–61, 117–118
Orthogonal intersection, 381
Oscillation, chemical
in isothermal system, 251–257
model, catalytic, 252–257

P

Parallel experimental procedure, 367, 368*f*, 369
Parametric analysis procedure, 257–262
Partial differential equation (PDE), 115–116
temporal analysis of products, 138–139
Pattern kinetics, 390
PDE. *See* Partial differential equation (PDE)
Periodic table, Mendeleev, 3
PFR. *See* Plug-flow reactor (PFR)
Physicochemical assumptions, 84–88
combining assumptions, 87–88

on experimental procedure, 87
on reactions and parameters, 85–86
on substances, 84–85
on transport-reaction characteristics, 86–87
Plug-flow reactor (PFR), 36, 41–43, 367, 371–372
differential, 42
nonideal, 145–146, 372–373
optimum conversion in continuous, 170–172
steady-state, 42
Poincaré-Bendixson theorem, 252–253, 255
Polymerization rate, 336–338
Power-law equations, 294
Power-law test
multiple reactions, 57*f*
single reaction, 57*f*
Pseudo-steady state (PSS). *See* Quasi-steady state (QSS)
Pulse reactor, 35–36
Python, reduced row echelon form (RREF) in, 31–33

Q

Quasi-equilibrium approximation, 92–94
Quasi-steady-state (QSS)
approximation, 94–98
assumption, 85, 210

R

Radial diffusion
in two and three dimensions, 139–141
Reaction rates, 357–361
bifurcation points, 358–361
Reaction reversibility
influence on hysteresis, 246–247
Reactor. *See also specific types of reactor*
batch, 35–36, 37*f*
conical, 147
continuous stirred-tank, 40–41
ideal, 39–44, 45*t*
for kinetic experiments, 35–36, 37*f*
moving bed, 159

Reactor (*Continued*)
  plug-flow, 41–42
  transport in, 37–39
Reactor-diffusion system, 110–115
Reactor model, 267–269, 280–283
Reduced Row Echelon Form
    (RREF), 13, 15*f*, 18, 25*f*
  augmented molecular matrix, 30*t*
  augmented stoichiometric matrix,
    25, 31*t*
  characteristics, 30–31*t*
  in Python, 31–33
Reversible adsorption, one-zone
    reactor, 124–126
Reversible first-order reaction,
    186–187
Reversible reaction, 50–52
  bipartite graph, 75*f*
  first-first order, 203, 204*f*
  first-order, 185–187
  with Langmuir-Hinshelwood-
    Hougen-Watson kinetics, 53–55
  second-first order, 203–205, 205*f*,
    207–208
  second-second order, 206–209,
    207*f*
  single, 203–209
Rosenbrock method, 288–290, 289*f*
RREF. *See* Reduced Row Echelon
    Form (RREF)

S

SAE. *See* State-altering experiment
    (SAE)
Schrödinger's equation, 3
SDE. *See* State-defining experiment
    (SDE)
Semibatch reactor, 35–36
Sequential experimental procedure,
    368, 368*f*, 370
Simplification, defined, 83–84
Simplification in chemical kinetics,
    88–101
  based on abundance, 89–90
  defined, 83–84
  and experimental observations,
    98–100
  lumping analysis, 100–101
  quasi-equilibrium approximation,
    92–94

quasi-steady-state
    approximation, 94–98
  rate-limiting step approximation,
    90–92
Single-particle TAP reactor,
    113–114, 113*f*
Single reaction
  vs. multiple reaction, 55, 56*f*
  power-law test for, 57*f*
Singular perturbation, zero
    approximation for, 354–355
Slow-fast system, 354
Solid catalyst, ideal reactor with,
    42–44
Solid-catalyzed reaction, 38
Stability, 221–231
  instability and, 221*f*
  local, 224–230
  of steady state, 239–242
Stability analysis, 361–363
State-altering experiment
    (SAE), 115
State-defining experiment (SDE),
    114–115
Steady-state(s)
  in isothermal heterogeneous
    catalytic systems, 242–251
  kinetic model, 62–63
  multiplicity of, 236–251, 243*f*,
    251*t*
  in nonisothermal system,
    236–242
  plug-flow reactor, 42
  for reaction mechanism, 246*t*
  reaction rate, 83, 247, 247–248*f*
  stability of, 239–242
Steady-state equation, 64–65
  for complex reaction, 62–68
  for multiroute mechanism,
    68–73, 69*f*, 71–72*f*
Steady-state reaction rates, 357
Stochastic modeling techniques,
    322–328
Stockmayer distribution, 339–340
Stoichiometric matrix, 21–26
Stoichiometric matrix, augmented,
    24–26
  enthalpy change and equilibrium
    coefficients, 26
  key and nonkey reactions, 25–26

reduced row echelon form
    (RREF) of, 25, 31*t*
Stoichiometry, 45–50
  elementary reaction, 45–46
  elementary *vs.* complex reaction,
    55–56
  heterogeneous reaction, 47–48
  homogeneous reaction, 46–47,
    48*t*
  rate expressions for single
    reactions, 49
  reaction rate *vs.* net rate of
    production, 49–50
  single *vs.* multiple reaction,
    55, 56*f*
Stokes friction, 106–107
Surface diffusion, 147–148
Suspension polymerization, 331–334
Symmetry relations
  for nonlinear reactions, 202–215
  between observables and initial
    data, 189–192

T

TAP. *See* Temporal analysis of
    products (TAP)
Temkin-Boudart mechanism,
    62–63, 62*t*, 210
Temporal analysis of products
    (TAP), 44, 84, 110–115, 132,
    267, 280–283, 367
  capacity zone, 127
  configuration, 110–114, 111*f*
  five-zone, 118*f*
  generalized boundary conditions,
    128–129
  infinite exit zone, 141–143
  internal injection, 143–145, 143*f*
  inverse problem solving in,
    132–133
  on Laplace and time domain,
    115–129
  matrix modeling, 152–154
  method, 5
  moment definitions, 115–121
  multiresponse, 129–132
  nonideal plug-flow reactor via
    transfer matrix, 145–146
  one-zone reactor, diffusion,
    121–123

one-zone reactor, irreversible adsorption, 123–124
one-zone reactor, reversible adsorption, 124–126
partial differential equation (PDE), 138–139
principles, 114–115
reactive thin zones, 126–127
reactor, 5
resistor zone, 128
single-particle, 113–114, 113f
theory, 148–149
thin-zone, 112–113, 112f
three-zone, 112, 112f
three-zone reactor, 126
Teorell's approach, 108
Thermodynamic Lyapunov function, 231–236
dissipativity of, 233–235
Thermodynamics, chemical, 159–166
equilibrium and principle of detailed balance, 165–166
reversibility and irreversibility, 164
steady state and equilibrium, 163
Thermodynamic state function, 160–163
Thermodynamic system, 159–160
ideal, 160
Thin-sandwiched TAP reactor (TSTR) matrix modeling, 152–154
Thin-zone TAP reactor (TZTR), 44–45, 112–113, 112f, 132, 132f, 151–152, 192–194
Three-dimensional modeling, 134–141
Three-zone reactor, 126
Three-zone TAP reactor, 112, 112f
Tobita-based Monte Carlo technique, 327–328
Transfer matrix
global, 120–122, 131, 145–146
zone, 120, 125–126, 140, 146
TSTR matrix modeling. *See* Thin-sandwiched TAP reactor (TSTR) matrix modeling
Two-dimensional modeling, 134–141
Two-step consecutive mechanism
one reversible and one irreversible step, 385–389
two irreversible steps, 374–385
TZTR. *See* Thin-zone TAP reactor (TZTR)

U

Univariate search method, 287–288, 287f

W

Water-gas shift (WGS) reaction, 62–63, 62t, 192–193
Wegscheider's analysis, 165

Y

Y procedure, 132–133